Arbuscular Mycorrhizal Fungi in Agriculture - New Insights

Edited by Rodrigo Nogueira de Sousa

Published in London, United Kingdom

Arbuscular Mycorrhizal Fungi in Agriculture - New Insights
http://dx.doi.org/10.5772/intechopen.104271
Edited by Rodrigo Nogueira de Sousa

Contributors
Addisu Fufa Ebbisa, YingWu Shi, Hongmei Yang, Ming Chu, XinXiang Niu, Huifang Bao, Ning Wang, Faqiang Zhan, Xuanqi Long, Rong Yang, Qing Lin, Kai Lou, Jamal Salih Alkobaisy, Muhammad Hanif, Zubairia Ashraf, Samar Bashir, Fatima Riaz, Rizwan Amanat, Samina Sarwar, Nousheen Yousaf, Malik A. Aziz, Shayesta Islam, Gousia Gani, Zaffar M. Dar, Amajad Masood, Syed H. Baligah, Stephen Larbi-Koranteng, Frederick Kankam, Joseph Adomako, Muntala Abdulai, Joanna F. Dames, Sílvia Natal David Maússe Sitoe, Yangrong Cao, Jiashan Wu, Weiyun Wang, Hui Zhu, Allal Douira, Amina Ouazzani Touhami, Zineb Sellal, Jamila Dahmani, Soukaina Maazouzi, Najoua Mouden, Mohamed Chliyeh, Karima Selmaoui, Rachid Benkirane, Cherkaoui El Modafar, Julio Alves Cardoso Filho, Nidhi Verma, Yeshveer Singh, Tanvi Singh, Anupam Patra, Sarah Otun, Ikechukwu Achilonu, Samah Ourras, Ismail El Aymani, Soumaya El Gabardi, Manal Adnani, Mariam Artib, Rodrigo Nogueira de Sousa

First published in London, United Kingdom, 2023 by IntechOpen
IntechOpen is the global imprint of INTECHOPEN LIMITED, registered in England and Wales, registration number: 11086078, 5 Princes Gate Court, London, SW7 2QJ, United Kingdom

British Library Cataloguing-in-Publication Data
A catalogue record for this book is available from the British Library

Additional hard and PDF copies can be obtained from orders@intechopen.com

Arbuscular Mycorrhizal Fungi in Agriculture - New Insights
Edited by Rodrigo Nogueira de Sousa
p. cm.
Print ISBN 978-1-83768-089-4
Online ISBN 978-1-83768-090-0
eBook (PDF) ISBN 978-1-83768-091-7

For EU product safety concerns:
IN TECH d.o.o., Prolaz Marije Krucifikse Kozulić 3, 51000 Rijeka, Croatia,
info@intechopen.com or visit our website at intechopen.com.

Meet the editor

Rodrigo Nogueira de Sousa obtained an undergraduate degree in Agronomic Engineering and a master's in Soil Science and Plant Nutrition from the Federal University of Viçosa, Brazil, in 2016 and 2018, respectively. He obtained a Ph.D. in Soil Science and Plant Nutrition from the University of São Paulo (USP). From 2014 to 2015, he studied at North Carolina Agricultural and Technical State University, USA. He also completed an internship at the Department of Crop and Soil Sciences, North Carolina State University, USA, in 2015. During his doctorate program, he studied the feasibility of alternative sources of phosphate fertilizers in tropical soils.

Contents

Preface **IX**

Section 1
The Role of Mycorrhizae Fungi in Plant Growth **1**

Chapter 1 **3**
Introductory Chapter: Mycorrhizal Fungi – A Current Overview on Agricultural Productivity and Soil Health
by Rodrigo Nogueira de Sousa

Chapter 2 **13**
Factors Affecting Mycorrhizal Activity
by Jamal Salih Alkobaisy

Chapter 3 **25**
Recent Advances in Plant: Arbuscular Mycorrhizal Fungi Associations and Their Application to Cassava Crops
by Sarah Otun and Ikechukwu Achilonu

Chapter 4 **45**
Mitigating Climate Change: The Influence of Arbuscular Mycorrhizal Fungi on Maize Production and Food Security
by Sílvia N.D. Maússe Sitoe and Joanna F. Dames

Chapter 5 **79**
Saffron Endomycorrhizae: Diversity and Effect on Plant Growth and Corm Formation
by Samah Ourras, Ismail El Aymani, Najoua Mouden, Soumaya El Gabardi, Manal Adnani, Karima Selmaoui, Mariam Artib, Rachid Benkirane, Cherkaoui El Modafar, Amina Ouazzani Touhami and Allal Douira

Section 2
The Effect of Mycorrhizae on Plant Nutrition and Protection **93**

Chapter 6 **95**
AM Fungi as a Potential Biofertilizer for Abiotic Stress Management
by Malik A. Aziz, Shayesta Islam, Gousia Gani, Zaffar M. Dar, Amajad Masood and Syed H. Baligah

Chapter 7 113
Mycorrhizal Association and Plant Disease Protection: New Perspectives
by Julio Alves Cardoso Filho

Chapter 8 131
Role of Mycorrhizae in Crop Protection
by Stephen Larbi-Koranteng, Frederick Kankam, Joseph Adomako and Muntala Abdulai

Chapter 9 149
Arbuscular Mycorrhizal Fungi (AMF) in Optimizing Nutrient Bioavailability and Reducing Agrochemicals for Maintaining Sustainable Agroecosystems
by Addisu Ebbisa

Chapter 10 171
Ectomycorrhizal Fungi as Biofertilizers in Forestry
by Muhammad Hanif, Zubaria Ashraf, Samar Bashir, Fatima Riaz, Rizwan Amanat, Nousheen Yousaf and Samina Sarwar

Section 3
Mycorrhizae Biology and Development 183

Chapter 11 185
Plant Microbiome and Mycorrhizal Fungi
by YingWu Shi, Hongmei Yang, Ming Chu, XinXiang Niu, Huifang Bao, Ning Wang, Faqiang Zhan, Xuanqi Long, Rong Yang, Qing Lin and Kai Lou

Chapter 12 203
Role of Non-Coding RNAs in Plant Nutrition through Mycorrhizal Interactions
by Nidhi Verma, Yeshveer Singh, Anupam Patra and Tanvi Singh

Chapter 13 225
Receptor Kinases and Signal Pathway in the Arbuscular Mycorrhizal Symbiosis
by Jiashan Wu, Weiyun Wang, Hui Zhu and Yangrong Cao

Chapter 14 247
Diversity of Arbuscular Mycorrhizal Fungi in the Rhizosphere of *Argania spinosa* in Morocco
by Zineb Sellal, Amina Ouazzani Touhami, Jamila Dahmani, Soukaina Maazouzi, Najoua Mouden, Mohamed Chliyeh, Karima Selmaoui, Rachid Benkirane, Cherkaoui El Modafar and Allal Douira

Preface

Microorganisms such as mycorrhizal fungi are of utmost importance because they contribute directly to plant growth and soil health. Mycorrhizae means a symbiotic, mutualistic association between specialized fungi living in the soil and plant roots. We say that it is a symbiotic association since the plants benefit from increased nutrient and water acquisition and at the same time they provide photo-assimilates to the fungi. Most of the known species are symbiotic with mycorrhizal fungi such as dicotyledonous, monocotyledonous, and magnoliids. In addition to all the benefits for plants, mycorrhizae in the soil play several important roles for the ecosystem, such as food source for other animals, cycling and conservation of nutrients through the fungal mycelium, and improvement of soil structure.

Ectomycorrhizae are formed mainly by basidiomycetes and ascomycetes fungi. The mycorrhizae belong to the phylum glomeromycota; they reproduce asexually and form glomerospores as reproductive structures. The mycorrhizae form the mantle and the Hartig net, which is a net of hyphae that extends into the root of the plant, penetrating between the epidermal and cortical cells.

There are several plant growth-promoting mechanisms linked to mycorrhizae. Their direct benefits include improved soil quality and health, better sustainability of ecosystems and efficiency in the use of natural resources, good plant development, and increased tolerance to biotic and abiotic stresses. Regarding nutritional effects in plants, mycorrhizae allow for increased water and nutrient uptake, fine root longevity, and higher mineralization and solubilization of nutrients from minerals. Indirect benefits include greater nutrient turnover and cycling leading to an accumulation of these elements in the plant and roots. Also, there are no nutritional effects such as increased tolerance to heavy metals and pathogens and enhancement of soil quality.

Fungi produce glomalin, which is a protein secreted by the hyphae of arbuscular mycorrhizal fungi; it works like a "super glue," binding soil particles together into aggregates. This protein helps fix soil particles, allowing the formation of stable aggregates. In addition, glomalin sequesters heavy metals, reducing the availability and risk of toxicity of these elements for organisms and plants growing in polluted soils.

Arbuscular Mycorrhizal Fungi in Agriculture - New Insights provides a comprehensive overview of the mechanisms and benefits of mycorrhizae in three sections : "The Role of Mycorrhizae Fungi in Plant Growth", "The Effect of Mycorrhizae on Plant Nutrition and Protection", and "Mycorrhizae Biology and Development".

Rodrigo Nogueira de Sousa
Department of Soil Science and Plant Nutrition,
University of São Paulo,
Luiz de Queiroz College,
Piracicaba, Brazil

Section 1

The Role of Mycorrhizae Fungi in Plant Growth

Chapter 1

Introductory Chapter: Mycorrhizal Fungi – A Current Overview on Agricultural Productivity and Soil Health

Rodrigo Nogueira de Sousa

1. Introduction

It is well known that near the roots of plants, at the zone called rhizosphere, there are several types of microorganisms, whether pathogenic or organic matter decomposing organisms (saprophytes,) or even those organisms that live in symbiosis with the plant as is the case of mycorrhizae [1].

The mycorrhiza is the positive association between the arbuscular mycorrhizal fungi (AMF) and roots of plants, which is named as mutualistic association. Mutualism is an ecological relationship between individuals of different species, in which both benefit from the interaction. Since it occurs among individuals of different species, it is a relationship called interspecific, and due to it benefits everyone involved, it is called a harmonic relationship. As its name says "mycorrhiza" comes from fungi and the "rrizo" that are the roots. Recently, these associations between fungi and roots have drawn attention as a scientific research topic when was discovered the ability of this mycorrhiza in increasing the absorption of nutrients by plants. As an example, the researchers have found a great ability of these fungi to increase the uptake of phosphorus by plants, especially in high-leached and poor soils. It is important to understand that the mycorrhiza will not increase the phosphorus content in the soil, as others solubilizing microorganisms do, these fungi will help enhance the absorption by roots as it will increase the specific surface area of that plant [2]. It is important to have not only the phosphorous (P) but also all essential elements (nutrients) available in the soil so that they can perform their effect in nourishing the plant. For this, and other facts, it is important to combine the biological with the chemical factors in soil for a healthier and more balanced environment.

The AMF are classified into two major groups. The first one is the ectomycorrhizae (ECM), which are those fungi that will not penetrate the plant or will not penetrate the roots of the plants. They will live there around the plant and over the roots, forming a kind of "ball" (called mantle) and Hartig net. This mantle will not only have the function of absorbing nutrients it also protects against some pathogens. This type of ectomycorrhizae association will happen mainly in forest species [2]. Another classification of mycorrhizae is endomycorrhizae. They have the ability to penetrate the roots of plants through the hyphae, which form structures within the cortical cells and also grow intercellularly. Inside the cell, those hyphae swell forming a kind

IntechOpen

of nutrient storage organ, which those elements will be used in the metabolism of that plant [1]. Therefore, the AMF receives carbohydrates provided by roots while the fungus transfers nutrients and water to the plant roots [3].

The endomycorrhizae occur mainly in crops of agricultural interest. These species are symbiotic obligatory since they need this lifestyle to survive. So, this kind of fungus will always have its host root being the connection between the plant's roots and soil. There are several species of fungi to establish this kind of connection and association, in which each species has its own ability to absorb a specific nutrient. Thus, depending on the species, they will absorb a particular nutrient or much more one kind of nutrient than others [1].

The soil system is interesting to have large varieties of these microorganisms, as each one will bring different benefits to the plants according to their specificities. It is important to emphasize that human being can directly or indirectly induce mycorrhizal survival and abundance in the crop (Hartmann et al., [4]). So, as we are working with microorganisms, which are living organisms, it is necessary to give specific conditions for their proper establishment in the soil. Inoculating various beneficial microorganisms such as mycorrhizal fungi in an attempt to bring benefits to the crop without appropriate and healthy soil management will be a waste of time and an expense. We cannot forget that soil health is the continued capacity of soil to function as a vital living ecosystem that sustains plants, animals and humans, and connects agricultural and soil science to policy, thus soil health is not only focused on crop production, but also includes the role of soil in water quality, climate change, and human health [5]. As we can see, the soil is multifunctional and quite complex, so we must seek its most complete balance.

The soil management needs to be adequate for full-plant growth. For example, the excess of fertilizer and pesticides, and low addition of organic matter will affect the soil health, since the microorganisms are directly affected. Also, the monocrop causes negative impacts on the soil microorganisms, since the input of organic residue is limited from a single crop. It is important to keep in mind that the soil is alive and its biological compartment is as important as the chemical and physical factors [2]. Thus, based on what has been discussed, this chapter covers the main positive points under the symbiotic association between mycorrhizal fungi and agricultural crops, in addition, we discussed the main agronomic practices that negatively and positively impact the soil quality.

2. Symbiotic association between mycorrhizal fungi and agricultural crops

Crop production is the first step to produce foods, so it is essential to study the factors that directly affect this production, such as the biological ones. Initially, it is important to mention that mycorrhizal fungi are one of the varieties of fungi of extreme importance for agricultural production due to their great contribution to the nutrient absorption by plants and several other functions. It is known that some fungi species can interact with others organisms leading to the establishment of symbiotic associations, and mycorrhizae are one of these associations [6].

When a fungus establishes an association with the roots of certain plant species, we have the so-called mycorrhizae, which are divided into two major categories: ectomycorrhizae and endomycorrhizae. This classification is based according to the morphological and anatomical aspects of the fungal colonization of plant roots [6]. We can observe this difference in **Figure 1**.

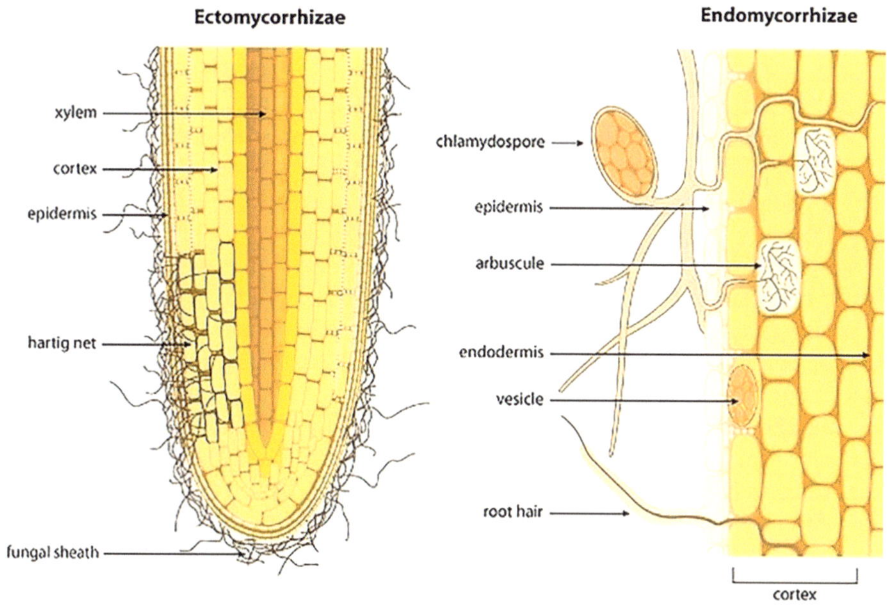

Figure 1.
Ectomycorrhizae and endomycorrhizae scheme of colonization of roots. Source: [7].

Ectomycorrhizae are formed mainly by basidiomycetes and ascomycete fungi, representing about 3% of phanerogams [8], which phanerogams is mainly a subkingdom of the plant kingdom which produce seeds to reproduce. Whereas, in colder regions with cooler temperate, around 90% of forest species present mutualism with this fungus [8].

In tropical regions like Brazil, ectomycorrhizae are more studied and found mainly on economically exploited species such as pine, eucalyptus, and acacia. In ectomycorrhizae, the fungi associated with the root do not penetrate the living cells of the root; and the hyphae grow between the cells of the root cortex, forming a characteristic structure, the Hartig's web (**Figure 1**). The roots of plants associated with ectomycorrhizae are devoid of hairs and their function is performed by fungal hyphae [9].

Endomycorrhizae are more common than ectomycorrhizae and occur in about 80% of vascular plants. The fungi that best represent this association are zygomycetes, which is a class of fungi with more than 1000 known species. The Endomycorrhizae penetrate the cortical cells of the plant roots, where they form very branched structures and chlamydospore (**Figure 1**), which is the defined as a thick-walled, non-deciduous, intercalary or terminal, asexual spore formed by the rounding of a cell or cells whose primary function is perennation, not dissemination [10]. Their hyphae extend for several centimeters in the soil, significantly increasing the amount of nutrients and phosphates essential for plant development.

Many studies have reported that root colonization by mycorrhizal fungi significantly increases the productivity of various plants in low-fertility soils [9]. This is due to the greater uptake of nutrients such as phosphorus, zinc, and copper which are essential for plant development. In turn, the fungus benefits greatly from this association since it can feed on the sugars, amino acids, and other organic sources (photoassimilates) produced by plants through photosynthesis [9].

As already well discussed, mycorrhiza is a symbiosis between plants and soil fungi in which both parties are mutually aided along the way. In this symbiotic relationship, first, organic carbon flows from the host plant to the fungi and then the inorganic elements flow from the fungi structure to the plant, so both parties benefit [11]. Mycorrhizal fungi can only survive when combined with plants because the fungi cannot produce its own food and although it decomposes organic compounds for energy acquisition, it is not enough, it needs to grow and develop together with plants [11].

In crops context, this symbiotic relationship is important as it determines the ability to stimulate plant growth due to higher nutrient acquisition and protection from pathogens attacks such as bacteria and fungi, protecting against diseases. As observed, microorganisms have great potential to increase crop productivity. Dark septate fungi, Pseudomonas bacteria, and bacteriocins are some of the studied microorganisms and substances that can lead to new agricultural bioproducts, as growth promoters or for the control of agricultural pests and diseases [11].

Researchers have demonstrated excellent results with the use of microorganisms in the soil to improve the agronomic efficiency of crops. Some selected bacteria have the ability to solubilize the phosphorus present in the soil and to promote it available to the plants, while others have the function of fixing nitrogen from the air into the roots. In general, soils in tropical regions are mostly deficient in nitrogen and phosphorus, and these fertilizers usually add up to more than 50% of the cultivation cost [11].

As we know, not all applied fertilizer is absorbed by plants, maize absorbs only 55–60% N [12], around 20% P [13], 50–70% potassium (K) [14], and 33% sulfur (S) [15]. One strategy to improve the plants' ability to absorb nutrients and consequently reduce fertilizer application, especially of P, is to inoculate AM fungi. The final response will be a maize plant well-nourished and presenting full growth and development.

Dark septate fungi are promising microorganisms that promote plant growth. At this fungus contact, rice plants grew shoots and roots 30% faster. In addition, tillers, which are new branches of the plant, increased by about 50%, which means greater nutrient uptake and grain production capacity [16]. As we observed, this fungus group also improves the nutritional state of the plants. In tests performed on tomatoes, nutrients such as nitrogen, phosphorus, and potassium accumulated to a greater extent compared to plants without the presence of microbes. In this way, it is possible to optimize plant nutrient use and obtain more vigorous crops [16].

When the farmer decides to use arbuscular mycorrhizal fungus, the plant must be inoculated, as it grows the fungus slowly develops and multiply, which can take many days for them to become fully established. Conversely, with the dark septum fungus use, the inoculum will be established in a few days. Another interesting feature of dark septum is its resistance to water stress. Researchers have observed that in the presence of this fungus, even with a lack of water, the plant developed similarly to another plant grown under normal conditions (without drought stress), confirming the full potential of this microorganism [16].

Research has shown that fungi and bacteria are effective and responsive in the restoration of highly degraded soils. Legume plants associated with these microorganisms have been successfully applied in several regions of mining and impoverished soils to lead to their recovery. As we have seen, microorganisms increase the absorption capacity of the roots, making them more resistant to environmental stress. This technique allows rapid revegetation, even when the subsoil has been exposed. While bacteria provide plants with the nutrient nitrogen, mycorrhizal fungi help plants uptake other nutrients, especially phosphorus and water.

The researchers selected the bacterial strains most effective at nitrogen fixation for each plant species and using microbiological multiplication techniques in the laboratory, it is possible to increase these more resistant and nitrogen-self-sufficient strains.

3. Agronomic practices that negatively impact mycorrhizal fungi

Agronomic practices such as intensive monocrop, soil use with recurrent harrowing, and subsoiling lead to a significant degradation and decrease in soil microbial biomass. There are other important impacts that we should highlight such as intensive grazing, indiscriminate use of pesticides and fertilizers, and mining activities. All these activities are responsible for sick, impoverished, and unproductive soil.

It is important to emphasize that fungi do not always act in a positive way, sometimes these microorganisms are the cause of a large number of diseases. Therefore, it is important to be careful with agronomic practices since if a person without chemical and biological knowledge recommends some wrong strain for inoculation, it may not help, but greatly hinder the production [16]. It is common that many field technicians with the intention of fertilizing the soil to invest in mycorrhizal fungi inoculation, however, it is necessary for a specific professional in the area to make such a symbiotic association between mycorrhizal fungi and agricultural crops.

It is important to say that the control of the disease caused by fungi requires mainly preventive measures since it is necessary to pay attention to the choice of the area to be cultivated until the harvests. This is due to the fact that for chemical control of diseases there are several products available on the market, from different companies.

4. Agronomic practices that positively impact mycorrhizal fungi

The different types of crop residues are very important for healthy soil, not only for supplying nutrients via biogeochemical cycling, but also for functions directly affecting physical, chemical, and biological properties.

Microorganisms are vital to agriculture, especially in two biological processes: (i) Biological efficiency – In this process, certain types of fungi act as solubilizers of soil nutrients, optimizing yields. (ii) Biocontrol – The process by which a group of fungi can do the work of bio-pesticides and biocides to facilitate the management. In addition to their use in agriculture, they can also be used in the medical field, for drug development [17] due to their bactericidal properties. They also can be used to make antibiotics, such as penicillin, whose function is to fight infectious diseases caused by bacteria. Another area where fungi excel is food, some are edible and can be used as part of human consumption [17].

The no-till system (NTS) and crop rotation are essential practices for maintaining good biological soil quality. The no-till farming system is considered a conservationist and sustainable practice, this agricultural practice is a form of soil management that involves techniques recommended to increase productivity while conserving or continuously improving the cropping environment. The main practices are: Absence or minimal soil disturbance. In relation to crop rotation, we can say that is the practice of alternating the plant species grown each season over the years for a more productive system. This conservation agriculture technique aims to reduce soil exhaustion. The crop rotation can benefit soil macro- and micro-fauna, since the richness and abundance of edaphic organisms are determined, among other factors, by the quantity and quality of aerial and root phytomass added to the soil [18].

The crop rotation system combined with no-till farming promotes an improvement in soil quality, as there are erosion reductions, increases in nutrient cycling, and, consequently, better biological activity. The straw has a fundamental function in the no-till farming principle and is highly effective in protecting the soil, avoiding the impact of raindrops, wind, and solar rays. It is also responsible for reducing temperature and water amplitudes, favoring biological activity. In addition, it increases the soil's chemical and physical characteristics such as organic matter content, cation exchange capacity (CEC), and water availability for plants and other organisms.

5. Yield increase X Arbuscular mycorrhizal fungi

Arbuscular mycorrhizal fungi (AMF) are the main microorganisms that compose the majority of the microbial biomass in cultivated soils. About 80% of plants are symbiotic with these fungi, and they are an important factor in improving soil quality and crop growth.

There is a great nutritional benefit for plants influenced by arbuscular mycorrhizal fungi. Phosphorus stands out among the nutrients that combine with the Arbuscular Mycorrhizal Fungi to facilitate uptake. This nutrient is very low in soil solutions due to sorption and precipitation reactions, also immobilization, this kind of behavior is common in heavily weathered soils. AMF hyphae help in accessing this nutrient because they utilize more soil and places that are difficult for plants' roots to access.

There are also reports in the literature that these fungi assist in the uptake of sulfur, potassium, nitrogen, zinc, copper, magnesium, iron, manganese, and calcium. Therefore, plants with high levels of mycorrhizal colonization are expected to be more tolerant to eventual nutrient unavailability and less dependent on the application of external inputs than plants with low symbiosis.

Indirectly, mycorrhizae also improve plant tolerance to nematode and disease attacks on coffee, increasing plant vigor. In addition, improving tolerance to pest and disease attacks is important because these organisms cause significant yield losses [17].

As seen earlier, AMF is an important stress reliever. These microorganisms increase drought resistance because they provide plants with increased water uptake. The increased hydration and stomatal conductance in plants associated with mycorrhizae promote transpiration, the main mechanism for cooling the leaves. Thus, hypothetically, coffee plants with a high degree of symbiosis are better able to withstand a high-temperature summer and suffer less scalding. These fungi also increased tolerance to metal toxicity such as copper, zinc, and aluminum and reduced damage caused by high osmotic pressure.

Topical application of amide and ammonia nitrogen fertilizers is also a major problem in these systems, as excessive acidification of the wetting bulb can increase the solubility of metal micronutrients to toxic levels [18]. Thus, mycorrhizae can be an important stress reliever in irrigated coffee farming improving soil quality the extensive development of extra-radicular hyphae and secretion of glomalin by these fungi are important mechanisms for improving soil physical, chemical, and biological properties [19].

Arbuscular mycorrhizal fungi improve soil structure by connecting particles and forming stabilized aggregates. This is due to the mechanical effects of extra-radicular hyphae growth and the secretion of glomalin. This glycoprotein exhaled by fungi has adhesive properties in the soil, and its content has been correlated with aggregate stability and organic matter. It increases the carbon content in the soil, stimulating biological activity because there is more food for the microorganisms.

6. Conclusions and Future perspectives with the adequate management of mycorrhizal fungi

Although several studies have shown that inoculation in young plants grown in nurseries works well, it is still not a common practice in the production of seedlings of coffee for example, or other perennial plants. Studies applied to plants in the field have shown inconsistent results because most cultivated soils naturally have sufficient inoculum of arbuscular mycorrhizal fungi. The best option to benefit from this partnership is to adopt conservation practices that stimulate microbial activity and endemic populations of AMFs.

The use of harrowing and subsoiling to constantly turn the soil should be avoided. These tools disrupt hyphae networks and expose soil surfaces to high temperatures, promoting the degradation of stable organic matter. Excessive use of herbicides and fungicides also negatively affects the populations of these fungi. Soils that have a high biological activity are usually intensively cultivated, producing large amounts of biomass per unit area. Therefore, it is not advisable to exploit its productive capacity to the maximum in order to preserve it.

Conservationist practices benefit bioactivity and the presence of mycorrhizal fungi in the rhizosphere of plants, this makes the production system more resilient and improves the quality of the soil for cultivation. A significant portion of producers is changing their production system to a model that sees and values the complexities of production systems with their particularities [9].

Abandoning the idea of soil as an inert, lifeless substrate, and that everything can be solved with agricultural chemical inputs, is a big step toward sustainable production, which is certainly necessary to enter an increasingly demanding market. Increasingly, farmers will have a real sense of the importance of keeping the soil environment healthy and balanced for the optimal propagation and establishment of microorganisms such as arbuscular mycorrhizal fungi.

Conflicts of interest

The author declares no conflict of interest.

Author details

Rodrigo Nogueira de Sousa
University of São Paulo, Luiz de Queiroz College (ESALQ/USP), Brazil

*Address all correspondence to: rodrigoagroufv@gmail.com

References

[1] Andrade S, Mazzafera P, Schiavinato M, Silveira A. Arbuscular mycorrhizal association in coffee. The Journal of Agricultural Science. 2009;**147**:1-11

[2] Hartmann A, Schmid M, Tuinen DV, Berg G. Plant-driven selection of microbes. Plant Soil. 2009;**321**:235-257

[3] Cogo FD et al. Arbuscular mycorrhiza in Coffea arabica L.: Review and meta-analysis. Coffee Science, Lavras. 2017;**12**(3):419-443

[4] Adholeya A, Das M. Biofertilizers: Potential for Crop Improvement under Stressed Conditions. 2012;**101**:183-200

[5] Lehmann J, Bossio DA, Kögel-Knabner I, et al. The concept and future prospects of soil health. Nature Reviews Earth & Environment. 2020;**1**:544-553. DOI: 10.1038/s43017-020-0080-8

[6] dos Santos FEFD, Carrenho R. Diversity of arbuscular mycorrizal fungi in an impacted forest remnant-Parque Cinquentenário, Maringá, Paraná, Brazil. Acta Botanica Brasilica. 2011;**25**(2):508-516

[7] Mcnear JR, David H. The rhizosphere-roots, soil and everything in between. Nature Education Knowledge. 2013;**4**(3):1

[8] Meyer FH. Distribution of ectomycorrhizae in native and man-made forests. In: Marks GC, Kozlowski TT, editors. Ectomycorrhizae. New York: Academic Press; 1973. pp. 79-105

[9] Feijen FA, Vos RA, Nuytinck J, Merckx VS. Evolutionary dynamics of mycorrhizal symbiosis in land plant diversification. Scientific Reports. 2018;**8**(1):1-7

[10] Griffiths DA. The origin, structure and function of chlamydospores in fungi. Nova Hewigia. 1974;**25**:503-540

[11] Fiorilli V, Vannini C, Ortolani F, Garcia-Seco D, Chiapello M, Novero M, et al. Omics approaches revealed how arbuscular mycorrhizal symbiosis enhances yield and resistance to leaf pathogen in wheat. Scientific Reports. 2018;**8**(1):1-18

[12] Jamal Z, Hamayun M, Ahmad N, Chaudhary F. Effects of soil and foliar application of different concentrations of NPF and foliar application of (NH4)2SO4 on different yield parameters in wheat. Agronomy. 2006;**5**:251-256

[13] Zhang W, Ma W, Ji Y, Fan M, Oenema O, Zhang F. Efficiency, economics, and environmental implications of phosphorus resource use and the fertilizer industry in China. Nutrient Cycling in Agroecosystems. 2008;**80**:131-144. DOI: 10.1007/s10705-007-9126-2

[14] Bukhsh M, Ahmad R, Ishaque M, Malik A. Response of maize hybrids to varying potassium application in Pakistan. Pakistan Journal of Agricultural Sciences. 2009;**46**:179-184

[15] Cooke G. Fertilizing for Maximum Yield. Granada, London: Granada; 1985

[16] Jobim K, Błaszkowski J, Niezgoda P, Kozłowska A, Zubek S, Mleczko P, et al. New sporocarpic taxa in the phylum Glomeromycota: Sclerocarpum amazonicum gen. et sp. nov. in the family Glomeraceae (Glomerales)

and Diversispora sporocarpia
sp. nov. in the Diversisporaceae
(Diversisporales). Mycological Progress.
2019;**18**(3):369-384

[17] Lammel DR et al. Microbiological
and faunal soil attributes of coffee
cultivation under different management
systems in Brazil. Brazilian Journal of
Biology, São Carlos. 2015;**75**(4):894-905

[18] Franchini JC, Costa JM,
Debiasi H, Torres H. Importância da
rotação de culturas para a produção
agrícola sustentável no Paraná. Londrina:
Embrapa Soja; 2011. p. 52

[19] Lira CRS, Alves MH, Baseia IG,
Berbara RLL, Goto BT, Gusmão LFP et al.
Fungos na Chapada do Araripe. Meiado,
MV Diversidade na Chapada do Araripe,
Ceará. Chapada do Araripe: Meiado; 2014

Chapter 2

Factors Affecting Mycorrhizal Activity

Jamal Salih Alkobaisy

Abstract

Mycorrhizal activity is affected by many factors, including biotic and abiotic factors, and these factors can affect the interaction between the plant and mycorrhizal. In short, these factors are as follows: First—abiotic factors include: 1. Soil fertility, 2. Heat, 3. The light, 4. pH, 5. Salinity. Secondly—biological factors. These factors may include some mycorrhizal effects on plant growth, as well as their effect on encouraging the absorption of nutrients, especially phosphorous and other nutrients, in addition to nonnutritive effects such as growth-stimulating effects, improving the water relationship and plant resistance to drought and plant resistance to salinity, and improving soil structure and biological control.

Keywords: mycorrhiza, mycorrhizal activity, abiotic factors, biological factors, fungi, arbuscular mycorrhizal (AM), active dissemination, passive dissemination

1. Introduction

Plant nutrition workers were interested in studying the effect of mycorrhizal fungi on plant growth, but what is the benefit of studying soil biology unless we have harnessed this biology to maximize production through symbotic relationships. At the same time, the plant host has a role that benefits the living organism so that the state of coexistence continues within the so-called microorganism plant society, which is formed from the state of coexistence between the plant host and the living organism. Some farmers in the world are directed to harness such mechanisms to increase production, such as India, New Zealand, Canada, Australia, and England, especially countries that suffer from a scarcity of energy sources, and the reason for this is to think about finding alternatives to compensate for the added fertilizers or reduce the cycle of fertilizer production, However, in this case, it is necessary to increase the pollen density, where the organisms are located within the population, which may affect the production process, but this pollen density may be weak to the extent that the effect of the organism on the plant does not appear, so the pollen density of a particular organism must be increased and then provided. This vaccine is more efficient in order to get what is called maximizing production.

2. Ecology of mycorrhiza

The mycorrhizal fungi are distinguished by their living in different environments, including tropical rainforest and sandy desert ones. Many factors affect the spread of mycorrhizal, as well as the processes of sporulation formation, and the way they coexist with mycoflora or other soil revivals, root colonization, colonization formation. We find ourselves in an urgent need to understand these stages of the presence of fungi as we approach the practical aspect of these fungi. Therefore, it is customary to express the presence of fungi with a term called dispersal. While concerning this process to such fungi, studies indicated that they were present in different continents before the flood and drought process occurred. As for the important study in mycorrhiza, the plant fossils proved that they contain a type of fungi and consist of structures similar to some extent to the mycorrhizal structures (vesicular—dendritic), which he led the researchers to say that these organisms were formed on the roots of forest plants and some natural plants. According to this, what Frank [1] made of his diagnosis of mycorrhizal fungi is a reality and a fact, but it did not attract the attention of scholars in that period or era.

2.1 Means of spreading mycorrhiza

Several ways to spread it have been suggested:

2.1.1 Active dissemination

It can appear from one area to another as a result of the growth of hyphae. It was found that the rate of mycorrhizal spread varies according to the density of these soils with these fungi, as well as the plant host that coexists with it. It has been proven that the rate of spread of these fungi reaches within 65 m per 150 years, which means 0.43 m/year, and it was clarified that the soils containing these fungi have a higher prevalence rate than the soils that do not contain these fungi. The type of plant and root density plays an important role in the spread of mycorrhizae. It was mentioned that the soils planted with subclover pasture plant in Australia have the prevalence of mycorrhizal *Glomus fasciculatum* up to 1 cm per week, but in another field when the vegetation cover prevails in it when one of the weeds was found. The spread of this type of mycorrhiza decreases and reaches about 0.07 cm, and from this we can show that the density of roots plays a key role in determining the spread of mycorrhizae, especially in the young stages of plant growth, where we find that the vegetation cover of weeds had a low prevalence rate, but the prevalence rate increased by changing the second type to clover because there is a wide difference in the root system of both plant hosts.

Poter [2] worked on the distribution of mycorrhizae with the type of soil when he took the mycorrhizal vaccines and added them to soil A. He noticed that the growth of hyphae as well as the infection and formation of spores differed significantly with other soil B, and at that time he attributed the reason to that there is a possibility that soil A in it encourages factors for the growth and formation of spores of this fungus compared to soil B, then he started transferring samples of mycorrhizae to soil C and noticed a strange observation than this, which is that mycorrhizae cause infection but do not form spores. As for soil B, after taking the inoculum from it and transferring it to soil C, the mycorrhizal system lost the ability to form spores, but it kept the infection, and it was found that there are three factors affecting the distribution of mycorrhizae in the soil, namely:

A. pH reaction number: it was found that the pH in soil A was within the biological pH number that enabled these fungi to grow, but the transfer of the inoculum from A to B and then to soil C became the biological reaction number eligible for infection and not eligible for spores to occur., Perhaps the process of spore formation is the reliable process in the production of the vaccine, so we can say that the soil must be chosen very carefully if I want to obtain mycorrhizal vaccines that contain significant amounts of spores per gram of soil.

B. light: the process of formation of spores in fungi is affected by a major interfacial factor, which is light. When the periods of illumination are compatible with the crop or plant host, we get the highest percentage of sporulation, but if the photoperiod is not compatible with the plant host, the process of sporulation formation may decrease or progress.

C. root density: the root density plays an important and major role in the spread of mycorrhizae. The most important studies on this subject are those of Poter [2] and Bolan et al. [3], who indicated that the plant host with a coarse root system is susceptible to infection with this type of fungus that is more than a soft root system.

2.2 Passive dissemination

It has been said that the dissemination within this method depends on the movement within the wind or water, but the surrounding environmental conditions will be a determining factor for the spread of the vesicular arbuscular mycorrhizae (VAM) fungi in this way. The plant organism, which is in the soil in many cases, has some characteristics that enabled it to spread, and one of these characteristics is that it can be the so-called spor capes structures, which are the structures of a group of spores present in one part. A spor cape is a manufacturer of large numbers of blackboards. There are indications from researchers about the free spread. It has been said that birds and insects have a role in the process of migration, through their movement within the remnants of these organisms and their spread from one area to another.

2.3 Factors affecting mycorrhizal activity

Mycorrhizal activity is affected by many factors, including biotic and abiotic factors, and these factors can affect the interaction between the plant and mycorrhizal.

In short, these factors are as follows:

2.4 First - Abiotic factors

There are many abiotic factors that affect the activity of mycorrhizal fungi, including soil fertility, temperature, pH, soil moisture and aeration, soil type, salinity, plant readiness for infection, soil organic matter, and others.

Abiotic factors include:

2.4.1 Soil fertility

Soil nutrients, especially phosphorous, are one of the most important abiotic factors that affect the mycorrhizal fungi. Both nitrogen and phosphorus significantly affect root colonies when they are present at high levels. Therefore, the state of balance for these two elements is a required condition, that is, the nutritional needs of the family. The plant and the nutritional needs of the living organism should not be within the critical sufficiency limit because this simply means entering into a kind of competition for the source of food, and this explains that some plants vaccinated with VAM may decrease their dry weight or growth rate because the VAM and the plant are included in this type from the competition so that the rate of net photosynthesis is insufficient for both the VAM and the plant, and thus the interference changes from Mutualism (+, +) to (−, +) Parasitism.

There is some information about the negative effects of nitrogen fertilizers on the formation of mycorrhizae, and it was found that nitrogen (ammonia nitrate) clearly reduces both the infection of mycorrhizae and the number of spores in wheat fields. As plants fertilized with high ratios of ammonium to nitrate have a higher phosphorous content in their tissues than plants fertilized with low ratios of (NH_4^+) to (NO_3^-), and these high concentrations of phosphorous in plant tissues inhibited infection. The reason for the inhibitory effect of ammonium may be attributed to the low pH in the area rhizosphere and to see the effect of fertilizers on mycorrhizae. The initial fertility of the soil must be known, because in poor soils the production of spores will be limited to the total quantity and not the percentage of infected roots derived from plant growth.

As for phosphate fertilizers, some studies have shown that these fertilizers have negative effects on the internal mycorrhizal fungi. Increasing the processed phosphorous may reduce the infection of mycorrhizae to levels that are insufficient to encourage the absorption of other elements. Phosphorus and zinc of the shoots of pollinated plants grown in low fertility soils were more than that of unpollinated plants. Increasing the level of ready phosphorus is an inhibitor of the growth of mycorrhizal, unlike insoluble forms such as rock phosphate, which is not considered an inhibitor. The results on wheat plants confirmed this, as it was found that the addition of phosphorous levels of 60, 120, and 240 kg phosphorus/ha led to a reduction in the percentage of infected roots in the fertilized treatments compared to the non-fertilized treatments, where the level led to 240 kg phosphorus/ha to the absence of infection significantly and the removal of the beneficial effect of the mycorrhizal infection.

The reason for the decrease in mycorrhizal infection as a result of the increase in phosphorous levels was shown by Cooper [4] that under conditions of phosphorus deficiency, the amount of phospholipids in the membranes of root cells decreases, leading to an increase in the permeability of these membranes, and this leads to an increase in the root secretion of reducing sugars. And amino acids lead to the formation of mycorrhizae, thus increasing the percentage of infected roots, but under conditions of availability of phosphorus, and the permeability of the membranes of the roots cells decreases due to the increase in phospholipids in them, and as a result, the secretions of the roots decrease from reducing sugars and amino acids, and this leads to a decrease in the percentage of infected roots. The decrease in the rate of infection may also be due to the increase in the concentration of phosphorus in the tissues of the plant, and the reason can be attributed to the fact that high levels of phosphorus may reduce the concentration of carbohydrates in the roots of plants, and as a result, the rate of infection is reduced. In general, high soil fertility leads to less mycorrhizal infection, so it is unlikely that we will find many mycorrhizae in densely cultivated soils. However, some crops are highly infested with fungi even in very fertile soils, as mycorrhizae are found in all poor and rich soils. Therefore, a low level of fertility is not always a condition for a significant development for mycorrhizal.

2.4.2 Temperature

Studies have shown that temperature has an effect on the formation of spores and colonies in greenhouse conditions, as the temperature usually affects the increase of colonies and thus increases the spores. The ideal temperature of mycelium on the surfaces of the roots will be 20–30°C. As for the formation of spores and the species that form the spore cyst, it will be at its strongest at 35°C. Studies have shown that the succession and decrease of temperatures increases the formation of colonies as well as the spores. Yoh-ichi Matsubara et al. [5] found that after 7 weeks of inoculation at a temperature of 20/25°C, the infection level of *Gigaspora margarita* in roots was 63.0 and 20% in *Glomus sp.* RIO, and the infected plants gave with fungi the highest values of plant height, dry weight of the vegetative part, and phosphorous concentration in the vegetative and root parts compared to unpollinated plants, as the effect was more pronounced in *Gigaspora margarita* than *Glomus sp.* RIO, but after 11 weeks from vaccination and when the temperature drops to 15°C, the infection level was recorded at 48.9% in *Gigaspora margarita* and 58.9% in *Glomus sp.* RIO. The plants infected with the fungus showed the highest values in all studied traits compared with the uninoculated ones. While after 11 weeks of insemination and at a temperature of 30°C, the infection level was 66.3% in *Gigaspora margarita* and 36.7% in *Glomus sp.* RIO.

2.4.3 Light

Light can indirectly affect soil microorganisms through its effects on plants, whose photosynthetic products are released from the roots [6]. The penetration of light through soil is important because of its effects on factors of ecological significance, such as spore germination, root growth, fungal growth, and formation of mycorrhizal and leguminous nodules. Light penetration can be affected by soil moisture content, soil type, cover material, and particle size. Phytochromes that are biliprotein photoreceptors enable some microorganisms to adapt to the light regime in the soil [7].

Fungi are unable to use light for photosynthesis; however, radiation plays a role in the biochemical and morphological responses of some fungi such as *Phycomyces blakeslleanus*, including their growth and differentiation. Physiologically and ecologically, a significant amount of light penetrates the soil approximately 4–5 mm from the surface, eliciting some phototrophic responses in plant roots. This information has led some VAM experts to hypothesize the function of LED on VAM formation. The induction of hyphal growth by light and chemicals, for example, the effect of blue light on hyphal branching, has been reported [8]. These authors demonstrated that blue light and some exudate components effectively stimulate hyphal branching, suggesting the involvement of a second messenger responsible for this synergism. The photo-induction caused by photo-mimetic compounds has been studied in many other fungi as well [9].

It is important to assess some environmental factors that stimulate hyphal growth and sporulation, such as root exudates, and LEDs applied individually and in combination. Lighting from red LED or red+blue LED could stimulate hyphal growth in *G. margarita* and *Glomus* spp. (R-10) *in vitro* [10]. Moreover, VAM colonization of corn roots was improved when the rhizosphere was exposed to light. The marginal VAM colonization of chalk false-brome [(*Brachypodium pinnatum* (L.) P.B.)] under shade conditions could show that when low light limits photosynthesis and thus growth of the plants, they dispense with the colonization of VAM in order to save the expenditure of organic carbon [11].

Previous studies have reported the effects of blue light on hyphae; however, its synergistic effect with root exudates on the production of new spores with minimum soil residues is still unclear. In this factor (light), two variables must be distinguished: (1) the period of illumination and (2) the intensity of illumination. The length of the illumination period is 32 hours, and the length of the illumination period is more important than its intensity in the formation of colonies. The growth of mycorrhizal onion plants was under light intensity of LUX 25,000 compared to LUX 13,000 and for 16 hours of light (23°C) with 8 hours of darkness (14°C) [12, 13].

2.4.4 The pH

Studies have shown that the reaction number pH has a significant effect on the mycorrhizal fungi, as well as the type of VAM plays a role in determining the appropriateness of the type of interaction, as it was found that *G. mosseae* excels in alkaline soils, most of the soils of Iraq work on this quality because it prevails in alkaline soils, and it can germinate its spores at a certain extent a broad range starts from 6 to 9, while we find that the species *Gigaspora coralloida*, which was isolated from acidic fluoride soils, its spores germinate at pH 6–4, and according to this we find that the *G.mosseae* is more suitable for alkaline media than the *Gigaspora coralloida*, and it was found that the type *G.epigaeum* has the ability to grow its spores in a wider range than the previous two species, which is from 6 to more than 8. On the other hand, it was found that both types of mushrooms *G. intaradices* and *G.mosseae* can grow in different types of soils, but they grow better in neutral and alkaline soils, while the fungus *Gigaspora margarita* prefers acidic soils. Therefore, it can be said that the pH affects the germination of VAM spores, and it is worth noting that there is a kind of fit between the pH appropriate for the growth of the plant host and the pH appropriate for the growth of the VAM.

To test the response of arbuscular mycorrhizal (AM) fungi to a difference in soil pH, the extraradical mycelium of *Scutellospora calospora* or *Glomus intraradices*, in association with *Plantago lanceolata*, was exposed to two different pH treatments, while the root substrate pH was left unchanged. Seedlings of *P. lanceolata*, colonized by one or other of the fungal symbionts, and nonmycorrhizal controls, were grown in mesh bags placed in pots containing pH-buffered sand (pH around 5 or 6). The systems were harvested at approximately 2-week intervals between 20 and 80 days. Both fungi formed more extra radical mycelium at the higher pH. *Glomus intraradices* formed almost no detectable extraradical mycelium at lower pH. The extraradical mycelium of *S. calospora* had higher acid phosphatase activity than that of *G. intraradices*. Total AM root colonization decreased for both fungi at the higher pH, and high pH also reduced arbuscule and vesicle formation in *G. intraradices*. In conclusion, soil pH influences AM root colonization as well as the growth and phosphatase activities of extraradical mycelium, although the two fungi responded differently [14].

2.4.5 Salinity

Several studies indicated that salinity has an important effect on the percentage of mycorrhizal infection and spore germination. These studies showed that there is a negative correlation between the incidence of infection, the number of spores, and the soil content of sodium when the Na concentration range is reached from 153 to 11,600 ppm, and it has been observed that 1 VAM fungi completely disappear when the concentration of Na increases to 3181 ppm. There are many studies that dealt with this subject, which showed that the salt concentrations are higher than 4 dSiemens. m2 in the root zone of the plant has caused a significant reduction in the infection rate for all mycorrhizal fungi, whether the addition of mycorrhizae with seeds or with seedlings, but the fungus *G. mosseae* recorded the highest rate of infection as an average at a higher salinity level than the rest of the species. The roots of tomato plants have caused a significant reduction in the incidence of infection with each type of mycorrhizal fungi and in the two ways of adding the inoculum (with seeds and with seedlings), area M. However, the fungus *G. mosseae* recorded the highest rate of infection as an average at each salinity level compared to the rest of the types and with both methods of additions to vaccines.

The process of spores of mycorrhizal fungi goes through four phases: the hydration phase, the activation phase, the emergence phase of the germination tube, and the hyphae growth phase. The failure of one or more of these phases due to high concentrations of dissolved salts in the soil solution may delay or stop the growth and development of the host plant. Three sources of salts (sodium chloride, sodium nitrate, and potassium chloride) with different concentrations were used to study their effect on spores' germination components of five types of endophytic mycorrhizae in three separate experiments. The fungus *G. mosseae* was significantly superior to the rest of the spores' germination components of other fungi, and the percentage of mycorrhizal spores germination decreased significantly when exceeding the critical concentrations of sodium and chlorine ions. Germination or inhibition of the germination process is by the toxic effect of sodium and chlorine ions. The toxic effect of sodium ion was more than the toxic effect of chlorine ion on the germination process, and in a study by Al-Khaliel [15], he used five levels of calcium carbonate (20, 30, 40, and 50%) and five levels of calcium sulfate (5, 15, 25, and 50%) to find out their effect on the components of spore germination of five types of mycorrhizal fungi. The results showed that the high levels of calcium carbonate (40 and 50%) and calcium sulfate (15, 25, and 50%) reduce the percentage of germination and other components of germination, but the effect differs according to the Mycorrhizae genus Glomus and Gigaspora, as well as between species within the genus Glomus, and the critical level of calcium carbonate is 20–30% and for calcium sulfate is 5–10% which after it is exceeded, the percentage of germination and other components of germination decrease, and they attributed the reason for the decrease in the percentage of germination to the lack of availability of nutritional needs of phosphorus and other elements due to the high percentage of calcium carbonate and sulfate, as well as due to the harmful effect of calcium and sulfate ions in the soil.

On the other hand, there are studies that showed that mycorrhizal fungi naturally appear in saline environments, despite the little affinity between mycorrhizae and halophytic plants such as hypophysis. The results of the research were that 21 of the 89 plants of the halophytic species were infected with VAM fungi, which indicates that the VAM works within saline concentrations, and 11 of the 89 plant species developed spores in the rhizosphere. This is sufficient evidence that different plants of 21% have an infection and 11% have an infection and spores. Al-Khaliel [15] found that the number of mycorrhizal spores did not decrease significantly with the increase of soil salinity, and the rate of spores was 100 per 100 g of dry soil, and that the majority of mycorrhizal species that were found in the soil of the plain whose salinity reaches 160 dSm are *G.etunicanum*, *G.versifform*, and *G.intraradices*, and this was attributed to stimulating the formation of spores under salt stress, meaning that the mycorrhizae is the reason for the production of spores at low levels of root infection under saline stress conditions, which inhibits the formation of mycorrhizae from spores and then the accumulation of spores in the soil.

2.4.6 Second: Biological factors

The biological interaction between mycorrhizal and other organisms took the space and thought of many researchers, and the matter of the fact is that the first to approach this topic and define the characteristics of the region in which this interference occurs is Hiltner [16] when he called the region surrounding the root hairs the rhizosphere region, which is the region affected by many factors, including the number of living organisms and growth and secretions of the roots, and he identified in an unequivocal way that the activity of organisms in this area is at its highest, and therefore it is not surprising that it is said that it is the key to microbiology to take its scope in maximizing plant production. And some are reduced, and some are free, and some are restricted. When there is an agreement between the plant and the living organism, the response is the greatest, and when there is no agreement between the plant and the living organism, the ability of the plant to grow and absorb is determined, and therefore the living organisms do not play the required role and mediate to facilitate the elements. In this area, there are mucilage, sloughed cells, and a wide quantitative production of enzymes, hormones, and growth regulators, in addition to antibiotics and multiple sugars, and all of these are diagnosed and studied, and thus the rhizosphere area has an applied importance from an agricultural point of view because it represents the true cradle of seeds and the roots in them and the increase in the activity of organisms in them. The interference between organisms, which is the effect of one organism on the activity, growth, and reproduction of another organism, and the interference is either:

1. Antagonistic: its results are inhibiting against the number of cells when the organism is bacterial and may be inhibiting the formation of spores in fungi.

2. Positive: when an organism stimulates some of the activities of an another living organism, such as mycorrhizalgia, it increases the number of spores of another fungus.

The unsatisfactory interactions that are of different types, including:

1. Simple overlap (monogamy), peaceful coexistence: it includes Micro- and Micro-plant, such as:

    a. Rhizobia and leguminous plants (symbiotic specialty).

    b. Mycorrhizae and plants (nonspecialized).

    c. Azotobacter or azospirilm and the plant (associative).

2. Unsatisfactory complex bilateral interference (Micro- and Micro-plant), such as Mycorrhizalgia + Azotobacter + Plant.

There are some organisms that are involved in the so-called Commelizan effect when two organisms are present with each other, so one of them stimulates the other. For example, the presence of mycorrhizae with Azotobacter increases the rate of cytoplasm release and thus enables plants to live in low levels of iron with no symptoms of deficiency. Many researches and studies indicated the effect of various microorganisms groups, especially bacteria, on the germination and growth of arboreal mycorrhizal fungi. Bacteria increase mycorrhizae by removing the inhibitor, such as self-inhibitors of the fungi spores or the production of chemicals that stimulate the growth of the spores, as well as an increase in plant secretions by increasing the permeability of the membranes and thus an increase in the growth of fungi and their ability to penetrate the host (we will address the relationships between mycorrhizae and other soil microbiota in other topics from this chapter). It was found that mung bean yellow mosaic virus reduced the rate of mycorrhizal infection and the production of spores by the fungi *Glomus constrictum*, *G. Fasciculatum*, and *Acaulospora morrowede*, but fungi *Gigaspora gilmorei* could not form any mycorrhizal roots in plants infected with the virus compared to healthy plants.

There have been many attempts for years that have been concerned with studying the state of competition between the mycorrhizae introduced into the growth medium (the rhizosphere) and the endemic mycorrhizae. Taking into account a number of variables, the most important of which is the rate of infection in the roots and then the calculation of the mycorrhizal roots, as well as knowledge of the formation of spores after adding VAM to the growth medium. Accordingly, the type of added vaccines will have two types of effects: either a positive effect or a negative effect (competition), the positive effect may increase the incidence of infection and spores and increase the production of spores due to the positive effect of one organism on another organism. Studies have shown that there is another type of competition that arises between the types of VAM introduced into the growth medium. And the technique that determines how this antagonism occurs between organisms is either the use of biochemistry and reliance on the metabolites secreted from living things and determining their structures, or the use of genetic engineering to determine this type of antagonism. Mycorrhizal fungi coexist with multiple plant families, and there are other families with a low tendency for this type of symbiosis, and most families have a high susceptibility to mycorrhizal infection, but some species such as cruciferous and ramiform do not infect mycorrhizal. Studies have shown that the history of vegetation cover in a region may participate in determining the density of mycorrhizal presence and in many times the supremacy of one sex over another.

3. Conclusions

The mycorrhizal fungi are distinguished by their living in different environments, including tropical rainforest and sandy desert ones. Many factors that affect the effectiveness of mycorrhiza, as well as the processes of sporulation formation, and the way they coexist with mycoflora or other soil revivals, root colonization, have been studied. We find ourselves in an urgent need to understand these stages of the presence of fungi as we approach the practical aspect of these fungi. Therefore, it is customary to express the presence of fungi with a term called dispersal. With regard to this process of such fungi, studies indicated that they were present in different continents before the flood and drought occurred. Since that time, there are many factors that limit the existence and effectiveness of mycorrhiza, and the most important of which are environmental factors and other biological factors (**Figure 1**).

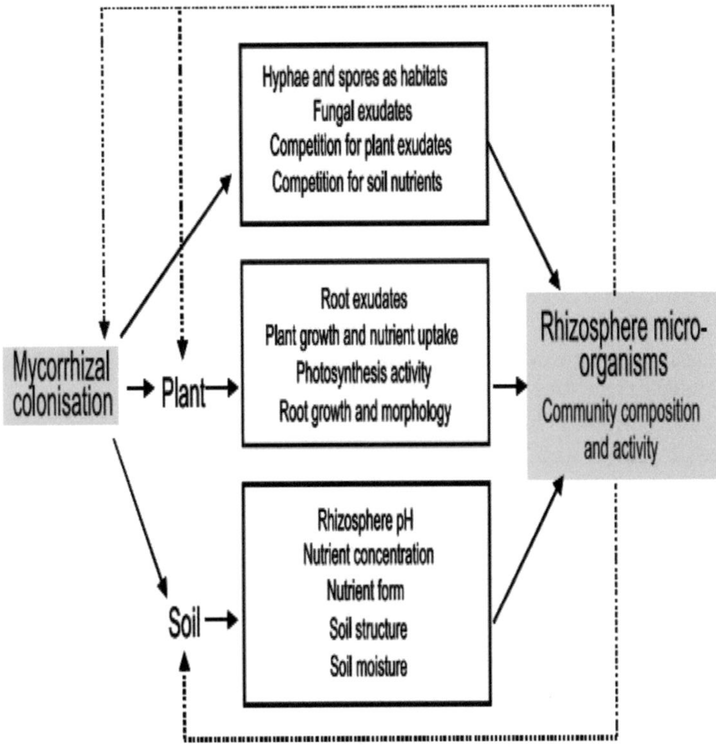

Figure 1.
The direct and indirect effect of mycorrhiza on the activity of other microbiology in the rhizosphere [17].

Author details

Jamal Salih Alkobaisy
Department of Soil and Water Sciences, College of Agriculture, University of Anbar, Anbar, Ramadi

*Address all correspondence to: ag.jamal.saleh@uoanbar.edu.iq

References

[1] Frank AB. Uber di auf werzelsymbiose beruhende Ernahrung gewisser Baume durch unterirdischeplize. Berichte der Deutschen Botanischen Gesellschaft. 1885;**3**:128-145

[2] Poter WM. The moet probable number method for enumerating infective propagules of VA- Mycorrhizal fungi in Soil. Australian Journal of Soil Research. 1979;**17**:515-519

[3] Bolan NS, Robson AD, Barrow NJ. Increasing phosphorus supply can increase the infection of plant roots by VA-mycorrhizal fungi. Soil Biology and Biochemistry. 1984;**16**:419-420

[4] Cooper KM. In: Powell CL, Bagyaraj DJ, editors. Physiology of VA-Mycorrhizal Association in VA Mycorrhiza. Boca. Raton, Florida: Press. Inc.; 1984. pp. 155-186

[5] Matsubara Y, Kayukawa Y, Fukui H. Temperature-stress tolerance of asparagus seedlings symbiosis with arbuscular mycorrhizal fungus through. Journal of Japanese Society for Horticultural Science. 2000;**69**(5):570-575

[6] Dehlin H, Peltzer D, Allison V, Yeates G, Nilsson M, Wardle D. Tree seedling performance and below-ground properties in stands of invasive and native tree species. New Zealand Journal of Ecology. 2008;**32**:67-79

[7] Rottwinkel G, Oberpichler I, Lamparter T. Bathy Phytochromes in Rhizobial Soil Bacteria. Journal of Bacteriology. 2010;**192**:5124-5133

[8] Nagahashi G, Douds D. Synergism between blue light and root exudate compounds and evidence for a second messenger in the hyphal branching response of Gigaspora gigantea. Mycologia. 2004;**96**:948-954

[9] He Q, Cheng P, Yang Y, Wang L, Gardner K, Liu Y. White collar-1, a DNA binding transcription factor and light sensor. Science. 2002;**297**:840-843

[10] Ishii T, Yachi Y, Oida K, Matsumura A, Horii S. Improvement of axenic culture of arbuscular mycorrhizal fungi. In: International Congress on Mycorrhizae, Montreal-Canada. 2003. p. 696

[11] Fuzy A, Bothe H, Molnar E, Biro B. Mycorrhizal symbiosis effects on growth of chalk false-brome (Brachypodium pinnatum) are dependent on the environmental light regime. Journal of Plant Physiology. 2014;**171**:1-6

[12] Akiyama K, Matsuzaki KI, Hayashi H. Plant sesquiterpenes induce hyphal branching in arbuscular mycorrhizal fungi. Nature. 2005;**435**: 824-827

[13] Cruz AF. Effect of light-emitting diodes on arbuscular mycorrhizal fungi associated with bahiagrass (Paspalum notatum Flügge) and millet [Pennisetum glaucum (L.) R.Br]. Bioagro. 2016;**28**(3):1-5

[14] Ingrid MA, Pål AO, Bengt S. Arbuscular mycorrhizal fungi respond to the substrate pH of their extraradical mycelium by altered growth and root colonization. New Phytologist. 2002;**155**:173-182

[15] Al-Khaliel AS. Effect of salinity stress on mycorrhizal association and growth response of peanut infected by Glomus mosseae. Plant Soil Environment. 2010;**56**(7):318-324

[16] Hiltner L. Uber neuere Erfahrungen and problem of dem Gebirt der Bonden bakteriology and unter besonderer Berucksichtigung der Grundungung and Brache. Arbeiten der Deutschen Landwirtschaftlichen Gesellschaft. 1904;**9**:59-78

[17] Mukerji KG, Manoharachary C, Singh J. Microbial Activity in the Rhizosphere, Soil Biology. Vol. 7. Berlin Heidelberg: Springer-Verlag; 2006, 2006 dual alteration to root longevity of poplar. Plant Soil 172: 327-329

Chapter 3

Recent Advances in Plant: Arbuscular Mycorrhizal Fungi Associations and Their Application to Cassava Crops

Sarah Otun and Ikechukwu Achilonu

Abstract

According to estimates, the world's population is growing at 0.96% yearly, meaning that there will be approximately 7.3 billion people on earth by the year 2050. Consequently, the agricultural sector is demanded to boost production and provide food security for the rising world's population. Unfortunately, almost 40% of the arable land has been damaged by several factors, such as industrialization, suburbanization, acidification, salinization, and erosion of the soil, environmental pollution, among others, resulting in a global agricultural and economical problem. However, several land recovery techniques have been developed over many years of research, such as the use of chemicals, cultural techniques, and Arbuscular Mycorrhizal Fungi (AMF). AMF forms a vital connection with the host plants and the soil nutrients and assists in the restoration of damaged agricultural lands. This reviews' objective includes (i) providing a brief overview of AMF; (ii) highlighting AMF's role in nutrient management; (iii) reviewing the roles of AMF in the regulation of plant (cassava) development; (iv) explaining the role of AMF in managing abiotic and biotic stressors; (vi) emphasizing the role of AMF in reducing greenhouse gas emissions, and (vi) highlighting significant areas within the study of AMF-cassava that has not yet been completely explored.

Keywords: arbuscular mycorrhizal fungi, beneficial soil microorganisms, cassava, soil nutrients, sustainable agriculture, nutrient-loss/uptake, abiotic/biotic stressors

1. Introduction

A symbiotic relationship between the root of the host plant and the fungi and spreads into the rhizosphere and the surrounding soil is known as mycorrhiza [1]. Mycorrhizal symbiosis has developed into a specialized area since Frank first used the term 136 years ago [2]. The scientific community aims to understand its characteristics and ramifications for both plants and fungi and other microorganisms like bacteria [1]. Mycorrhizae are now recognized as a type of "biological fertilization", mostly because of their presence in almost all healthy plant roots [3]. Biofertilization has been presented as the bio-sustainable substitute for chemical fertilization, because of its variety of possible benefits, including fertilizing the host plants with nutrients, protection from biotic (pathogens), abiotic (drought, unfavorable temperature, amongst others) [1, 3].

Mycorrhiza, as a crucial functional group of the soil biota, provides various nutrients and aids the host plant in battling against unfavorable soil conditions (for instance, in drought settings it aids in increasing the surface of the roots) [4]. Endomycorrhizal fungi and ectomycorrhizal fungi are two different forms of mycorrhizal fungi that form a root biotrophic relationship [5]. In contrast to ectomycorrhizal fungi, which develop around the surrounding root cells, endomycorrhizal fungi colonize the root cells of plants [5].

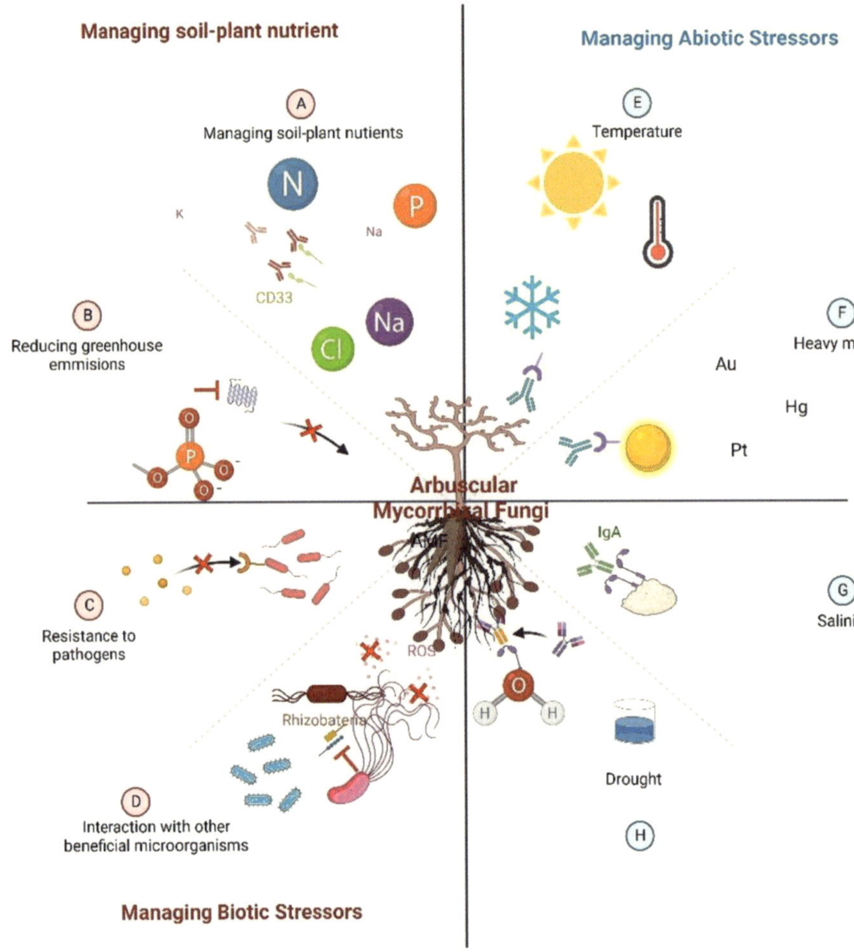

Figure 1.
An illustration of the potential for arbuscular mycorrhizal fungal colonization in managing soil–plant nutrients, biotic stressors, abiotic stressors, and reduction of greenhouse emissions.

AMF has been reported to be present in 80% of terrestrial plant species, including economically important crops like wheat, tomatoes, cassava, etc. [6]. Cassava is a crucial staple crop that produces starch-rich, tuberous roots, especially in sub-Saharan African smallholder farming systems [7]. Partly due to its resilience to drought, it has earned the title "drought', 'war', and famine' crop" [8]. Although cassava yield decreases noticeably during dry seasons, its relationship with and reliance on AMF may be partially responsible for its relatively strong resistance to drought [9]. Approximately 80% of angiosperms have connections made by AMF with their roots and exchange carbon molecules for water and vital nutrients including phosphorus and nitrogen for the host plant [10]. Additionally, they give increased resistance to both biotic and abiotic stressors [9, 10].

There are reports that AMF plays a significant role in improving cassava productivity [8–10]. Hence, cassava growers could considerably benefit from this technique because cassava is a crop that relies heavily on mycorrhizal fungi for the extraction of nutrients. On the symbiotic interaction between AMF and cassava, the most recent research trends were covered in this review. In addition, we examined several recent studies and extrapolated findings to improve the uses, advantages (**Figure 1**), and limitations of using AMF in cassava cultivation. The objective of this review is to highlight recent discoveries in a plant-AMF relationship that are associated with cassava crop production. This will contribute to developing a lasting solution to the global problem associated with poor quality and quantity in cassava production, hence solving the problem of hunger and poverty globally, especially in Southern Africa.

2. Database retrieval strategy

The literature review was done electronically utilizing 'Google Scholar', 'PubMed', 'Scientific Electronic Library Online (SciELO)', 'cassavabase', and the 'Scopus databases. The articles to be examined were initially chosen using the following keyword combinations: Arbuscular mycorrhizal fungi, plant-pathogen relationship, cassava mycorrhizae; plant pathogen; abiotic factors; biotic factors, salinity; heavy metals. The majority of the chosen research articles (80% of all references) were published between 2018 and 2022. Given its relevance in the history of mycorrhizae, an old publication from 1885 was added.

3. The significance of arbuscular mycorrhizal fungi (AMF)

AMF is the most common type of endophytic fungus to colonize the root of its host [6]. Their positive impacts on the survival of their host are widely documented, and these effects include but are not limited to nutrient uptake, assisting with soil aggregation, and resilience to biotic and abiotic stress [6–9]. As a result, AMF symbioses have a major impact on the yield of the host and the health of the surrounding soil in both natural and laboratory settings [10]. According to Salomon et al. [11], certain AMF species can colonize a host plant and increase the number of its mycorrhizae in organic soils. The AMF's hyphae graze on the soil's water and mineral deposits before passing them on to the host plant, ensuring plant production and diversity even in soils low in micronutrients like phosphorus (P) [12]. The root system then passively refluxes these nutrients to the fungus [11]. Consequently, utilizing the soil microbial communities can result in the most sustainable and healthy crop production method. This will help to protect the biosphere by enhancing not just the soil's fertility even in adverse weather, as well as the plants' nutrition and health. When these potentials are combined, they would support agriculture and boost global food security [12, 13].

Studies on the interaction of AMF and cassava plants in responding to abiotic and biotic stressors are scarce, despite the extensive study on AMF symbiosis. However, due to the rising and unavoidable stress conditions (with biotic and abiotic factors), it is challenging to meet the global food needs. This study consequently discusses the role of AMF in controlling plant nutrition and growth in response to biotic stresses (plant diseases) as well as abiotic stresses including salinity, water scarcity, floods, high and low temperatures, soil acidity, and soil management, in addition to the involvement of AMF in the sustainable soil management nutrients.

4. The role of AMF in nutrients management

The arbuscular mycorrhizal fungus can enhance the health of the soil and the plants by synthesizing metabolites and plant growth hormones, increasing the availability of essential vitamins to the plants in nutrient-deficient soils, and providing other ecological services. All these factors are further discussed.

4.1 The function of arbuscular mycorrhizal fungi in soil nutrient uptake

Numerous macro- and micronutrients are required by plants, including nitrogen (N), sodium chloride [NaCl], potassium (K), copper (C), calcium (Ca), iron (Fe), and zinc (Zn), amongst others [14]. Plants often take up these nutrients in an inorganic or fixed form from the soil or the air [6], before transferring them to the host plant. Conventionally, chemically produced fertilizer has been used to meet all the needs of plants because the quality of the soil's nutrients had decreased owing to overuse and pollution [15]. Unfortunately, the continuous application of these chemicals on farms has led to issues with pollution and the degradation of soil quality [14].

Therefore, the use of AMF in connection with other nutrients-solubilizing or -fixing microorganisms has recently been taken into consideration as a sustainable alternative to soil nourishment [9, 10, 16]. The soil's nutrients were mobilized to the host plant by AMF, through the mycorrhizal hyphae, which connect the plant roots to the soil. Several investigations into the mechanism of action of nutrient absorption and translocation with AMF have found that:

i. The mycorrhizal hyphae can stretch and explore a larger region of the ground soil than the roots of the host plant can achieve, thereby enabling it to gain access to both micro and macronutrients that the plant alone is unable to absorb [17]. Hence, AMF can cross the depletion zone caused by the plant's quick uptake of nutrients in the region around its root system and provide the appropriate nutrient elements to the plant [12].

ii. The fungal hyphae are favorable for nutrient intake because of their small (<10 μm diameter) sizes, which enable them to enter tiny pores that are inaccessible to plant roots, having significant effects on water and micronutrient absorption [12]. Similarly, Püschel [18], reported that the ability of mycorrhizal to modify their hyphal diameter by the size of the soil pore enables them to provide nourishment for plants regardless of soil texture.

Similarly, several researchers documented the beneficial impact of mycorrhizae on cassava plants with nutrient intake [8, 19–22]. For instance, AMF's contribution to Phosphate uptake was examined by Ndeko et al. [23]. They discovered that using AMF is suitable for enhancing cassava's phosphate nutrition in various soil types. The results of their experiment demonstrated that the root abundance and dry weight were increased by the inoculation of an unusual fungal strain (*Rhizophagus iregularis*). However, following AMF inoculation in unsterilized soil, the root dry weight dropped. They concluded that the *Rhizophagus iregularis* strain, particularly when the soil is not treated with phosphorus, increases the Phosphate uptake of the cassava plants [23].

Furthermore, the advantages of phosphorus nutrition were revealed using cassava and *Rhizophagus irregularis* inoculum. To determine whether the paradigm holds in tropical field settings, field tests were carried out in three areas utilizing varied AMF and cassava cultivars in both Kenya and Tanzania at varied Phosphorus fertilization levels. It was discovered that contrary to what the paradigm would have us believe, Cassava's ability to colonize AMF and respond to inoculation does not necessarily decrease as phosphorus availability rises. The obtained results showed that cassava genotypes and fungal availability play a role in maximizing inoculation responsiveness, which is not always the case in low Phosphorus availability settings [24].

Also, Poku [21], investigated how AMF could help cassava absorb more phosphorus from the soil. Phosphorus-fertilizer, AMF, Phosphorus + AMF, and Untreated (Control) were the four treatments used. In comparison to Phosphorus-fertilizer treatment plots, the Phosphorus + AMF were significantly (p0.05) taller, and the Control and AMF-treated plots were significantly (p0.05) identical. At all four locations, the percentage of leaf Phosphorus was statistically comparable, with a grand mean was 0.4%. The content of Phosphorus in the leaves was significantly raised to 0.5% by adding AMF and Phosphorus + AMF to the soil. In comparison to control plots, tubers taken from Phosphorus + AMF-treated plot lines were meaningfully longer (p 0.05). when compared to the control samples, tuber length rose in plots treated with Phosphorus and AMF. The tuber yields on all soil treatments were higher than on control-treatment plots by a substantial amount (p 0.05). Phosphorus+AMF treated plots and AMF treated plots, however, had considerably higher values than P than the control plots. According to this study, cassava yield can be increased by utilizing AMF or Phosphorus+AMF in comparison to Phosphorus alone or untreated control plants and this can be used to maximize tuber yield [21].

Furthermore, in the work of Lopes et al. [25], they sought to ascertain whether co-inoculating micro-propagated cassava with AMF (*Glomus clarum*) and PGPBs (Plant growth promoting bacteria) improved greenhouse growth. Inoculated PGPB strains in the cassava variety "*BRA Pretinha III*" affected the number of glomerospores and mycorrhizal colonization, while *Glomus clarum* and PGPBs had synergistic interactions, the *Glomus clarum* and PGPBs combined inoculation promoted higher performance in cassava development with time like all the variants examined. Hence, Co-inoculating PGPBs and AMF can meet cassava's need for nitrogen, therefore, minimizing the need for nitrogen fertilizer [25].

4.2 How arbuscular mycorrhizal fungi contribute to soil aggregation

Sand, silt, and clay particles are bonded together to form aggregates of different sizes, and this arrangement is referred to as the soil structure [26]. Soil aggregation is essential to the health of the entire ecosystem because it serves as a major site for the exchange of water, gaseous, and nutrient flows as well as a significant source of carbon storage [27]. It is believed that fungal hyphae are one of the key binding agents involved in maintaining micro aggregates. However, intensive agricultural practices used today have significantly impacted soil structure by lowering aggregation stability [26, 27], via the following steps.

The first step is that the extraradical hyphae compress the soil physically as they ramify around plant roots, causing clay particles to reorient and ramification in macroaggregate pores [28], thereby affecting the plant water status, and contributing to the soils' cohesion and strength, particularly in drought conditions [29]. Also, the production of glomalin, which is a hydrophobic glycoprotein generated by AMF hyphae enables the hydrophilic fungal wall to stick to hydrophobic surfaces found on soil particles [30]. Glomalin production also increases carbon storage and availability, which has an impact on the microbial community, aggregate stability, and soil structure. Based on the design of the plant's roots and how they are connected to the fungus, glomalin promotes aggregation to varying degrees; the greatest impact on macro aggregation was observed with thin host plants' roots (0.2–1 mm in diameter) [28].

Furthermore, numerous authors have positively confirmed AMF's ability to lessen soil aggregation's detrimental impact on plant growth [28–31], but few studies have explicitly focused on cassava crops. One of these uncommon investigations was conducted by Morris et al. [31]. They were able to assess how AMF altered aggregate turnover durations. They demonstrated how AMF accelerated the production of large macroaggregates and slowed the dissolution of both big and small macroaggregates. In the presence of AMF, macroaggregates turnover increased. The internal aggregate organization suggested that although the accretion of soil to organic materials in the form of micro aggregates is a prevalent process, it is not the only one at work [31].

5. How arbuscular mycorrhizal fungi manages abiotic stressors

Numerous abiotic stress studies have demonstrated how human activities associated with agriculture (such as irrigation, overuse of chemical fertilizers and pesticides deforestation, and waste material diffusion) have hurt the plant's growth, health, and output, leading to major production losses [12, 26, 27]. A general route is involved in how plants respond to stress; it begins with the membrane receptor acquiring the stress signal that culminates in the creation of genes, whose byproducts may defend the plant either directly or indirectly [32]. However, numerous investigations on AMF symbiosis have demonstrated that the contributing fungus typically uses several strategies to help the plant resist some abiotic stressors, including but not limited to salinity, heavy metal pollution, and drought [33]. Abiotic stressors are the main obstacle to achieving global food security since they significantly reduce crop production quality and quantity [30, 32, 33].

5.1 Salinity stress

High salt concentrations in the soil make it difficult for roots to draw water from the ground and can be damaging to crops, with adverse effects such as ethylene production, plasmolysis, an unbalanced diet, the inhibition of photosynthesis, and the creation of reactive oxygen species (ROS) [34]. However, one of the ways by which plants manage salinity stress is via osmotic adaptation. A physiological technique employed by crops to sustain a variety of water movements between cells without experiencing turgor or growth decreases is called osmotic adaptation [35]. The buildup of appropriate solutes in plant cells, including proline and glycine betaine, serves as an illustration of this [34]. Furthermore, the existence of salt-tolerant AMF species was demonstrated by several recent scientific studies to alleviate the plant's salinity stress. The following four AMF properties are specifically mentioned by the researchers as having the capacity to reduce salt stress:

i. Increased water intake: Mycorrhizal hyphae possesses a stronger capability to penetrate the ground soil, allowing it to absorb more water and lessen the two primary consequences of salinity that jeopardize a plant's water status—turgor loss and dehydration [36].

ii. Increased mineral uptake: Under osmotic stress, sodium (Na^+) levels in the soil are frequently quite high, which has a detrimental effect on some other transporters present in the roots, like the potassium ion (K^+) selective channels [37]. However, it was noted that plants connected to AMF showed a rise in total nutrient availability and a large K^+ build-up, which assisted the plants in preserving a low Na^+/K^+ ratio and avoiding harm to their biological processes [37].

iii. Abundant synthesis of suitable organic solutes: AMF-plants produce more proline, glycine, betaine, and sugars, and these chemicals appear to be positively associated with the invasion of fungi [38]. Because they are necessary for AMF's function in ROS detoxification, regulating membrane structure, and sustaining enzymes and proteins, their synthesis can aid in cellular osmotic adjustment [38].

iv. Increased antioxidant enzyme activity: Research has demonstrated that AMF symbiosis enhances enzyme system performance in ROS detoxification, particularly that of peroxides, superoxide, hydroxyl radicals, and alpha oxygen, whose generation in plants is significantly influenced by stress conditions like salt. In AMF plants, which normally show less oxidative damage, the effects of these chemicals on cell metabolisms, such as DNA damage, the oxidation of polyunsaturated fatty acids in lipids and amino acids in proteins, and the inhibition of enzymes, appear to be less severe [36, 39].

The benefits of arbuscular mycorrhiza fungi (AMF) on cassava plant growth under salinity have rarely been studied in field settings but in controlled settings. An example is a study conducted by Carretero et al. [20]. This study examined the effects of *G. intraradices* colonization on three cassava cultivars' biomass and salt tolerance (measured as growth) (SOM-1, 05, and 50). In both AMF-inoculated and non-inoculated cassava cultivars, the survival rate of the root, stem, and leaf development, as well as nutrient accumulation, were assessed in the presence of different sodium chloride concentrations (0, 68.4, or 136.8 mM) in the medium. It was reported that at 136.8 mM of salt, the AMF colonization boosted plant survival and encouraged growth. The SOM-1 cultivars outperformed the other two in terms of salt tolerance. In addition to promoting growth, the *G. intraradices*-inoculation proved essential for protecting salt-sensitive cassava cultivars (especially in salt-sensitive cultivars). When compared to non-mycorrhizal cultivars growing in the absence of salt, the AMF cultivars grew in 136.8 mM of $NaCl_2$ and produced more dry weight. The results show that AMF-colonization offers a biological process whereby cassava cultivars can increase their salt tolerance and biomass, which, in both low- and high-stress settings, is essential for optimal cassava development [20].

5.2 Drought stress

Drought is another frequent abiotic factor that negatively influences plant growth, survival, and development [40]. The main symptoms of drought stress include wilting as well as a decline in the net rate of photosynthesis, transpiration rate, water usage effectiveness, relative moisture content, and overall chlorophyll content [41]. Also, drought compromises the electron transport system, which results in the generation of activated oxygen and the shutting of plant stomata, consequently reducing CO_2 absorption [42].

AMF play the following crucial functions in the response to drought stress; several studies have demonstrated how crucial plant symbiosis is for reducing the detrimental effects of drought:

i. Regulation of water uptake: To assist in keeping the plant moist, AMF hyphae delve deeper into the soil and scour a large area for water [43].

ii. Osmotic adjustment: The activity of the AMF sustains several activities, including stomatal opening, cellular expansion, and development, which enable the cells to maintain their turgor [44].

iii. Trehalose biosynthesis: Trehalose is a sugar produced by AMF that assists the plant to resist drought stress, maintaining biological nitrogen fixation, and protecting it from a shortage of water [45].

iv. An increase in antioxidant levels: Catalase, peroxidase, and superoxide dismutase, which decrease ROS, and hydroxyl radicals, appear to be present in higher amounts in plants colonized by AMF [46].

v. Gene expression: AMF typically encourages the expression of some drought-resistant genes such as *1-pyrroline-5-carboxylate synthetase*, and *9-cis-epoxy-carotenoid dioxygenase genes*. Consequently, the stomata are close to stopping water loss, and water movement either within or outside the cell [47].

It was reported that cassava growth could be stimulated under drought stress by inoculating with an AMF. For instance, according to the experiment by Ekanayake et al. [48] the inoculation with *G. clarum*, and *G. mosseae* significantly increased the photosynthetic photochemical efficiency of the photosystem 11 light reactions in cassava. The PS11 photochemistry's maximal quantum yield (Fv/Fm) was similarly significantly decreased (p 0.05) by water stress, whereas AMF inoculation significantly (p 0.05) decreased the negative effects of water stress on cassava cultivars grown under a water deficit regime.

Similarly, Pea et al. [49] investigated whether intraspecific variation in *R. irregularis* affects the physiological responses of cassava to water stress because it frequently experiences seasonal drought where it is cultivated. Two genetically distinct *R. irregularis* isolates were inoculated into cassava to promote recovery, which was then subjected to a drought situation before being re-watered. Cassava samples treated with the two distinct fungi had considerably different physiological stress reactions to drought. However, after being re-watered, both plants recovered but at different rates. They concluded that intraspecific genetic variation in AMF considerably affects the physiological reactions of cassava under water stress. This shows the opportunity to increase cassava resistance to water stress by utilizing naturally occurring polymorphism in AMF [49].

5.3 Heavy metal stress

For several enzyme-catalyzed or redox reactions, as well as the metabolism of nucleic acids, plants require specific mineral elements, such as copper (Cu), iron (Fe), nickel (Ni), and zinc (Zn) amongst others [50]. However, high concentrations of these heavy metals, can alter the protein structure or cause mutations in the plant's genetic makeup [51]. This may cause indications of deficiency such as chlorosis, diminished germination, and slow growth. Diffusion of these heavy metals into the plant's roots after a groundwater and soil surface deposition is mostly caused by anthropogenic activities including fertilizer application and the diffusion of industrial waste, which pose a significant stressor for these plants [50, 51].

However, AMF had been reported in supporting plants in heavy metal-contaminated environments using these mechanisms.

i. The AMF hyphae have strong soil exploration abilities and function as a great adsorbent position for the buildup of cations that inhibit the buildup of dangerous metals [52]. Heavy metals are typically precipitated in the extraradical hyphae by AMF-produced proteins termed 'glomalin' [51].

ii. The symbiosis also has a localized benefit in the soil where AMF generates exudates that contain citric acid, lactic acid, etc. By forming compounds with the metals, these organic acids lower their intensity in the soil [53].

Recently, the benefits of AMF as reviewed by Riaz et al. [52], highlighted their potential as plant-based remediation techniques for extremely heavy metal-contaminated soils [52]. Additionally, according to Dushenkove et al. [54], certain plants, like cassava, can remove heavy metals from soils that have been contaminated by crude oil by using high biomass crops in tandem with a system of soil amendments using the rhizofiltration approach. Rhizofiltration is a technique that uses plants to remove toxins from aqueous streams [54].

For the first time, the capacity of cassava to phytoextract mercury (Hg) and gold (Au) from biosolids and mine tailings that contain these metals was successfully proven [55]. Pre-rooted cassava cuttings with 5–7 nodes were grown in a blend of 25% mine tailings and 75% biosolids. Plant cuttings were additionally grown in hydroponics samples containing Hg and/or Au to gauge the two metals' root uptake. Up to 12.59 g/kg of Hg and 1.89 g/kg of Au were discovered in the cassava's fibrous roots. Hence, due to the cultivation simplicity, cassava provides a sustainable choice for Hg removal and Au recuperation.

5.4 Temperatures stress

Increased temperature due to several factors including global warming has a significant impact on the cycling of nutrients because temperature regulates both the decomposition of organic materials and soil microbial activity [56]. In AMF, reduced spore diameter, root colonization, species variety, and soil glomalin content have all been linked to higher temperatures, resulting in varying water and N levels impacting the plants' growth [57]. Reduced mycorrhizal reliance had an impact on the growth of mycobionts, particularly in AMF even while plants could obtain the necessary nutrients. AMF has been reported to act as a buffer for plants that were affected by climate change by widening their "niche," [58].

Also, low soil temperature inhibits AMF root growth much like high soil temperature does [33]. Even though the ideal temperature for AMF development varies according to the fungal isolates, most species prefer temperatures between 18 and 30OC [59]. At 15°C, mycorrhization of roots declines, and a further drop in temperature (10°C or 5°C) prevents AMF production [33]. AMF sporulation was impacted by low temperatures [12]. For instance, *R. intraradices'* sporulation slows down at 15°C, although spores' metabolic processes are not affected until temperatures of 10°C or lower [60]. Organic additions may improve AMF's effectiveness in reducing cold stress. An example is the development of *Lolium perenne* L. on salty soils at low temperatures was enhanced by the synergistic inoculation of *R. intraradices* and biochar [61].

6. How arbuscular mycorrhizal Fungi manages biotic stressors

6.1 Resistance to pathogens

In the agricultural industry, AMF can be utilized as a biological agent against many plant diseases, providing an effective substitute for chemical pesticides by fostering sustainability and lowering the hazards to the general populace's health [62]. It is known that AMF symbiosis helps to lessen the harm done by a variety of soil-borne diseases, such as nematodes and fungi that are responsible for significant production losses [63]. The following are the biological control mechanisms of AMF against pathogens:

i. Modifications in root development and morphological characteristics: AMF colonization alters the dynamics of pathogens and microbial populations and may stimulate microbiota elements that have antagonistic activity toward specific root illnesses [64].

ii. Modifications in the nutrition of the plant host: The pathogens' reduction of root biomass or function is made up for by the plant's increased vigor and the ensuing rise in resilience brought on by the AMF symbiosis [33].

iii. Competition for colonization spots and photosynthates: Root diseases and AMF compete for the carbon molecules that get to the root. Both are dependent on host photosynthates. The increasing carbon demand, though, might prevent pathogen growth since AMF have main access to photosynthates [63].

iv. The activation of the host defense mechanisms: AMF colonization causes the activation of the host defense mechanisms. Examples of these defense mechanisms are the synthesis of Phytoalexins, phenylpropanoid pathway enzymes, chitinases, b-1,3-glucanases, peroxidases, pathogenesis-related (PR) proteins, callose, hydroxyproline-rich glycoproteins (HRGP), and phenolics are amongst the biological controls that the host plant produces in response to AMF [65].

For instance, the capacity of two local AMF species (*A. Colombiana* and *A. appendicula*) to promote growth and enhance resistance to root-knot nematode and water stresses in the cassava Yayo cultivar was examined using both single and multiple inoculations in a greenhouse setting. It was determined that of the two AMF species, *A. colombiana* greatly boosted cassava growth and drought tolerance. However, *A. colombiana* and *A. appendicula* both gave cassava plants bioprotective properties against the nematode, including tolerance or resistance [22]. Finally, Investigations into the effects of the ACMV (African Cassava Mosaic Virus) on AMF root colonization and leaf symptoms were conducted. The results revealed that both mycorrhized and non-mycorrhized plants had comparable colonization parameters, and ACMV infection after mycorrhization establishment, had no effect on the AMF root colonization [66].

6.2 The advantages of arbuscular mycorrhizal Fungi interaction with other beneficial soil microorganisms

Several soil microorganisms engage in interactions with AMF [67]. The interactions of AMF with other microorganisms could either be positive, neutral, or negative [68]. They could help plants acquire nutrients, biologically regulate pathogens that cause root infections, and increase soil quality and plant resistance to abiotic stress [65, 68]. Similar to how AMF and plant growth-promoting Rhizobacteria (PGPR) collaborate to aid the growth of the host plant [67]. While AMF and nitrogen-fixing bacteria provide plants with essential soil nutrients, More et al.'s [69] hypothesis states that co-inoculating PGPR and AMF will have the greatest synergistic effects. This notion was confirmed by the interactions in their experiment between AMF and modifying rhizobial bacteria [69, 70].

7. The function of arbuscular mycorrhizal fungi in reducing greenhouse gas emissions

Reduced emissions of greenhouse gases and boosted carbon sequestration are the two primary goals the Food and Agriculture Organization (FAO) has developed for climate-smart agriculture [71]. N_2O is a potent greenhouse gas with a larger global warming potential (280–310) than CO_2 and longer persistence (118–131 years) in the atmosphere and sadly, agriculture is a significant source of N_2O emissions [72]. Dissimilatory nitrate reduction to ammonium and ammonia oxidation are amongst the main sources of N_2O emissions in farms through a variety of denitrification processes. Slow-growing nitrifiers are outcompeted by AMF hyphae for ammonium, which hinders nitrification and, ultimately, the production of N_2O [73].

Under varying moisture conditions, the emissions of greenhouse gases from soil can also be controlled by AMF (as well as the water relationship between plants) [11]. The impact of AMF in controlling N_2O emissions has not been studied in cassava, but it has been studied in legume systems. A field experiment with AMF on legumes revealed a significant reduction in yield-scaled N_2O emissions due to greater biological nitrification inhibition [74]. The term "biological nitrification inhibition" (BNI) refers to the ability of the host plant's roots to lower the activity of soil nitrifiers by generating and releasing nitrogen removal inhibitors. To manage the soil nitrifier activities, help reduce emissions of greenhouse gases and generally make agriculture more ecologically friendly and effective, BNI-enabled plants and pastures have been suggested for use in agriculture [72, 73].

8. Limitations

Given the projected rise in population, it will be necessary to enhance the production of economically important crops like cassava. Hence, increasing soil fertility in an environmentally sustainable approach may be the greatest strategy to accomplish the expected crop yield. Given the effects of climate change, nutrient depletion, drought, salt, and metal toxicity, it has been well established that AMF may sustainably improve plant growth, production, and crop nutritional quality for this purpose, thereby ensuring food security for both present and future generations.

However, most studies on AMF-mediated advances in plant health, nutrient uptake, and regulation of the impacts of biotic and abiotic stress conditions have thus far used in vitro studies on laboratory scales, and under controlled temperature and low nutrients profiles trials studies. However, if more farm field experiments are done, the results may offer a more in-depth understanding of the intricate mechanisms underpinning Plant-nutrient connections that are mediated by AMF and AMF interactions with both the biotic and abiotic stressors. These understandings may have an advantage to the effective application of AMF in boosting the production of more important commercial crops, such as cassava, regardless of the region.

Also, the manipulation of AMF in both artificial and natural systems has proven to be a particularly difficult barrier to overcome. It is anticipated that future tests would use cutting-edge techniques to understand the mystery surrounding plants-AMF community ecology, what patterns do plants and AMF communities follow? what does this mean in terms of ecosystem management, recovery, and restoration? To get answers to these issues, we both hope and anticipate that we will not have to wait another 13 years.

Additionally, the glycoproteins (glomalin) produced by AMF are not currently considered in soil health assessment programs even though they are linked to several ecosystem processes and act as a low-cost proxy for soil quality, the efficiency of agricultural management and restoration methods, and AMF biomass [75]. Controlled studies using these methods (ELISA and advanced spectroscopic classification) and AMF quantification (quantitative PCR assays, AMF-signature lipids, microscopic measurements) are necessary to inform researchers and farmers about the potential of these glycoproteins as a C-sequestration and quality indicator, along with field demonstrations.

9. Conclusions

Mycorrhizae and their usage in the agronomic field have gained popularity in numerous scientific studies conducted all over the world in recent years. Most of the research has been on the advantages that arbuscular fungus provides for host plants regarding its productivity and resistance to all biotic and abiotic challenges. Numerous studies of the fungus' metabolic processes and pathways have been conducted to enhance the fungus' capacity for nutrient and water absorption as well as its defense against infections, salt, and heavy metals.

The cassava plant, one of the most significant food crops in the world, has been the centre of numerous studies on mycorrhizal inoculation. However, choosing cassava varieties with very effective mycorrhizal symbiosis may serve as the cornerstone to produce foods in low-impact agricultural systems. The useful rhizosphere microbial inoculants are now widely used in agricultural techniques for many crops than cassava. However, there might also be negative interactions, including rivalry and hostility, between various rhizosphere microbe kinds. It is counterproductive when specific indigenous soil microbes interact negatively with externally applied microorganisms. Hence, choosing the most contagious and effective mycorrhizal fungi and cassava together will make it easier to employ them as biofertilizers to replace the lost biological fertility of the soil, reduction of chemical usage and mitigate the effects of biotic and abiotic stress.

Acknowledgements

Dr. Otun acknowledges the University of Witwatersrand, the SARChi programme of the Department of Science and Technology and the South African National Research Foundation for her postdoctoral fellowship funding.

Conflict of interest

The authors declare no conflict of interest.

Author details

Sarah Otun* and Ikechukwu Achilonu
Faculty of Science, Protein Structure-Function and Research Unit, School of Molecular and Cell Biology, University of the Witwatersrand, Johannesburg, South Africa

*Address all correspondence to: sarahholabamiji@gmail.com

References

[1] Pathan SI, Ceccherini MT, Sunseri F, Lupini A. Rhizosphere as hotspot for plant-soil-microbe interaction. Carbon and Nitrogen Cycling in Soil. Springer; 2020. [cited 2022 Jul 14]:17-43. DOI: 10.1007/978-981-13-7264-3_2

[2] Frank AB. Ber Dent. Bot Ges. 1885. [cited 2022 Jul 14];**3**:128. Available from: https://scholar.google.com/scholar_lookup?title=&journal=Uber%20Die%20Auf%20Wurzelsymbiose%20Beruhende%20Ern%C3%A4hrung%20Gewisser%20Ba%C3%BCme%20Durch%20Unterirdische%20Pilze&volume=3&pages=128-145&publication_year=1885&author=Frank%2CA

[3] Anand K, Kumar Pandey G, Kaur T, Pericak O, Olson C, Mohan R, et al. Arbuscular mycorrhizal fungi as a potential biofertilizers for agricultural sustainability. Journal of Applied Biology and Biotechnology. 2022. [cited 2022 Jul 14];**10**(1):90-107. Available from: http://creativecommons.org/licenses/by-nc-sa/3.0/

[4] Mahmud K, Missaoui A, Lee K, Ghimire B, Presley HW, Makaju S. Rhizosphere microbiome manipulation for sustainable crop production. Current Plant Biology. 2021;**27**:100210

[5] Pandey D, Kehri HK, Zoomi I, Akhtar O, Singh AK. Mycorrhizal Fungi: Biodiversity, ecological significance, and industrial applications. Recent Advancement in White Biotechnology Through Fungi. Springer; 2019. [cited 2022 Jul 14]:181-199. DOI: 10.1007/978-3-030-10480-1_5

[6] Gupta MM, Chourasiya D, Sharma MP. Diversity of arbuscular mycorrhizal fungi in relation to sustainable plant production systems. In: Microbial Diversity in Ecosystem Sustainability and Biotechnological Applications: Volume 2 Soil & Agroecosystems. Springer; 2019. [cited 2022 Jul 14]. pp. 167-186. DOI: 10.1007/978-981-13-8487-5_7

[7] Noort MWJ, Renzetti S, Linderhof V, du Rand GE, Marx-Pienaar NJMM, de Kock HL, et al. Towards sustainable shifts to healthy diets and food security in sub-Saharan Africa with climate-resilient crops in bread-type products: A food system analysis. Food. 2022;**11**(2):135. [cited 2022 Jul 14]. Available from: https://www.mdpi.com/2304-8158/11/2/135/htm

[8] Thanni B, Merckx R, de Bauw P, Boeraeve M, Peeters G, Hauser S, et al. Spatial variability and environmental drivers of cassava—Arbuscular mycorrhiza fungi (AMF) associations across southern Nigeria. Mycorrhiza. 2022. [cited 2022 Jul 14];**32**(1):1-13. DOI: 10.1007/s00572-021-01058-x

[9] Nahuelcura J, Ruiz A, Gomez F, Cornejo P. The effect of arbuscular mycorrhizal fungi on the phenolic compounds profile, antioxidant activity and grain yields in wheat cultivars growing under hydric stress. Journal of the Science of Food and Agriculture. 2022. [cited 2022 Jul 14];**102**(1):407-416. DOI: 10.1002/jsfa.11370

[10] Bhantana P, Rana MS, Cheng SX, Moussa MG, Saleem MH, Syaifudin M, et al. Arbuscular mycorrhizal fungi and its major role in plant growth, zinc nutrition, phosphorous regulation and phytoremediation. Symbiosis. 2021. [cited 2022 Jul 14];**84**(1):19-37. DOI: 10.1007/s13199-021-00756-6

[11] Salomon MJ, Demarmels R, Watts-Williams SJ, McLaughlin MJ,

Kafle A, Ketelsen C, et al. Global evaluation of commercial arbuscular mycorrhizal inoculants under greenhouse and field conditions. Applied Soil Ecology. 2022;**169**:104225

[12] Thangavel P, Anjum NA, Muthukumar T, Sridevi G, Vasudhevan P, Maruthupandian A. Arbuscular mycorrhizae: Natural modulators of plant–nutrient relation and growth in stressful environments. Archives of Microbiology. 2022. [cited 2022 Jul 14];**204**(5):1-22. DOI: 10.1007/s00203-022-02882-1

[13] Gupta MM. Arbuscular mycorrhizal Fungi: The potential soil health indicators. Soil Health. Springer; 2020. [cited 2022 Jul 14]:183-195. DOI: 10.1007/978-3-030-44364-1_11

[14] Yeatman F, Yeatman K. Decline in nutrients in soils and foods, and the role of nutrients a summary and analysis of the literature. Advanced Agricultural Science Oasis Agri (PTY) LTD. 2020. [cited 2022 Jul 14]; Available from: www.oasis-agri.co.za2020

[15] Nadarajan S, Sukumaran S. Chemistry and toxicology behind chemical fertilizers. Controlled Release Fertilizers for Sustainable Agriculture. Elsevier; 2021:195-229

[16] de Aza CH, Armenteros S, McDermott J, Mauceri S, Olaizola J, Hernández-Rodríguez M, et al. Fungal and bacterial communities in tuber melanosporum plantations from northern Spain. Forests. 2022. [cited 2022 Jul 14];**13**(3):385 Available from: https://www.mdpi.com/1999-4907/13/3/385/htm

[17] Li Q, Umer M, Guo Y, Shen K, Xia T, Xu X, et al. Karst soil patch heterogeneity with gravels promotes plant root development and nutrient utilization associated with arbuscular mycorrhizal Fungi. Agronomy. 2022. [cited 2022 Jul 14];**12**(5):1063 Available from: https://www.mdpi.com/2073-4395/12/5/1063/htm

[18] Püschel D, Bitterlich M, Rydlová J, Jansa J. Facilitation of plant water uptake by an arbuscular mycorrhizal fungus: A Gordian knot of roots and hyphae. Mycorrhiza. 2020. [cited 2022 Jul 14];**30**(2-3):299-313. DOI: 10.1007/s00572-020-00949-9

[19] Kuyper TW, Wang X, Muchane MN. The interplay between roots and arbuscular mycorrhizal Fungi influencing water and nutrient acquisition and use efficiency. The Root Systems in Sustainable Agricultural Intensification. Wiley Online Library; 2021. [cited 2022 Jul 14]:193-220. DOI: 10.1002/9781119525417.ch7

[20] Carretero CL, Cantos M, García JL, Azcón R, Troncoso A. Growth responses of micropropagated cassava clones as affected by Glomus Intraradices colonization. Journal of Plant Nutrition. 2008. [cited 2022 Jul 15];**31**(5):959-971. DOI: 10.1080/01904160802608601

[21] Poku E. Improving phosphorus uptake by cassava (Manihot esculanata Crantz) using Arbuscular Mycorrhizal Fungi (AMF) [Doctoral dissertation]. Durban: University of KwaZulu-Natal; 2014. [cited 2022 Jul 14]. Available from: https://ukzn-dspace.ukzn.ac.za/handle/10413/10855

[22] Séry DJM, Kouadjo ZGC, Voko BRR, Zézé A. Selecting native arbuscular mycorrhizal fungi to promote cassava growth and increase yield under field conditions. Frontiers in Microbiology. 2016;**7**(DEC):2063

[23] Byamungu Ndeko A, Chuma Basimine G, Mukengere Bagula E,

Mugumaarhahama Y, Nabintu Ndusha B, Rehema P, et al. Comparative effect of Rhizophagus irregularis strain on cassava root development and phosphorus uptake under acidic soils conditions of Walungu territory. Journal of Applied Biosciences. 2019. [cited 2022 Jul 14];**148**:15167-15175. Available from: http://www.m.elewa.org/Journals/wp-content/uploads/2020/04/1.Ndeko_-5.pdf

[24] Peña Venegas RA, Lee SJ, Thuita M, Mlay DP, Masso C, Vanlauwe B, et al. The phosphate inhibition paradigm: Host and fungal genotypes determine arbuscular mycorrhizal fungal colonization and responsiveness to inoculation in cassava with increasing phosphorus supply. Frontiers in Plant Science. 2021;**12**:1163

[25] EAP L, ADA d S, ACDES M, EVN d S, Santiago AD, MDVB F. Co-inoculation of growth promoting Bacteria and *Glomus clarum* in micropropagated cassava plants. Revista Caatinga. 2019. [cited 2022 Jul 14];**32**(1):152-166. Available from: http://www.scielo.br/j/rcaat/a/dNTvSCHJ59HkbkFcGymTrYH/abstract/?lang=en

[26] Mukherjee S. Soil structure. Current Topics in Soil Science: An Environmental Approach. Springer Nature; 2022. [cited 2022 Jul 14]:67-75. DOI: 10.1007/978-3-030-92669-4_7

[27] Prasad S, Malav LC, Choudhary J, Kannojiya S, Kundu M, Kumar S, et al. Soil microbiomes for healthy nutrient recycling. Current Trends in Microbial Biotechnology for Sustainable Agriculture. Springer; 2021. [cited 2022 Jul 14]:1-21. DOI: 10.1007/978-981-15-6949-4_1

[28] Ganugi P, Masoni A, Pietramellara G, Benedettelli S. A review of studies from the last twenty years on plant–arbuscular mycorrhizal Fungi associations and their uses for wheat crops. Agronomy. 2019. [cited 2022 Jul 14];**9**(12):840. Available from: https://www.mdpi.com/2073-4395/9/12/840/htm

[29] Chen XW, Wong JTF, Wang JJ, Ng CWW, Wong MH. Effects of mycorrhizal Bermuda grass on low-range soil matric suction. Journal of Soils and Sediments. 2021. [cited 2022 Jul 14];**21**(2):990-1000. DOI: 10.1007/s11368-020-02839-1

[30] Coban O, de Deyn GB, van der Ploeg M. Soil microbiota as game-changers in restoration of degraded lands. Science (1979). American Association for the Advancement of Science; 2022. [cited 2022 Jul 14];**375**(6584):abe0725. DOI: 10.1126/science.abe0725

[31] Morris EK, Morris DJP, Vogt S, Gleber SC, Bigalke M, Wilcke W, et al. Visualizing the dynamics of soil aggregation as affected by arbuscular mycorrhizal fungi. The ISME Journal. 2019. [cited 2022 Jul 14];**13**(7):1639-1646. Available from: https://www.nature.com/articles/s41396-019-0369-0

[32] Meisrimler CN, Allan C, Eccersall S, Morris RJ. Interior design: How plant pathogens optimize their living conditions. New Phytologist. 2021. [cited 2022 Jul 15;**229**(5):2514-2524. DOI: 10.1111/nph.17024

[33] Diagne N, Ngom M, Djighaly PI, Fall D, Hocher V, Svistoonoff S. Roles of arbuscular mycorrhizal Fungi on plant growth and performance: Importance in biotic and abiotic stressed regulation. Diversity. 2020. [cited 2022 Jul 15];**12**(10):370. Available from: https://www.mdpi.com/1424-2818/12/10/370/htm

[34] Bensidhoum L, Ait Bessai S, Nabti E. Restoration of plant growth under saline soil by halotolerant plant growth-promoting Rhizobacteria (PGPR). Saline Soil-based Agriculture by

Halotolerant Microorganisms. Springer; 2019. [cited 2022 Jul 15]:23-51. DOI: 10.1007/978-981-13-8335-9_2

[35] Jaramillo Roman V, den Toom LA, Castro Gamiz C, van der Pijl N, Visser RGF, van Loo EN, et al. Differential responses to salt stress in ion dynamics, growth and seed yield of European quinoa varieties. Environmental and Experimental Botany. 2020;**177**:104146

[36] Jansson JK, Hofmockel KS. Soil microbiomes and climate change. Nature Reviews Microbiology. 2019. [cited 2022 Jul 15];**18**(1):35-46. Available from: https://www.nature.com/articles/s41579-019-0265-7

[37] Selvakumar G, Shagol CC, Kim K, Han S, Sa T. Spore associated bacteria regulates maize root K+/Na+ ion homeostasis to promote salinity tolerance during arbuscular mycorrhizal symbiosis. BMC Plant Biology. 2018. [cited 2022 Jul 15];**18**(1):1-13. DOI: 10.1186/s12870-018-1317-2

[38] Gupta S, Schillaci M, Walker R, Smith PMC, Watt M, Roessner U. Alleviation of salinity stress in plants by endophytic plant-fungal symbiosis: Current knowledge, perspectives and future directions. Plant and Soil. 2020. [cited 2022 Jul 15];**461**(1):219-244. DOI: 10.1007/s11104-020-04618-w

[39] Ganugi P, Pè M, Pietramellara G. Genetic Characterization of Tetraploid Wheats and Evaluation of their Mycorrhizal Affinity. flore.unifi.it. PhD thesis; 2020. [cited 2022 Jul 15]; Available from: https://flore.unifi.it/bitstream/2158/1239668/1/PhD_thesis_GanugiPaola.pdf

[40] Khan N, Ali S, Tariq H, Latif S, Yasmin H, Mehmood A, et al. Water conservation and plant survival strategies of Rhizobacteria under drought stress. Agronomy. 2020. [cited 2022 Jul 15];**10**(11):1683. Available from: https://www.mdpi.com/2073-4395/10/11/1683/htm

[41] Ye Q, Wang H, Li H. Arbuscular mycorrhizal Fungi improve growth, photosynthetic activity, and chlorophyll fluorescence of Vitis vinifera L. cv. Ecolly under drought stress. Agronomy. 2022. [cited 2022 Jul 15];**12**(7):1563. Available from: https://www.mdpi.com/2073-4395/12/7/1563/htm

[42] Kapoor D, Bhardwaj S, Landi M, Sharma A, Ramakrishnan M, Sharma A. The impact of drought in plant metabolism: How to exploit tolerance mechanisms to increase crop production. Applied Sciences. 2020. [cited 2022 Jul 15];**10**(16):5692. Available from: https://www.mdpi.com/2076-3417/10/16/5692/htm

[43] Bouskout M, Bourhia M, al Feddy MN, Dounas H, Salamatullah AM, Soufan W, et al. Mycorrhizal Fungi inoculation improves Capparis spinosa’s yield, nutrient uptake and photosynthetic efficiency under water deficit. Agronomy. 2022. [cited 2022 Jul 15];**12**(1):149. Available from: https://www.mdpi.com/2073-4395/12/1/149/htm

[44] Ben-Laouane R, Baslam M, Ait-El-mokhtar M, Anli M, Boutasknit A, Ait-Rahou Y, et al. Potential of native arbuscular mycorrhizal Fungi, rhizobia, and/or green compost as alfalfa (Medicago sativa) enhancers under salinity. Microorganisms. 2020. [cited 2022 Jul 15];**8**(11):1695. Available from: https://www.mdpi.com/2076-2607/8/11/1695/htm

[45] Kosar F, Akram NA, Sadiq M, Al-Qurainy F, Ashraf M. Trehalose:

A key organic Osmolyte effectively involved in plant abiotic stress tolerance. Journal of Plant Growth Regulation. 2018. [cited 2022 Jul 15];**38**(2):606-618. DOI: 10.1007/s00344-018-9876-x

[46] Hu S, Hu B, Chen Z, Vosátka M, Vymazal J. Antioxidant response in arbuscular mycorrhizal fungi inoculated wetland plant under Cr stress. Environmental Research. 2020;**191**:110203

[47] Xie W, Hao Z, Zhou X, Jiang X, Xu L, Wu S, et al. Arbuscular mycorrhiza facilitates the accumulation of glycyrrhizin and liquiritin in Glycyrrhiza uralensis under drought stress. Mycorrhiza. 2018. [cited 2022 Jul 15];**28**(3):285-300. DOI: 10.1007/s00572-018-0827-y

[48] Ekanayake World Bank IJ. The relationship between water deficit, Arbuscular mycorrhizal fungi (AMF) inoculation and chlorophyll fluorescence in cassava (Manihot esculenta Crantz) Foundations for a Modern Food System in Myanmar View project Myanmar Agriculture Development Support Project View project. 2007. [cited 2022 Jul 15]. Available from: https://www.researchgate.net/publication/322593900

[49] Peña R, Robbins C, Corella JC, Thuita M, Masso C, Vanlauwe B, et al. Genetically different isolates of the arbuscular mycorrhizal fungus Rhizophagus irregularis induce differential responses to stress in cassava. Frontiers in Plant Science. 2020;**11**:1904

[50] Maurya AK, Sinha D, Kamzakshi, Mukherjee S, et al. Plant response to heavy metals (at the cellular level). In: Heavy Metals in Plants Physiological to Molecular Approach. CRC Press; 2022. [cited 2022 Jul 15]. pp. 125-148. DOI: 10.1201/9781003110576-7

[51] Ghori NH, Ghori T, Hayat MQ, Imadi SR, Gul A, Altay V, et al. Heavy metal stress and responses in plants. International journal of Environmental Science and Technology. 201. [cited 2022 Jul 15];**16**(3):1807-1828. DOI: 10.1007/s13762-019-02215-8

[52] Riaz M, Kamran M, Fang Y, Wang Q, Cao H, Yang G, et al. Arbuscular mycorrhizal fungi-induced mitigation of heavy metal phytotoxicity in metal contaminated soils: A critical review. Journal of Hazardous Materials. 2021;**402**:123919

[53] Adedeji AA, Häggblom MM, Babalola OO. Sustainable agriculture in Africa: Plant growth-promoting rhizobacteria (PGPR) to the rescue. Scientific African. 2020;**9**:e00492

[54] Dushenkov S, Kapulnik Y, Blaylock M, Sorochisky B, Raskin I, Ensley B. Phytoremediation: A novel approach to an old problem. Studies in Environmental Science. 1997;**66**(C):563-572

[55] Alcantara HJP, Doronila AI, Kolev SD. Phytoextraction potential of Manihot esculenta Crantz. (cassava) grown in mercury- and gold-containing biosolids and mine tailings. Minerals Engineering. 2017;**114**:57-63

[56] AminiTabrizi R, Dontsova K, Graf Grachet N, Tfaily MM. Elevated temperatures drive abiotic and biotic degradation of organic matter in a peat bog under oxic conditions. Science of the Total Environment. 2022;**804**:150045

[57] Yang X, Mariotte P, Guo J, Hautier Y, Zhang T. Suppression of arbuscular mycorrhizal fungi decreases the temporal stability of community productivity under elevated temperature and nitrogen addition in a temperate meadow. Science of the Total Environment. 2021;**762**:143137

[58] Bennett AE, Ee A, Classen T. Climate Change Influences Mycorrhizal Fungal-Plant Interactions, but Conclusions Are Limited by Geographical Study bias. Ecology. Wiley Online Library; 2020;**101**(4):e02978

[59] Püschel D, Kolaříková Z, Šmilauer P, Rydlová J. Survival and long-term infectivity of arbuscular mycorrhizal fungi in peat-based substrates stored under different temperature regimes. Applied Soil Ecology. 2019;**140**:98-107

[60] Agnihotri R, Maheshwari HS, Sharma AK, Sharma MP. Biobank for Conservation of Arbuscular Mycorrhizal (AM) Fungi. Microbial Resource Conservation. Springer; 2018. [cited 2022 Jul 15]. pp. 199-221. DOI: 10.1007/978-3-319-96971-8_7

[61] Yan P, Li G, Sun H, Zhang Z, Yang R, Sun J. Can arbuscular mycorrhizal fungi and biochar enhance plant resistance to low-temperature stress? Agronomy Journal. 2021. [cited 2022 Jul 15];**113**(2): 1457-1466. DOI: 10.1002/agj2.20520

[62] Verma PP, Shelake RM, Das S, Sharma P, Kim JY. Plant growth-promoting rhizobacteria (PGPR) and fungi (PGPF): Potential biological control agents of diseases and pests. In: Microbial Interventions in Agriculture and Environment: Volume 1: Research Trends, Priorities and Prospects. Springer; 2019. [cited 2022 Jul 15]. pp. 281-311. DOI: 10.1007/978-981-13-8391-5_11

[63] Dowarah B, Gill SS, Agarwala N. Arbuscular mycorrhizal Fungi in conferring tolerance to biotic stresses in plants. Journal of Plant Growth Regulation. 2022;**41**(4):1429-1444

[64] Lucini L, Colla G, Miras Moreno MB, Bernardo L, Cardarelli M, Terzi V, et al. Inoculation of Rhizoglomus irregulare or Trichoderma atroviride differentially modulates metabolite profiling of wheat root exudates. Phytochemistry. 2019;**157**:158-167

[65] Himaya S, Sivasubramaniam N, Afreen S. A review on role of mycorrhizal Fungi in plant disease management. Faculty of Technology, South Eastern University of Sri Lanka; 2021. [cited 2022 Jul 15]; Available from: http://ir.lib.seu.ac.lk/handle/123456789/5470

[66] Wasnaire J, Declerck S, Hols P. Mycorrhizal association dynamics between Rhizophagus irregularis and Manihot esculenta in a context of African Cassava Mosaic Virus infection. dial.uclouvain.be. 2022. [cited 2022 Jul 20]; Available from: https://dial.uclouvain.be/downloader/downloader.php?pid=thesis%3A23010&datastream=PDF_02&cover=cover-mem

[67] Sangwan S, Prasanna R. Mycorrhizae helper Bacteria: Unlocking their potential as bioenhancers of plant–arbuscular mycorrhizal fungal associations. Microbial Ecology. Springer; 2021:1-10

[68] Berger F, Gutjahr C. Factors affecting plant responsiveness to arbuscular mycorrhiza. Current Opinion in Plant Biology. 2021;**59**:101994

[69] More N, Verma A, Bharagava RN, Kharat AS, Gautam R, Navaratna D. Sustainable development in agriculture by revitalization of PGPR. Bioremediation. CRC Press; 2022. [cited 2022 Jul 15]:127-142. DOI: 10.1201/9781003181224-8

[70] Sagar A, Rathore P, Ramteke PW, Ramakrishna W, Reddy MS, Pecoraro L. Plant growth promoting Rhizobacteria, arbuscular mycorrhizal Fungi and their synergistic interactions to counteract the negative effects of saline soil on

agriculture: Key macromolecules and mechanisms. Microorganisms. 2021. [cited 2022 Jul 15];**9**(7):1491 Available from: https://www.mdpi.com/2076-2607/9/7/1491/htm

[71] Saha ND, Chakrabarti B, Bhatia A, Jain N, Sharma A, Gurjar DS. Greenhouse gas (GHG) emission mitigation options: An approach towards climate smart agriculture. Innovative Approaches for Sustainable Development. Springer; 2022. [cited 2022 Jul 15]:43-63. DOI: 10.1007/978-3-030-90549-1_3

[72] Walling E, Vaneeckhaute C. Greenhouse gas emissions from inorganic and organic fertilizer production and use: A review of emission factors and their variability. Journal of Environmental Management. 2020;**276**:111211

[73] Reimer J. Microbial Regulation of Soil Greenhouse Gas Emissions in a Non-Bloat Legume Grazing System. 2021. [cited 2022 Jul 15]. Available from: https://128.233.230.70/handle/10388/13209

[74] Grant R, Dyck M, Puurveen D. Soil DPCJ of, 2020 undefined. Nitrogen and phosphorus control carbon sequestration in agricultural ecosystems: Modelling carbon, nitrogen, and phosphorus balances at the Breton plots with ecosys under historical and future climates. cdnsciencepub.com. Canadian Journal of Soil Science. Canadian Science Publishing; 2020. [cited 2022 Jul 15];**100**(4):408-429. DOI: 10.1139/CJSS-2019-0132

[75] Agnihotri R, Sharma MP, Prakash A, Ramesh A, Bhattacharjya S, Patra AK, et al. Glycoproteins of arbuscular mycorrhiza for soil carbon sequestration: Review of mechanisms and controls. Science of the Total Environment. 2022;**806**:150571

Chapter 4

Mitigating Climate Change: The Influence of Arbuscular Mycorrhizal Fungi on Maize Production and Food Security

Sílvia N.D. Maússe Sitoe and Joanna F. Dames

Abstract

Anthropogenic activities have contributed to the increased atmospheric concentration of greenhouse gases, which are an important contributor to climate change. From 1940 to 2004, global emissions increased by 70%, and projections suggest a continual increase by 2050 due to agriculture, forestry, and other land uses. Arbuscular mycorrhizal (AM) fungi are ubiquitous in undisturbed soils and form a symbiotic relationship with various plants. The relationship that enhances nutrient uptake and plant growth, among other benefits, is well known. Several soil management practices employed in agriculture adversely affect the symbiosis. *Zea mays* (maize) provides 30% of total caloric intake to 4.5 billion people worldwide and is an important staple crop, vulnerable to climate change. Higher temperatures can result in increased water demand, while changes in precipitation can result in crop failure. AM fungi can be applied as inoculants to maize. Resulting in improved plant growth, yield, and nutrient uptake and providing superior food quality properties, such as increased antioxidants, vitamins, and minerals. AM fungi are considered a crucial biotechnological tool in crop production. This review illustrates their essential role in sustainable maize production and emphasizes the need to maintain AM fungal communities in the soil to mitigate the effects of climate change.

Keywords: *Zea may*, symbiotic benefits, drought and salinity, soil management, sustainable agriculture

1. Introduction

Crop production is the primary source of staple foods and is particularly sensitive to climate change. Activities such as synthetic fertilizer use, livestock rearing, change in land-use patterns, deforestation, waste disposal, burning of fossil fuels, industrial manufacturing and transportation have contributed to the increased atmospheric concentration of greenhouse gases, contributing to climate change [1]. Of the global anthropogenic greenhouse gas emissions, approximately 13% are attributed to agricultural practices [2]. Predictions indicate that by 2030 these emissions will rise almost 40%, primarily due to increasing demand from a growing population and changing consumption patterns for food, including increasing demand for ruminant meats [3].

Between 1940 (pre-industrial) and 2004, global greenhouse emissions (GHG) increased by 70% [2]. Average temperatures have increased by 1.5–2°C. Urgent climate change mitigation is required to halt further increases. The effect on agriculture is severe, with adverse growing conditions such as drought and extreme weather events already evident [4]. Agriculture can reduce these negative effects, increasing carbon storage by improving cropping and land management practices [5].

Mitigation technologies may not be cheap and easy, but the cost and benefits will be less than the losses caused by climate change [6, 7]. The expected impacts of climate change will be most adverse in low- and middle-income countries, where millions of people depend on agriculture and are vulnerable to food insecurity [8, 9]. The impact on global food security will relate to food supply and food quality, food access and utilization, and the stability of food security resulting in reduced per capita calorie availability, childhood malnutrition, and child deaths [10]. Climate change may affect the nutritional properties of some crops. The concentrations of minerals in some crops (e.g., wheat, rice, and soybeans) were up to 8% lower under conditions of elevated carbon dioxide (CO_2) levels [11]. Studies on yields (primarily wheat, maize, rice and soybeans) under different climate change conditions indicate that climate change may significantly reduce these in the long run [12].

Maize (*Zea mays* L.) is known globally as the queen of cereals because it can be grown throughout the year, is photo-thermo insensitive and has the highest genetic yield potential among cereal crops. Maize provides 30% of total caloric intake to 4.5 billion people worldwide [13]. The grain contains about 72% starch, 10% protein, 4.8% oil, 5.8% fiber, 3.0% sugar, and 1.7% ash [14]. The major maize-producing countries are the USA, China, Brazil, Argentina, Ukraine, Indonesia, India and Mexico [15]. Globally, maize is grown mainly for livestock feed, food, and industrial materials [16]. Regarded by most people as a breakfast cereal, maize is also a source of fuel (ethanol) and starch in processed form. Enzymatically maize starch is converted into products such as sorbitol, dextrin, sorbic and lactic acid and appears in household items such as beer, ice cream, syrup, shoe polish, glue, fireworks, ink, batteries, mustard, cosmetics, aspirin, and paint [17]. In South Africa, maize is the most important grain crop for feed and is a staple of most of the population [17]. Of the total maize produced in South Africa, approximately 60% is white (for human consumption), and 40% is yellow (mainly for animal feed) [18]. Adverse effects of climate change on agricultural production will be most severe in the global south, particularly South Asia and Sub-Saharan Africa, many regions of which remain underdeveloped [19]. Finding ways to improve the productivity of agriculture, particularly of staple crops such as maize, is one means of meeting these future challenges.

2. Improving nutrient uptake

Management technologies, crop species, and soil type play a significant role in soil microbial diversity [20]. Arbuscular mycorrhizal (AM) fungal diversity positively contributes to nutrient and water use efficiency [20]. The frequent use of soil amendments such as fertilizers, organic residues, and pH adjustments to improve crop yields can change the soil properties, which leads to variations in plant and fungal responses that can modify the outcome of the symbiosis [21].

In many soils worldwide, low phosphorus (P) availability can limit plant growth [22]. Availability and acquisition of P in terms of plant root architecture and mycorrhizal association is controlled by rhizospheric chemical and biological processes which compensate for this limitation [23]. Maize has a high requirement for nitrogen (N) and P nutrients, and these are applied in large quantities to realize high yields [24]. Not all applied fertilizer is absorbed. For example, maize absorbs only 55–60% N [25], around 20% P [26], 50–70% potassium (K) [27] and 33% sulfur (S) [28]. One strategy to reduce fertilizer usage, especially P, is to use AM fungi to improve maize nutritional status and growth [29].

AM colonized plants absorb more P (nearly 80% of the plant P uptake) at lower concentrations in the soil solution than non-mycorrhizal plants [30–32]. The AM fungi forms an extraradical mycelium (ERM) network that effectively exploits the soil environment in the search of nutrients [33, 34], providing access to more soluble forms of phosphate [35, 36], and promotes plant N uptake [37]. Acquisition of other soil nutrients such as copper (Cu), iron (Fe), K, zinc (Zn), calcium (Ca) and S, especially when plants are grown in nutrient deficient soils are also improved by the symbiosis [38–41].

Different AM fungal species exhibit different levels of effectiveness. A study by Bi et al. [42] on maize's growth and nutrient uptake showed that the effect of two AM fungi, *Funneliformis mosseae* and *F. versiforme*, increased plant growth when compared with non-inoculated controls. *F. mosseae* being the most effective. In a field experiment, commercial AM fungal inoculants applied to maize improved plant growth, yield, and P uptake under both un-fertilized and P-fertilized treatments [43]. Inoculation of maize with *Glomus caledonium* increased soil organic C content in maize straw-amended soils, likely due to enhanced rhizosphere acidification and increased nutrient (notably P) uptake in the seedling period.

Intercropping can also improve plant nutrient uptake [44] because intercropping systems can better use one or more agricultural resources both in time and in space [45]. He et al. [46] reported N transfer from N_2 fixing legumes to maize and other crop species. Marzban et al. [47] demonstrated that intercropping maize and green bean could increase root growth of maize (about 7.2% more than in monocropping). Biological nitrogen fixation is heavily dependent on P, which provides energy for converting atmospheric N to useable N compounds. Arbuscular mycorrhizal fungi form a tripartite association with N-fixing rhizobia and legumes assisting in acquiring P [48].

Plant nutrient uptake involves two pathways, the direct pathway (DP) involves uptake of nutrients from the rhizosphere by the root epidermis and root hairs [49]. The mycorrhizal pathway (MP) develops behind the root hair zone and involves uptake of nutrients by the ERM, rapid translocation over many centimeters, delivery to the symbiotic interfaces, and transfer to the plants [34]. The two pathways are potentially independent and involve different cell types and different nutrient transporters, providing capacity for independent and coordinated regulation and nutrient access from different regions and volumes of soil [39, 41]. The conserved P-sensing pathway regulates the direct and indirect P acquisition pathways, centered on phosphate starvation response (PHR) transcription factors [50]. PHR transcription factor was characterized in many plant species [51, 52]. Xu et al. [53] identified 18 ZmPHR genes involved in relocating inorganic P across different maize plant tissues.

The mycorrhizal pathway involves the uptake of orthophosphate (Pi) by AM fungal high-affinity Pi transporters in the ERM, followed by translocation of P along the hyphae to intracellular structures in the root cortex and then transfers P to the root [34, 54–56]. Polyphosphate (polyP: linear chains of Pi residues linked by phosphoanhydride bonds) accumulate in the hyphae, where it buffers cytoplasmic Pi concentration, providing temporary P storage, and translocate P along hyphae [50, 57] resulting in rapid, long distance P translocation from sites of uptake in the ERM to sites of transfer to the plant [58]. Pi and polyP are known to carry negative charge, which cations must balance. In soil-grown plants, K^+ and Mg^{2+} may play this role [59, 60]. H^+-ATPases, which energize perifungal membranes surrounding arbuscules, are involved in all Pi-uptake steps [61–63].

The advantage of the AM symbiosis for plants in acquiring P is that AM fungi provide a very effective pathway, the AM pathway, for scavenging P from large volumes of soil and rapidly delivering to cortical cells within the root bypassing direct uptake [34]. Research by Smith and colleagues [34, 64] revealed that the AM pathway plays a significant role in P uptake, regardless of how an AM plant benefits in terms of increased growth or P uptake.

In addition, AM fungi can deliver substantial amounts of N to the host plant [65]. Researchers have different opinions on the mechanisms of N transfer to the host plant. Smith et al. [66] suggested that N might be transferred from the fungus to the host in the form of amino acids (AA) or amides. While Kaldorf et al. [67] suggested transfer in the form of NO_3^-, Bago et al. [68] postulated that N is transferred to the host as NH_4^+. The allocation of C to the AM fungi has been reported to depend on the N status of the mycorrhizal root, which indicates that the stoichiometry of C and N regulates the nutrient exchange between the fungus and the host plant [69]. The ERM absorbs various forms of N with 21% of total N taken up by the fungal ERM transferred to roots in root organ culture [70]. Additionally it was demonstrated that 74% of the total N in the leaves of maize was derived from the slow-release urea added to the hyphal compartment [71].

Studies showed that the AM fungus *Glomus hoi* enhanced degradation of organic residues and N uptake by the host plant [72, 73]. Jin et al. [74] demonstrated that urea or NH_4^+ were absorbed more rapidly than NO_3^-, amino acids (AA), and proteins when supplied as N sources for AM fungal uptake. Assimilation of NH_4^+ is the principal means of N absorption in AM fungal systems mediated by a specific carrier [75]. GintAMT1, which encodes the high-affinity NH_4^+ transporter in the AM fungus *Rhizophagus intraradices* [76] is one such carrier. The uptake of NO_3^- is linked to an H^+ symporter that alkalinizes the external mycelium. It has been shown that mycorrhizal roots growing in a NO_3^- amended soil induced an initial (30 d) alkalinization of the mycorrhizosphere, which was followed (at 60 d) by strong acidification. This acidification of the mycorrhizosphere would be the consequence of the unmasking of other cation/anion balances involved in different nutrient uptake processes once the nitrate is depleted [77, 78].

3. Carbon sequestration

The AM symbiosis represents a significant link between atmospheric and soil-contained carbon (C). Soil is one of the planet's largest C sinks. It stores at least twice as much C as currently occurs in the world's vegetation plus atmosphere. The estimated total soil organic carbon (SOC) to 2-meter depth is 2400 Pg, which is three times the amount of CO_2 currently in the atmosphere (~830 Pg C) and 240 times current annual fossil fuel emissions (~10 Pg) [79]. Carbon storage depends on the balance between carbon sequestration by plant photosynthesis and carbon release to the atmosphere through soil respiration [79, 80]. Atmospheric CO_2 concentrations will reach 550 ppm by 2100, accompanied by an increase in global average annual temperature of 4.4 ± 0.5°C for 2070–2099 [81]. One of the consequences of this increase may be increased carbon availability to fungi to develop the mycorrhizal mycelium [82]. AM fungi receive increased levels of photosynthates under elevated CO_2 before other soil microbes [83, 84], increasing AM fungal colonization [85, 86]. Treseder and Allen [87] reported that mycorrhizal dependent plants allocate 5–20% of the net photosynthate to maintain symbiosis [88] increasing the sink effect and movement of photo-assimilates from aerial parts to the roots [89]. CO_2 enhancement of AM fungi might alter terrestrial ecosystem C dynamics by stimulating the decomposition of soil organic C in AM fungal active zones [90]. Globally forest soils release approximately 24Pg carbon per year into the atmosphere via CO_2 efflux and generate CO_2 from a wide variety of belowground organisms, with AM fungi as the dominant carbon source [91]. Elevated CO_2 increases allocation to AM hyphae in the soil outside plant roots [86, 92] due to carbon sequestration [93]. Research conducted in a German grassland community reported that AM fungi stimulated soil respiration of pasture soil, leading to elevated CO_2 levels and temperature, with most carbon sequestered in belowground parts [94, 95].

High-temperature stress negatively affects plant morphological, physiological, and biochemical growth, leading to reduced plant productivity [96, 97]. Hatfield et al. [98] showed that temperatures above 35°C affected maize vegetative and reproductive growth, from germination to grain filling. Temperature also regulates mycorrhizal fungal growth and metabolic activity [99], with colonization peaking during the growing season when temperatures are warm [100]. Changes in atmospheric CO_2 concentration and temperature and resulting changes in soil physicochemical properties and microbial activity [101] can influence mycorrhizal symbiosis on various scales [102].

Crop growth and development are critical factors in determining the impact of a changing environment. Increasing temperatures affect all major grain crops [103]. Maize is one of the most important crops grown in tropical countries and tolerates temperatures up to 32–33°C. Beyond this range, crop growth and yield starts to decline [103]. Photosynthesis is one of the most heat-sensitive processes in plants and is essential to maintaining the mycorrhizal symbiosis [104, 105]. The AM fungal network provides the host plant with nutrients and water from the soils and enhances the plant's tolerance to various abiotic stresses [106] and can thus alleviate additional stress placed on crop plants due to climate change.

4. Tolerance to drought, temperature and salinity

Environmental factors are stressors that impact plant growth [107]. Drought is a significant challenge [108] resulting from low precipitation and a high rate of evapotranspiration causing reduction in plant cell division and proliferation of roots, closure of stomatal pores, changes in plant and water uptake efficiency, and high production of abscisic acid, which further decreases evapotranspiration by controlling stomatal pores [109]. Plants have developed various mechanisms to retain water under these conditions.

The ability of AM fungi to exploit soil resources assists the host plants' ability to grow under drought stress [110]. The AM fungal hyphae can absorb water by entering soil pores that are too small for root hairs to access [111, 112]. AM fungal mycelia can also improve soil moisture retention by enhancing soil aggregation [113]. Glomalin is a hydrophobic protein [113] present in AM fungal hyphal walls [114], which forms an insoluble glue with a strong cementing capacity that stabilizes soil aggregates [92, 115–119]. Glomalin is quantified by measuring several glomalin related soil-protein (GRSP) pools [92]. The glomalin compound contains 30–40% C, which protects the soil from drying out [120], 0.9–7.3% of N, 0.03–0.1% of P and metal ions [121]. Hyphae and glomalin contributed up to 15% of soil organic C in a grassland [122]. As a result, a considerable amount of C allocated to AM fungi is used in glomalin production, governed by plant productivity [122]. The regulation of plant nutrient uptake, stomatal conductance, leaf water potential, photosynthesis, and transpiration [123, 124] help plants to produce significantly higher yields under stressful conditions. The AM fungi modify the root hairs allowing plants to overcome drought [123]. They assist in maintaining high relative water content of the leaf, improving water use efficiency [125], increasing leaf area, delaying senescence [126] and maintaining ion balance [127]. Li et al. [128] observed that in the arbuscule-enriched cortical cells and ERM of maize roots, the expression of two functional aquaporin AM fungal genes i.e., GintAQPF1 and GintAQPF2, were enhanced under drought stress. Maize is sensitive to drought and heat stress, particularly at the reproductive stages of development reducing grain yield [129]. Some AM fungal species, such as *Funneliformis mosseae*, minimize the adverse effects of drought by the accumulation of AA, increase in trehalose content and higher trehalase activity [130]; *Rhizophagus intraradices* increased plant dry weight, uptake of P, N, K, and Mg in shoot, and water use efficiency [131].

In many parts of the world, maize production occurs in semi-arid environments where high temperatures and water scarcity [132, 133] are common challenges. In China 60% of crops in maize growing regions are often subjected to spells of heat and drought, resulting in 30% yield losses per year [133]. These climate change-induced stresses will significantly threaten maize yields and decrease world maize production by 15–20% annually [133, 134]. AM fungi play an essential role in improving drought tolerance. They mediate the increase in nutritional status by increasing the developing root surface area, enhancing the uptake of P [131, 135, 136], increasing resistance to withering [137], increasing proline accumulation levels in roots [138, 139] and by increasing photosynthetic activity as detected by the increase in chlorophyll [140, 141]. Temperature is another important environmental factor that determines the growth and productivity of crops [142]. Temperature stress (low and high temperature) can occur during the growing season [97, 143] resulting in the disruption of physiological and biochemical processes and functions. This results in injuries such as damage of cell membrane structure and lipid composition, cellular leakage of electrolytes and amino acids, peroxidation of membrane lipids, a diversion of electron flow to alternate pathways, denaturation and aggregation of proteins, redistribution of intracellular calcium ions, inactivation of enzymes in chloroplast and mitochondria, and production of toxic compounds and reactive oxygen species (ROS) [97, 136, 144, 145]. AM symbiosis can alter plant physiology to deal with these stress conditions [146].

Zhu et al. [136] reported that *Claroideoglomus etunicatum* could be used to reduce high temperature effects on maize by reducing membrane lipid peroxidation, membrane permeability and increasing accumulation of osmotic adjustment compounds and antioxidant activity. At 35°C and 40°C, positive net photosynthesis rate, transpiration rate, stomatal conductance, chlorophyll and carotenoid contents, relative water content and negative intercellular CO_2 concentrations were recorded [136, 142].

Soil, water and the environment can influence crop salt tolerance [89]. Among abiotic stresses, soil salinization is probably one of the most important in the world [147]. Globally, arid and semi-arid soils are significant factors limiting agricultural productivity due to the high soil salinity [148]. More than 800 million hectares of land worldwide is affected by either salinity (397 million hectares) or sodicity (434 million hectares) [149]. The accumulation of soluble salts in the rhizosphere can reduce water potential and, consequently, water availability to plants. Under such circumstances, the uptake of these salts can affect the physiological processes of plants growing in these environments [150, 151] and poses the biggest challenge to food security [152]. *Rhizophagus intraradices* and *Funneliformis geosporum* increased leaf length, plant height, leaf number, chlorophyll a content, photosynthetic rate, stomatal conductance, and transpiration rate in maize [106]. Farooq et al. [153] found that AM fungal colonization and symbiosis improved salt resistance in maize due to better nutrient availability, increased potassium/sodium ratios in plant tissues, and better osmotic adjustment.

Increased temperatures and resultant drought and salinity are not the only legacies of climate change. Anthropogenic activities, which include industrial activities

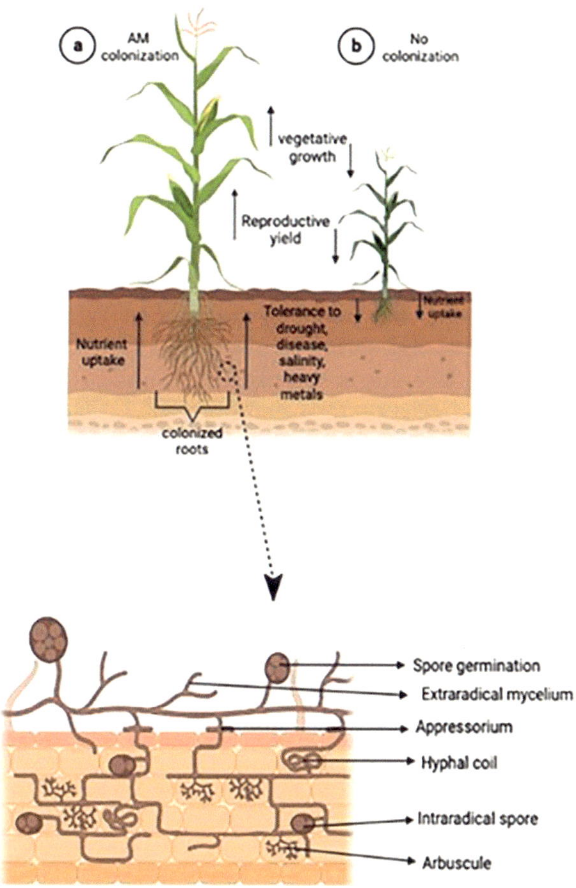

Figure 1.
Schematic representation of the effect of AM fungi on maize crop: (a) extensive roots on plant colonized by AM fungi, (b) reduced roots on plant with no colonization. (17/02/2021 - Biorender.com).

(mining, metal processing, fossil fuel combustion) and agricultural practices (application of fertilizers, fungicides, and sewage sludge disposal), have been described as the primary contaminants of the environment with heavy metals and are not included in this review [154, 155], however these and other growth benefits to maize are illustrated in **Figure 1**.

5. Soil management for enhanced maize production

Severely disturbed land is a global concern because changes in land use are one of the biggest threats to biodiversity and ecosystem services worldwide. This is exacerbated by increased demand for agricultural production. Studies conducted have shown that disturbance not only reduces AM fungal abundance, diversity and infectivity but can also result in drastic shifts in the AM fungal community [156]. AM fungal hyphae and root litter are the most abundant carbon source in the soil [157], providing energy for other soil microbes to flourish [92]. They may increase the diversity and abundance of microorganisms beneficial to plant growth and health [157]. Hyphae are highly susceptible to disturbance and disturbance results in reduced infective potential of AM fungi [157]. As the scale of degradation increases, the abundance and diversity of AM fungi reduces [158].

Tillage management plays a central role in ecological and biological stability, which is closely related to soil quality, by influencing the activities of soil microbial communities [159–161]. Soil disturbance, caused by tillage or plowing, decrease AM fungal colonization, disrupt AM hyphal networks [162–166], reducing spore numbers [167, 168], AM fungal species richness [156, 169] and glomalin production [170].

Conventional or conservation tillage are soil management methods employed [171]. Conservation tillage results in less disturbance when compared with conventional tillage and tends to benefit the soil by conserving aggregate stability and organic matter content [172]. Other reported benefits include higher microbial biomass in conservation tillage due to less disruption and preservation of the hyphal network, contributing to aggregate stability [173, 174]. Reduced tillage increased the abundance of AM fungal and saprotrophic fungal lipids in shallow soil layers [175].

Soil disturbance reduced P uptake from the soil by maize plants, while the uptake of P by canola was not affected [176]. Canola, a non-mycorrhizal host plant, supports the hypothesis that soil disturbance reduces the effectiveness of the mycorrhizal association. McGonigle et al. [177] found that tillage reduced colonization of maize, and the P and Zn contents of maize shoots. The alteration of AM fungal communities by tillage has been reported under field conditions [178]. They demonstrated that colonization by AM fungi from the genus *Scutellospora* was depressed by intensive tillage, while members of the genus *Glomus* were not affected. Fairchild and Miller [179] demonstrated improved AM colonization of maize growing in the undisturbed soil compared to the disturbed soil when amended with P.

Globally, the prevalence of low fertility soils requires amendment with large amounts of inorganic fertilizers and application of pesticides to achieve maximum plant growth and crop yield [180, 181]. The excessive and inappropriate application of chemical fertilizers can cause a series of environmental problems and soil degradation. Balzergue et al. [182] found that the high P concentrations in plants induced by high P fertilization inhibited mycorrhizal symbiosis. Symbiosis modulating compounds in root exudates such as strigolactone are also reduced under high P conditions [183]. Under these conditions plants are less reliant on mycorrhizal mediated P uptake and reduce carbohydrate sharing [184–186], this results in a decreased supply of soluble carbohydrate in roots reducing appressorial formation and new colonization [183, 184]. These phenomenal impact AM colonization, arbuscule formation and active P transfer to plants [187]. Some AM fungi can be relatively susceptible to fungicides, particularly when applied to the seed or the soil [188] while other fungicides can also stimulate mycorrhizal growth [189]. Fungicides such as flutolanil, azoxystrobin, fenpropimorph and fenhexamid can inhibit spore germination of *Rhizophargus irregularis* [190, 191]. The insecticide, oxamyl reduced root colonization by a commercial *Funneliformis mosseae* inoculum [192], and azadirachtin inhibiting *Claroideoglomus etunicatum* in the field causing a significant shift in the AM fungal community [193].

The response of AM fungi to agrochemicals is both substance- and dose-dependent. A field experiment showed that most AM fungi belonging to the *Glomus* group were sensitive to high levels of herbicide nicosulfuron which accumulated in soil due to repeated applications in later culture cycles [194]. Atrazine has been used as an agricultural herbicide worldwide, mostly on maize, sorghum, and sugarcane. Studies on maize showed a significant reduction in AM fungal spores [181] and AM colonization [195]. Makarian et al. [196] found a significant effect of the herbicide (metribuzin) on maize dry weight where an increase in herbicide concentration resulted in a decrease in the maize dry weight. Low herbicide concentrations resulted in increased shoot height of AM plants than when applied at high concentrations suggesting that mycorrhizal fungi can alleviate crop stress under lower doses of the herbicide [196].

The extensive use of agrochemicals reduces ecosystem functioning, contributing to soil and water degradation [197]. It also exerts deleterious effects on human health, mainly through the exposure of workers [198–200] and the intake of contaminated food crops [201]. Therefore, increasingly, the enhancement of more biologically based cultivation for safer and healthier food is a rising need, along with finding alternatives to replace agrochemicals in plant production [202–204]. The use of biofertilizers appears to be a natural option, particularly in low agrochemical input systems, because of their capacity to maintain long term soil fertility and sustainability by improving the uptake efficiency and availability of nutrients to plants [205]. Plants inoculated with AM fungi not only have improved growth but also have superior food quality properties, such as increased antioxidants, vitamins, and minerals [206]. AM fungal benefits related to maize are summarized in **Table 1**.

Monoculture is the cultivation of a single crop over a large area over consecutive years [241] and was adopted as a means to increase production [242]. Cultivated crops usually have identical genetic similarities, uniform growth patterns, and resistance to certain common diseases in monoculture. This system includes crop varieties uniquely suited to the specific conditions of a particular location [243]. This approach is criticized for its environmental impacts and is known as one of the major causes of soil degradation due to nonrotational cropping [244].

Hijri et al. [243] found that in continuous maize monoculture diversity of AM fungi decreased but found high diversity in long term field experiments where low-input agriculture involving crop rotation provided better conditions for their preservation. Sangabriel-conde et al. [245] investigated the AM fungal symbioses in native maize landraces at different levels of phosphorus fertilization. They showed a high diversity of AM fungi, most of which colonized several maize varieties, was best achieved at a moderate P level.

Parameter (p/f)*	Benefits	Reference
Nutrient uptake (p)	Increased - N; P; K; Ca; Mg; Na	[37, 186, 207–223]
Nutrient uptake (f)	Increased - P; K; Ca; Mg; Fe	[221, 224–226]
Vegetative growth (p)	Increased - shoot and root biomass; root length; plant, leaf and tassel length; stem girth	[186, 203, 205, 206–209, 211, 213, 214, 216, 219, 221–229]
Vegetative growth (f)	Increased - shoot and root biomass; root length; plant height; leaf mass and area	[217, 221, 224, 226, 230–234]
Yield (p)	Increased - cob and grain yield; number of grains per cob	[215, 221, 235–237]
Yield (f)	Increased - cob and grain yield; number of grains per cob; silage yield	[226, 231, 233, 234, 238]
Salinity tolerance (p)	Increased - shoot and root biomass; K^+; Na^+; root volume and diameter; Reduced Cl and Na in shoots;	[208, 211, 239]
Salinity tolerance (f)	Increased – soil macroaggregates; soil bacterial diversity; nutrient uptake; photosynthesis and chlorophyll; K^+; Na^+	[217, 240]

**p/f indicates whether the experiment was conducted under pot (p) trial or field (f) trial conditions.*

Table 1.
Compendium of studies showing effect of arbuscular mycorrhizal (AM) fungi on maize.

One of the most important soil properties is its structure [246]. Soil structure results from the iterations of the soil's chemical, physical and biological factors [247, 248]. Soil management practices, especially tillage systems, affect almost all soil properties, including AM fungi activity, diversity, and glomalin production [248].

Arbuscular mycorrhizal fungi have essential functions in the construction of the soil structure by acting on the formation and stabilization of the aggregates [247, 249–251]. The effect of AM fungi on soil aggregation is a result of ERM growth into the soil matrix creating the skeletal structure that physically entangles soil particles which along with roots enable microaggregates to form in the soil. Microaggregates form larger aggregated [251] via the production of a soil glycoprotein, glomalin [247, 252–254].

AM fungi account for 5–50% of the biomass of soil microbes [255]. Approximately 10–100 m mycorrhizal mycelium per cm root has been estimated [164], the biomass of the ERM may amount to 54–900 kg/ha [256]. Rilling et al. [120] estimated that pools of organic carbon such as glomalin produced by AM fungi might even exceed soil microbial biomass by a factor of 10–20. Glomalin is present in the soil in large amounts. The concentration of glomalin in soil depends on the vegetation cover and the manner of soil management [257] and ranges from 1.6 to 2.3 mg/g soil [258]. Some examples of glomalin concentration in diverse ecosystems include; Agricultural land 0.3–0.7 mg/g [114, 259]; Boreal forest 1.1 mg/g [260]; Desert 0.003–0.13 mg/g [170, 261]; Temperate forest 0.60–5.8 mg/g [170, 262, 263]; Temperate grassland 0.23–2.5 mg/g [263–265]; Tropical rainforest 2.6–13.5 mg/g [170, 266] and Antarctic region 0.007–0.15 mg/g [267]. For example, in the top 10 cm of a tropical rain forest in Costa Rica up to 12.5 mg of glomalin $cm^{-3}$was reported [266] and up to 60 mg of glomalin cm^{-3} in a chrono sequence of Hawaiian soils [120]. Glomalin has a longer residence time in soil than hyphae, allowing for a long persistent contribution to soil aggregate stabilization. For hyphae, the residence time varies from days to months [86, 266], while for glomalin, it varies from 6 to 42 years [120]. The effects of AM fungi on soil aggregation are probably more easily detected in nutrient-poor soils with neutral or alkaline soil pH [268]. The management of mycorrhizal fungi and diversity in the soil can be considered a biological approach to improving soil structure [269, 270]. Improved soil structure results in improved water infiltration and can mitigate raindrop impact through higher soil stability, increasing resistance to slaking and reduced particles detachment [271]. Significant decreases in AM fungi hyphae and GRSP concentrations have been correlated to losses of C and N protected in macroaggregates as a result of reduced aggregate stabilization [272].

Maize is an obligatory mycorrhizal species (**Table 1**) readily colonized by many non-host-specific AM fungi [273]. Agricultural techniques employing direct sowing and reduced tillage interfere as little as possible with the soil structure and do not cause tearing of the trunks of mycorrhizal fungi [273, 274] resulting in an increase and activity of soil microorganisms and enzymes, especially in the top 20 cm layer. Roldán et al. [275] examined the effect of different management practices on the soil profile distribution of organic matter and physical and microbiological soil quality indicators in a maize field under subtropical conditions. They concluded that the tillage system significantly affected aggregate stability and glomalin. The increases in glomalin suggested that the proliferation of AM fungi could have mediated the improvement in soil aggregate stability under no-tillage. Investigating the influence of tillage and no-tillage on the mycorrhizal status of a field cultivated with maize or bean [276] revealed that GRSP was greater under no-tillage Maize plants (with a root mass of 450 g m^{-3}) had a more marked effect on improving soil aggregate stability than bean plants (with a smaller root mass of 42 g m^{-3}).

6. Conclusions and recommendations

Adoption of good agriculture management practices can increase productivity, reduce erosion, increase soil fertility, and increase the soil's water-holding capacity. For farmers, it is often easier to use the agricultural systems that they are familiar with and that are supported by existing research and existing industries rather than search for the necessary solutions. Some of these systems do not consider the differences of the agroecological zones, cultures and resource limitations which tend to fail in most of the areas where they are applied. Improved soil fertility is critical. Therefore, strategies must include biologically based systems to rebuild soil fertility. Populations and space pressures are forcing farmers to use land more intensely. Reducing land degradation and replenishing soil fertility requires an integrated sustainable approach that promotes agricultural management practices that enhance AM fungal diversity. AM fungi are a key ecosystem partner that relates to sustainable management in their activity contributing to many ecosystem functions, including soil aggregation, reduced nutrient losses, and improved plant nutrient acquisition, which may reduce the amounts of fertilizer required to achieve elevated yields. Because mycorrhizal networks can create indefinitely large numbers of fungal linkages connecting many plants in a community, AM fungal formation could be a critical element in the plant succession of ecosystems and reducing greenhouse gases. The distribution pattern of AM fungi and glomalin are helpful components in monitoring desertification and soil degradation [277].

Mitigating of climate change cannot ignore the role of AM fungi and their symbiotic interaction with important crops such as maize. Adoption of soil management approaches that sustain AM fungal populations whether indigenous or introduced are therefore essential.

Acknowledgements

This research was funded by Foundational Biodiversity Information Programme – National Research Foundation, German Academic Exchange Service and Rhodes University. We sincerely thank Mrs. Margot Brooks for the English language review of this manuscript.

Author details

Sílvia N.D. Maússe Sitoe and Joanna F. Dames*
Rhodes University, Grahamstown, South Africa

*Address all correspondence to: J.dames@ru.ac.za

References

[1] Houghton J, Meira Pilho L, Callander B, Harris N, Kattonberg A, Maskeel K. In: Lakeman J, editor. Climate Change 1995: The Science of Climatic Change. First ed. Cambridge: Cambridge University Press; 1996

[2] Metz B, Davidson O, Bosch P, Dave R, Meyer L. Climate Change 2007: Mitigation of Climate Change. Vol. 9780521880. Cambridge; 2007

[3] Smith P, Martino D, Cai Z, Gwary D, Janzen H, Kumar P, et al. Agriculture. In: Metz B, Davidson OR, Bosch PR, Dave R, Meyer LA, editors. Climate Change 2007: Mitigation. Contribution of Working Group III to the Fourth Assessment Report of the Intergovernmental Panel on Climate Change. Cambridge: Cambridge University Press; 2007. pp. 498-540

[4] Tubiello F, Salvatore M, Golec CR, Ferrara A, Rossi S, Biancalani R, et al. Agriculture, Forestry and Other Land Use Emissions by Sources and Removals by Sinks: 1990-2011 Analysis. Vol. 2. Italy: FAO, ESS/14-02; 2014

[5] IRRI Rice farming: Saving water through Alternate Wetting Drying (AWD) method, Indonesia; 2013. pp. 1-3

[6] Kasterine A, Vanzetti D. The effectiveness, efficiency and equity of market-based and voluntary measures to mitigate greenhouse gas emissions from the Agri-food sector. Trade and Environment Review. 2010;**2010**:1-37

[7] Uprety D, Dhar S, Hongmin D, Kimball B, Garg A, Upadhyay J. Technologies for Climate Change Mitigation: Agriculture Sector. In: Haggar J, editor. New Delhi: Magnum Custom; 2012. ISBN 2013206534

[8] FCCC Adoption of the Paris Agreement. In: Proceedings of the Conference of the Parties, Twenty-first session. Durban: United Nations; 2015

[9] FAO. The future of food and agriculture Trends and challenges. Rome: Food Agriculture Organization of the United Nations (FAO); 2017. ISBN 9789251095515

[10] Porter J, Xie L, Challinor A, Cochrane K, Howden S, Iqbal M, et al. Food security and food production systems. In: Field C, Barros V, Dokken D, Mach K, Mastrandrea M, Bilir P, Chatterjee M, Ebi K, Estrada Y, Genova R, Girma B, Kissel E, Levy A, MacCracken S, Mastrandrea P, White L, editors. Climate Change 2014 Impacts, Adaptation and Vulnerability: Part a: Global and Sectoral Aspects. United Kingdom and New York: Cambridge University Press; 2014. pp. 485-533 ISBN 9781107415379

[11] FAO. Climate change and food systems: Global assessments and implications for food security and trade. In: Elbehri A, editor. Rome: Food Agriculture Organization of the United Nations (FAO); 2015. ISBN 9789251086995

[12] FAO. Climate change and food security: Risks and responses. Rome: Food Agriculture Organization of the United Nations (FAO); 2015. ISBN 9789251089989

[13] Zhao J, Yang X. Distribution of high-yield and high-yield-stability zones for maize yield potential in the main growing regions in China. Agricultural and Forest Meteorology. 2018;**248**:511-517. DOI: 10.1016/j.agrformet.2017.10.016

[14] Chaudhry A. Maize in Pakistan. Faisalabad: Punjab Agricultural Research Coordination Board; 1983

[15] Erenstein O, Jaleta M, Sonder K, Mottaleb K, Prasanna B. Global maize production, consumption and trade: Trends and R&D implications. Food Security. 2022:1-25. DOI: 10.1007/s12571-022-01288-7

[16] Lv Y, Liang Z, Ge M, Qi W, Zhang T, Lin F, et al. Genome-wide identification and functional prediction of nitrogen-responsive intergenic and intronic long non-coding RNAs in maize (Zea mays L.). BMC Genomics. 2016;**17**:1-15. DOI: 10.1186/s12864-016-2650-1

[17] du Plessis J. Maize Production. Pretoria: Department of Agriculture and ARC-Grain Crops Institute; 2003

[18] DAFF. Trends in the Agricultural Sector 2011. Pretoria: Department of Agriculture; 2012

[19] Ludwig F, van Scheltinga C, Verhagen J, Kruijt B, van Ierland E, Dellink R, et al. Climate change impacts on Developing Countries - EU Accountability. Netherlands: Wageningen University and Research Centre; 2007

[20] Brussaard L, de Ruiter PC, Brown GG. Soil biodiversity for agricultural sustainability. Agriculture, Ecosystems and Environment. 2007;**121**:233-244. DOI: 10.1016/j.agee.2006.12.013

[21] Carrenho R, Trufem SFB, Bononi VLR, Silva ES. The effect of different soil properties on arbuscular mycorrhizal colonization of peanuts, sorghum and maize. Acta Botanica Brasilica. 2007;**21**:723-730. DOI: 10.1590/S0102-33062007000300018

[22] Lynch JPT, review no. Roots of the second green revolution. Australian Journal of Botany. 2007;**55**:493-512. DOI: 10.1071/BT06118

[23] Rai A, Rai S, Rakshit A. Mycorrhiza-mediated phosphorus use efficiency in plants. Environmental and Experimental Biology. 2013;**11**:107-117

[24] Bojtor C, Illés Á, Mohammad S, Mousavi N, Széles A, Oth BT, et al. Evaluation of the nutrient composition of maize in different NPK fertilizer levels based on multivariate method analysis. International Journal of Agronomy. 2021;**2021**:1-13. DOI: 10.1155/2021/5537549

[25] Jamal Z, Hamayun M, Ahmad N, Chaudhary F. Effects of soil and foliar application of different concentrations of NPF and foliar application of (NH4)2SO4 on different yield parameters in wheat. Agronomy. 2006;**5**:251-256

[26] Zhang W, Ma W, Ji Y, Fan M, Oenema O, Zhang F. Efficiency, economics, and environmental implications of phosphorus resource use and the fertilizer industry in China. Nutrient Cycling in Agroecosystems. 2008;**80**:131-144. DOI: 10.1007/s10705-007-9126-2

[27] Bukhsh M, Ahmad R, Ishaque M, Malik A. Response of maize hybrids to varying potassium application in Pakistan. Pakistan Journal of Agricultural Sciences. 2009;**46**:179-184

[28] Cooke G. Fertilizing for Maximum Yield. London: Granada Publishing Lmt5; 1985

[29] Almagrabi O, Abdelmoneim T. Using of arbuscular mycorrhizal fungi to reduce the deficiency effect of phosphorous fertilization on maize plants (Zea mays L.). Life Science Journal. 2012;**9**:1648-1654

[30] Lynch JP, Brown KM. Topsoil foraging - an architectural adaptation of plants to low phosphorus availability.

Plant and Soil. 2001;**237**:225-237. DOI: 10.1023/A:1013324727040

[31] Douds D, Nagahashi G, Reider C, Hepperly P. Inoculation with arbuscular mycorrhizal fungi increases the yield of potatoes in a high P soil. Biological Agriculture and Horticulture. 2007;**25**:67-78. DOI: 10.1080/01448765.2007.10823209

[32] Rakshit A, Bhadoria PS. Role of VAM on growth and phosphorus nutrition of maize with low soluble phosphate fertilization. Acta Agronómica. 2010;**59**:119-123

[33] Plenchette C, Clermont-Dauphin C, Meynard J, Fortin J. Managing arbuscular mycorrhizal fungi in cropping systems. Canadian Journal of Plant Science. 2005;**85**:31-40. DOI: 10.4141/P03-159

[34] Smith S, Read D. Mycorrhizal Simbiosos. Third ed. Great Britain: Elsevier Ltd.; 2008 ISBN 9780123705266

[35] Barea J, Azcón R, Azcón-Aguilar C. Mycorrhizosphere interactions to improve plant fitness and soil quality. Antonie van Leeuwenhoek. International Journal of General and Molecular Microbiology. 2002;**81**:343-351. DOI: 10.1023/A:1020588701325

[36] Lecomte J, St-Arnaud M, Hijri M. Isolation and identification of soil bacteria growing at the expense of arbuscular mycorrhizal fungi. FEMS Microbiology Letters. 2011;**317**:43-51. DOI: 10.1111/j.1574-6968.2011.02209.x

[37] Corkidi L, Rowland D, Johnson N, Allen E. Nitrogen fertilization alters the functioning of arbuscular mycorrhizas at two semiarid grasslands. Plant and Soil. 2002;**240**:299-310. DOI: 10.1023/A:1015792204633

[38] Ortas I, Ortakçi D, Kaya Z. Various mycorrhizal fungi propagated on different hosts have different effect on citrus growth and nutrient uptake. Communications in Soil Science and Plant Analysis. 2002;**33**:259-272. DOI: 10.1081/CSS-120002392

[39] Ortas I, Varma A. Field trials of bioinoculants. In: Varma A, Oelmuller R, editors. Advanced techniques in soil microbiology. New York: Springer Berlin Heidelberg; 2007. pp. 397-412. ISBN 9783540708643

[40] Smith F, Smith S. What is the significance of the arbuscular mycorrhizal colonisation of many economically important crop plants? Plant and Soil. 2011;**348**:63-79. DOI: 10.1007/s11104-011-0865-0

[41] Ortas I, Akpinar C. Response of kidney bean to arbuscular mycorrhizal inoculation and mycorrhizal dependency in P and Zn deficient soils. Acta Agriculturae Scandinavica Section B Soil and Plant Science. 2006;**56**:101-109. DOI: 10.1080/09064710510029196

[42] Bi Y, Li X, Christie P, Hu Z, Wong M. Growth and nutrient uptake of arbuscular mycorrhizal maize in different depths of soil overlying coal fly ash. Chemosphere. 2003;**50**:863-869. DOI: 10.1016/S0045-6535(02)00231-X

[43] Lu K, Yang X, Gielen G, Bolan N, Ok Y, Niazi N, et al. Effect of bamboo and rice straw biochars on the mobility and redistribution of heavy metals (Cd, Cu, Pb and Zn) in contaminated soil. Journal of Environmental Management. 2017;**186**:285-292. DOI: 10.1016/j.jenvman.2016.05.068

[44] Li L, Zhang F, Li X, Christie P, Sun J, Yang S, et al. Interspecific facilitation of nutrient uptake by intercropped maize and faba bean. Nutrient Cycling in Agroecosystems. 2003;**65**:61-71. DOI: 10.1023/A:1021885032241

[45] Rodrigo VHL, Stirling CM, Teklehaimanot Z, Nugawela A. Intercropping with banana to improve fractional interception and radiation-use efficiency of immature rubber plantations. Field Crops Research. 2001;**69**:237-249. DOI: 10.1016/S0378-4290(00)00147-7

[46] He X, Xu M, Qiu G, Zhou J. Use of 15N stable isotope to quantify nitrogen transfer between mycorrhizal plants. Journal of Plant Ecology. 2009;**2**:107-118. DOI: 10.1093/jpe/rtp015

[47] Marzban Z, Faryabi E, Torabian S. Effects of arbuscular mycorrhizal fungi and Rhizobium on ion content and root characteristics of green bean and maize under intercropping. Acta Agriculturae Slovenica. 2017;**109**:79-88. DOI: 10.14720/aas.2017.109.1.08

[48] Xavier L, Germida J. Bacteria associated with Glomus clarum spores influence mycorrhizal activity. Soil Biology and Biochemistry. 2003;**35**:471-478. DOI: 10.1016/S0038-0717(03)00003-8

[49] Gordon-Weeks R, Tong Y, Davies TGE, Leggewie G. Restricted spatial expression of a high-affinity phosphate transporter in potato roots. Journal of Cell Science. 2003;**116**:3135-3144. DOI: 10.1242/jcs.00615

[50] Ezawa T, Smith S, Smith F. P metabolism and transport in AM fungi. Plant and Soil. 2002;**244**:221-230. DOI: 10.1023/A:1020258325010

[51] Wang J, Sun J, Miao J, Guo J, Shi Z, He M, et al. A phosphate starvation response regulator Ta-PHR1 is involved in phosphate signalling and increases grain yield in wheat. Annals of Botany. 2013;**111**:1139-1153. DOI: 10.1093/aob/mct080

[52] Zhou J, Jiao F, Wu Z, Li Y, Wang X, He X, et al. OsPHR2 is involved in phosphate-starvation signaling and excessive phosphate accumulation in shoots of plants. Plant Physiology. 2008;**146**:1673-1686. DOI: 10.1104/pp.107.111443

[53] Xu Y, Liu F, Han G, Cheng B. Genome-wide identification and comparative analysis of phosphate starvation-responsive transcription factors in maize and three other gramineous plants. Plant Cell Reports. 2018;**37**:711-726. DOI: 10.1007/s00299-018-2262-0

[54] Bucher M. Functional biology of plant phosphate uptake at root and mycorrhiza interfaces. The New Phytologist. 2007;**173**:11-26. DOI: 10.1111/j.1469-8137.2006.01935.x

[55] Ferrol N, Barea J, Azcón-Aguilar C. The plasma membrane H+-ATPase gene family in the arbuscular mycorrhizal fungus Glomus mosseae. Current Genetics. 2000;**37**:112-118. DOI: 10.1007/s002940050017

[56] Javot H, Pumplin N, Harrison MJ. Phosphate in the arbuscular mycorrhizal symbiosis: Transport properties and regulatory roles. Plant, Cell and Environment. 2007;**30**:310-322. DOI: 10.1111/j.1365-3040.2006.01617.x

[57] Hijikata N, Murase M, Tani C, Ohtomo R, Osaki M, Ezawa T. Polyphosphate has a central role in the rapid and massive accumulation of phosphorus in extraradical mycelium of an arbuscular mycorrhizal fungus. The New Phytologist. 2010;**186**:285-289. DOI: 10.1111/j.1469-8137.2009.03168.x

[58] Viereck N, Hansen P, Jakobsen I. Phosphate pool dynamics in the arbuscular mycorrhizal fungus Glomus intraradices studied by in

vivo 31P NMR spectroscopy. The New Phytologist. 2004;**162**:783-794. DOI: 10.1111/j.1469-8137.2004.01048.x

[59] Ryan MH, McCully ME, Huang CX. Location and quantification of phosphorus and other elements in fully hydrated, soil-grown arbuscular mycorrhizas: A cryo-analytical scanning electron microscopy study. The New Phytologist. 2003;**160**:429-441. DOI: 10.1046/j.1469-8137.2003.00884.x

[60] Ryan MH, McCully ME, Huang CX. Relative amounts of soluble and insoluble forms of phosphorus and other elements in intraradical hyphae and arbuscules of arbuscular mycorrhizas. Functional Plant Biology. 2007;**34**:457-464. DOI: 10.1071/FP06242

[61] Gianinazzi-Pearson V, Arnould C, Oufattole M, Arango M, Gianinazzi S. Differential activation of H+-ATPase genes by an arbuscular mycorrhizal fungus in root cells of transgenic tobacco. Planta. 2000;**211**:609-613. DOI: 10.1007/s004250000323

[62] Krajinski F, Hause B, Gianinazzi-Pearson V, Franken P. Mtha1, a plasma membrane H+-ATPase gene from Medicago truncatula, shows Arbuscule-specific induced expression in mycorrhizal tissue. Plant Biology. 2002;**4**:754-761

[63] Rosewarne G, Smith F, Schachtman D, Smith S. Localization of proton-ATPase genes expressed in arbuscular mycorrhizal tomato plants. Mycorrhiza. 2007;**17**:249-258. DOI: 10.1007/s00572-006-0101-6

[64] Smith S, Smith F. Roles of arbuscular mycorrhizas in plant nutrition and growth: New paradigms from cellular to ecosystem scales. Annual Review of Plant Biology. 2011;**62**:227-250. DOI: 10.1146/annurev-arplant-042110-103846

[65] Jin HR, Liu J, Liu J, Huang XW. Forms of nitrogen uptake, translocation, and transfer via arbuscular mycorrhizal fungi: A review. Science China. Life Sciences. 2012;**55**:474-482. DOI: 10.1007/s11427-012-4330-y

[66] Smith S, Gianinazzi-Pearson V, Koide R, Cairney J. Nutrient transport in mycorrhizas: Structure, physiology and consequences for efficiency of the symbiosis. Plant and Soil. 1994;**159**:103-113. DOI: 10.1007/BF00000099

[67] Kaldorf M, Schmelzer E, Bothe H. Expression of maize and fungal nitrate reductase genes in arbuscular mycorrhiza. Molecular Plant-Microbe Interactions. 1998;**11**:439-448. DOI: 10.1094/MPMI.1998.11.6.439

[68] Bago B, Pfeffer P, Shachar-Hill Y. Could the urea cycle Be translocating nitrogen in the arbuscular mycorrhizal Symbiosis ? The New Phytologist. 2001;**149**:4-8

[69] Olsson P, Burleigh S, Van Aarle I. The influence of external nitrogen on carbon allocation to Glomus intraradices in monoxenic arbuscular mycorrhiza. The New Phytologist. 2005;**168**:677-686. DOI: 10.1111/j.1469-8137.2005.01532.x

[70] Toussaint J, St-Arnaud M, Charest C. Nitrogen transfer and assimilation between the arbuscular mycorrhizal fungus Glomus intraradices Schenck & Smith and Ri T-DNA roots of Daucus carota L. in an in vitro compartmented system. Canadian Journal of Microbiology. 2004;**50**:251-260. DOI: 10.1139/w04-009

[71] Tanaka Y, Yano K. Nitrogen delivery to maize via mycorrhizal hyphae depends on the form of N supplied. Plant, Cell and Environment. 2005;**28**:1247-1254. DOI: 10.1111/j.1365-3040.2005.01360.x

[72] Leigh J, Hodge A, Fitter AH. Arbuscular mycorrhizal fungi can transfer substantial amounts of nitrogen to their host plant from organic material. The New Phytologist. 2009;**181**:199-207. DOI: 10.1111/j.1469-8137.2008.02630.x

[73] Hodge A, Fitter AH. Substantial nitrogen acquisition by arbuscular mycorrhizal fungi from organic material has implications for N cycling. Proceedings of the National Academy of Sciences of the United States of America. 2010;**107**:13754-13759. DOI: 10.1073/pnas.1005874107

[74] Jin H, Pfeffer P, Douds D, Piotrowski E, Lammers P, Shachar-Hill Y. The uptake, metabolism, transport and transfer of nitrogen in an arbuscular mycorrhizal symbiosis. The New Phytologist. 2005;**168**:687-696. DOI: 10.1111/j.1469-8137.2005.01536.x

[75] Botton B, Chalot M. Nitrogen assimilation: Enzymology in Ectomycorrhizas. In: Varma A, Hock B, editors. Mycorrhiza: Structure, Function, Molecular Biology and Biotechnology. Berlin: Springer-Verlag; 1995. pp. 325-363 ISBN 9783662088999

[76] López-Pedrosa A, González-Guerrero M, Valderas A, Azcón-Aguilar C, Ferrol N. GintAMT1 encodes a functional high-affinity ammonium transporter that is expressed in the extraradical mycelium of Glomus intraradices. Fungal Genetics and Biology. 2006;**43**:102-110. DOI: 10.1016/j.fgb.2005.10.005

[77] Subramanian KS, Charest C. Acquisition of N by external hyphae of an arbuscular mycorrhizal fungus and its impact on physiological responses in maize under drought-stressed and well-watered conditions. Mycorrhiza. 1999;**9**:0069-0075. DOI: 10.1007/s005720050289

[78] NSAC. Agriculture and Climate Change: Policy Imperatives and Opportunities to Help Producers Meet the Challenge. Washington D.C.; 2019

[79] Hu H, Wang Z, Liu G, Fu B. Vegetation carbon storage of major shrublands in China. Journal of Plant Ecology. 2006;**30**:539-544

[80] Zhang F, Zhao H, Song Y, Chen L. The effect of coal-mining subsidence on water environment in the shenfu-dongsheng mining area. Acta Geoscientia Sinica. 2007;**28**:521-527

[81] Beyene T, Lettenmaier DP, Kabat P. Hydrologic impacts of climate change on the Nile River basin: Implications of the 2007 IPCC scenarios. Climatic Change. 2010;**100**:433-461. DOI: 10.1007/s10584-009-9693-0

[82] Rouhier H, Read DJ. Plant and fungal responses to elevated atmospheric CO2 in mycorrhizal seedlings of Betula pendula. Environmental and Experimental Botany. 1999;**42**:231-241. DOI: 10.1016/S0098-8472(99)00039-8

[83] Drigo B, Pijl A, Duyts H, Kielak A, Gamper H, Houtekamer M, et al. Shifting carbon flow from roots into associated microbial communities in response to elevated atmospheric CO2. Proceedings of the National Academy of Sciences of the United States of America. 2010;**107**:10938-10942. DOI: 10.1073/PNAS.0912421107

[84] Drigo B, Kowalchuk G, Knapp B. Impacts of 3 years of elevated atmospheric CO 2 on rhizosphere carbon flow and microbial community dynamics. Global Change Biology. 2013;**19**:621-636. DOI: 10.1111/gcb.12045

[85] Alberton O, Kuyper T, Gorissen A. Takingmycocentrismseriously:Responses mycorrhizal fungal and elevated CO. The New Phytologist. 2005;**167**:859-868

[86] Treseder K. A meta-analysis of mycorrhizal responses to nitrogen, phosphorus, and atmospheric CO2 in field studies. The New Phytologist. 2004;**164**:347-355. DOI: 10.1111/j.1469-8137.2004.01159.x

[87] Treseder K, Allen M. Mycorrhizal fungi have a potential role in soil carbon storage under elevated CO2 and nitrogen deposition. The New Phytologist. 2000;**147**:189-200. DOI: 10.1046/j.1469-8137.2000.00690.x

[88] Luo Y, Zhou X. Soil Respiration and Environment. Burlington: Elsevier; 2006 ISBN 9780120887828

[89] Begum N, Qin C, Ahanger M, Raza S, Khan M, Ashraf M, et al. Role of arbuscular mycorrhizal Fungi in plant growth regulation: Implications in abiotic stress tolerance. Frontiers in Plant Science. 2019;**10**:1-15. DOI: 10.3389/fpls.2019.01068

[90] Yu CH, Huang CH, Tan CS. A review of CO2 capture by absorption and adsorption. Aerosol and Air Quality Research. 2012;**12**:745-769. DOI: 10.4209/aaqr.2012.05.0132

[91] Qi Y, Xu M, Wu J. Temperature sensitivity of soil respiration and its effects on ecosystem carbon budget.Pdf. Ecological Modelling. 2002;**153**:131-142

[92] Rillig M. Arbuscular mycorrhizae, glomalin, and soil aggregation. Canadian Journal of Soil Science. 2004;**84**:355-363. DOI: 10.4141/S04-003

[93] Wang Z, Bi Y, Jiang B, Zhakypbek Y, Peng S, Liu W, et al. Arbuscular mycorrhizal fungi enhance soil carbon sequestration in the coalfields, Northwest China. Scientific Reports. 2016;**6**:1-11. DOI: 10.1038/srep34336

[94] Friedlingstein P, Cox P, Betts R, Bopp L, von Bloh W, Brovkin V, et al. Climate-carbon cycle feedback analysis: Results from the C4MIP model intercomparison. Journal of Climate. 2006;**19**:3337-3353. DOI: 10.1175/JCLI3800.1

[95] Bahn M, Rodeghiero M, Anderson-Dunn M, Dore S, Gimeno C, Drösler M, et al. Soil respiration in European grasslands in relation to climate and assimilate supply Europe PMC funders group. Ecosystems. 2008;**11**:1352-1367. DOI: 10.1007/s10021-008-9198-0

[96] Rykaczewska K. Impact of heat and drought stresses on size and quality of the potato yield. Plant, Soil and Environment. 2017;**63**:40-46. DOI: 10.17221/691/2016-PSE

[97] Wahid A, Gelani S, Ashraf M, Foolad MR. Heat tolerance in plants: An overview. Environmental and Experimental Botany. 2007;**61**:199-223. DOI: 10.1016/j.envexpbot.2007.05.011

[98] Hatfield JL, Boote KJ, Kimball BA, Ziska LH, Izaurralde RC, Ort D, et al. Climate impacts on agriculture: Implications for crop production. Agronomy Journal. 2011;**103**:351-370. DOI: 10.2134/agronj2010.0303

[99] Gavito M, Schweiger P, Jakobsen I. P uptake by arbuscular mycorrhizal hyphae: Effect of soil temperature and atmospheric CO2 enrichment. Global Change Biology. 2003;**9**:106-116. DOI: 10.1046/j.1365-2486.2003.00560.x

[100] Soudzilovskaia N, Douma C, Akhmetzhanova A, van Bodegom P, Cornwell W, Moens E, et al. Global patterns of plant root colonization intensity by mycorrhizal fungi explained by climate and soil chemistry. Global Ecology and Biogeography. 2015;**24**:371-382. DOI: 10.1111/geb.12272

[101] Sun H, Santalahti M, Pumpanen J, Köster K, Berninger F, Raffaello T, et al. Fungal community shifts in structure and function across a boreal forest fire chronosequence. Applied and Environmental Microbiology. 2015;**81**:7869-7880. DOI: 10.1128/AEM.02063-15

[102] Leppälammi-Kujansuu J, Ostonen I, Strömgren M, Nilsson LO, Kleja DB, Sah SP, et al. Effects of long-term temperature and nutrient manipulation on Norway spruce fine roots and mycelia production. Plant and Soil. 2013;**366**:287-303. DOI: 10.1007/s11104-012-1431-0

[103] Hatfield JL, Prueger JH. Temperature extremes: Effect on plant growth and development. Weather and Climate Extremes. 2015;**10**:4-10. DOI: 10.1016/j.wace.2015.08.001

[104] Bagyaraj D, Sharma M, Maiti D. Phosphorus nutrition of crops through arbuscular mycorrhizal fungi. Current Science. 2015;**108**:1288-1293. DOI: 10.18520/cs/v108/i7/1288-1293

[105] Millar N, Bennett A. Stressed out symbiotes: Hypotheses for the influence of abiotic stress on arbuscular mycorrhizal fungi. Oecologia. 2016;**182**:625-641. DOI: 10.1007/s00442-016-3673-7

[106] Mathur S, Sharma M, Jajoo A. Improved photosynthetic efficacy of maize (Zea mays) plants with arbuscular mycorrhizal fungi (AMF) under high temperature stress. Journal of Photochemistry and Photobiology B: Biology. 2018;**180**:149-154. DOI: 10.1016/j.jphotobiol.2018.02.002

[107] Schluter D. Ecology and the origin of species. Trends in Ecology & Evolution. 2001;**16**:372-380. DOI: 10.1016/S0169-5347(01)02198-X

[108] da Silva E, Nogueira R, da Silva M, de Albuquerque M. Drought stress and plant nutrition. Plant Stress. 2011;**5**:32-41

[109] Yamaguchi-Shinozaki K, Shinozaki K. Transcriptional regulatory networks in cellular responses and tolerance to dehydration and cold stresses. Annual Review of Plant Biology. 2006;**57**:781-803. DOI: 10.1146/annurev.arplant.57.032905.105444

[110] Tian C, Feng G, Li X, Zhang F. Different effects of arbuscular mycorrhizal fungal isolates from saline or non-saline soil on salinity tolerance of plants. Applied Soil Ecology. 2004;**26**:143-148. DOI: 10.1016/j.apsoil.2003.10.010

[111] Hardie K. The effect of removal of Extraradical hyphae on water uptake by vesicular-arbuscular mycorrhizal plants. The New Phytologist. 1985;**101**:677-684. DOI: 10.1111/j.1469-8137.1985.tb02873.x

[112] Hammer E, Balogh-Brunstad Z, Jakobsen I, Olsson P, Stipp S, Rillig M. A mycorrhizal fungus grows on biochar and captures phosphorus from its surfaces. Soil Biology and Biochemistry. 2014;**77**:252-260. DOI: 10.1016/j.soilbio.2014.06.012

[113] Bedini S, Pellegrino E, Avio L, Pellegrini S, Bazzoffi P, Argese E, et al. Changes in soil aggregation and glomalin-related soil protein content as affected by the arbuscular mycorrhizal fungal species Glomus mosseae and Glomus intraradices. Soil Biology and Biochemistry. 2009;**41**:1491-1496. DOI: 10.1016/j.soilbio.2009.04.005

[114] Wright S, Anderson R. Aggregate stability and glomalin in alternative crop rotations for the central Great Plains. Biology and Fertility of Soils. 2000;**31**:249-253. DOI: 10.1007/s003740050653

[115] Wright S, Upadhyaya A. Extraction of an abundant and unusual protein from soil and comparison with hyphal protein of arbuscular mycorrhizal fungi. Soil Science. 1996;**161**:575-586. DOI: 10.1097/00010694-199609000-00003

[116] Wright S, Upadhyaya A. A survey of soils for aggregate stability and glomalin, a glycoprotein produced by hyphae of arbuscular mycorrhizal fungi. Plant and Soil. 1998;**198**:97-107. DOI: 10.1023/A:1004347701584

[117] Rillig M, Wright S, Eviner V. The role of arbuscular mycorrhizal fungi and glomalin in soil aggregation. Plant and Soil. 2002;**238**:325-333

[118] Driver J, Holben W, Rillig M. Characterization of glomalin as a hyphal wall component of arbuscular mycorrhizal fungi. Soil Biology and Biochemistry. 2005;**37**:101-106. DOI: 10.1016/j.soilbio.2004.06.011

[119] Borie F, Rubio R, Morales A. Arbuscular mycorrhizal Fungi and soil aggregation. Revista de la Ciencia del Suelo y Nutrición Vegetal. 2008;**8**:9-18. DOI: 10.4067/s0718-27912008000200003

[120] Rillig M, Wright S, Nichols K, Schmidt W, Torn M. Large contribution of arbuscular mycorrhizal fungi to soil carbon pools in tropical forest soils. Plant and Soil. 2001;**233**:167-177. DOI: 10.1023/A:1010364221169

[121] Huang Y, Wang D, Cai J, Zheng W. Review of glomalin-related soil protein and its environmental function in the rhizosphere. Chinese Journal of Plant Ecology. 2011;**35**:232-236. DOI: 10.3724/sp.j.1258.2011.00232

[122] Treseder and Turner 2007 Glomalin in Ecosystems

[123] Li T, Lin G, Zhang X, Chen Y, Zhang S, Chen B. Relative importance of an arbuscular mycorrhizal fungus (Rhizophagus intraradices) and root hairs in plant drought tolerance. Mycorrhiza. 2014;**24**:595-602. DOI: 10.1007/s00572-014-0578-3

[124] Yang Y, Tang M, Sulpice R, Chen H, Tian S, Ban Y. Arbuscular mycorrhizal Fungi Alter fractal dimension characteristics of Robinia pseudoacacia L. seedlings through regulating plant growth, leaf water status, photosynthesis, and nutrient concentration under drought stress. Journal of Plant Growth Regulation. 2014;**33**:612-625. DOI: 10.1007/s00344-013-9410-0

[125] Meddich A, Jaiti F, Bourzik W, Asli A, El; Hafidi, M. Use of mycorrhizal fungi as a strategy for improving the drought tolerance in date palm (Phoenix dactylifera). Scientia Horticulturae (Amsterdam). 2015;**192**:468-474. DOI: 10.1016/j.scienta.2015.06.024

[126] Beltrano J, Ronco M. Improved tolerance of wheat plants (Triticum aestivum L.) to drought stress and rewatering by the arbuscular mycorrhizal fungus Glomus claroideum: Effect on growth and cell membrane stability. Brazilian. Journal of Plant Physiology. 2008;**20**:29-37. DOI: 10.1590/S1677-04202008000100004

[127] Giri B, Kapoor R, Mukerji K. Improved tolerance of Acacia nilotica to salt stress by arbuscular mycorrhiza, Glomus fasciculatum may be partly related to elevated K/Na ratios in root and shoot tissues. Microbial Ecology. 2007;**54**:753-760. DOI: 10.1007/s00248-007-9239-9

[128] Li T, Hu Y, Hao Z, Li H, Chen B. Aquaporin genes GintAQPF1 and GintAQPF2 from Glomus intraradices contribute to plant drought tolerance. Plant Signaling & Behavior. 2013;**8**:1-3. DOI: 10.4161/psb.24030

[129] Hussain H, Men S, Hussain S, Chen Y, Ali S, Zhang S, et al. Interactive effects of drought and heat stresses on morpho-physiological attributes, yield, nutrient uptake and oxidative status in maize hybrids. Scientific Reports. 2019;**9**:1-12. DOI: 10.1038/s41598-019-40362-7

[130] Schellenbaum L, Müller J, Boller T, Wiemken A, Schüepp H. Effects of drought on non-mycorrhizal and mycorrhizal maize: Changes in the pools of non-structural carbohydrates, in the activities of invertase and trehalase, and in the pools of amino acids and imino acids. The New Phytologist. 1998;**138**:59-66. DOI: 10.1046/j.1469-8137.1998.00892.x

[131] Zhu X, Song F, Xu H. Arbuscular mycorrhizae improves low temperature stress in maize via alterations in host water status and photosynthesis. Plant and Soil. 2010;**331**:129-137. DOI: 10.1007/s11104-009-0239-z

[132] Zhao F, Zhang D, Zhao Y, Wang W, Yang H, Ta F, et al. The difference of physiological and proteomic changes in maize leaves adaptation to drought, heat, and combined both stresses. Frontiers in Plant Science. 2016;**7**:1-19. DOI: 10.3389/fpls.2016.01471

[133] Lobell DB, Bänziger M, Magorokosho C, Vivek B. Nonlinear heat effects on African maize as evidenced by historical yield trials. Nature Climate Change. 2011;**1**:42-45. DOI: 10.1038/nclimate1043

[134] Chen J, Xu W, Velten J, Xin Z, Stout J. Characterization of maize inbred lines for drought and heat tolerance. Journal of Soil and Water Conservation. 2012;**67**:354-364. DOI: 10.2489/jswc.67.5.354

[135] Karti PDMH. Pengaruh Pemberian Cendawan Mikoriza Arbuskula terhadap Pertumbuhan dan Produksi Rumput Setaria splendida Stapf yang Mengalami Cekaman Kekeringan. Media Peternak. 2004;**27**:51-55

[136] Zhu X, Song F, Xu H. Influence of arbuscular mycorrhiza on lipid peroxidation and antioxidant enzyme activity of maize plants under temperature stress. Mycorrhiza. 2010;**20**:325-332. DOI: 10.1007/s00572-009-0285-7

[137] Pebriansyah A, Karti P, Permana A. Effect of drought stress and addition of Arbuscula mycorrhizal fungi (AMF) on growth and productivity of tropical grasses (Chloris gayana, Paspalum dilatatum, and Paspalum notatum). Pastura: Journal of Tropical Forage Science. 2012;**2**:41-48. DOI: 10.24843/Pastura.2012.v02.i01.p10

[138] Porcel R, Ruiz-Lozano J. Arbuscular mycorrhizal influence on leaf water potential, solute accumulation, and oxidative stress in soybean plants subjected to drought stress. Journal of Experimental Botany. 2004;**55**:1743-1750. DOI: 10.1093/jxb/erh188

[139] Bhaskara GB, Yang TH, Verslues PE. Dynamic proline metabolism: Importance and regulation in water limited environments. Frontiers in Plant Science. 2015;**6**:1-7. DOI: 10.3389/fpls.2015.00484

[140] Zai X, Zhu S, Qin P, Wang X, Che L, Luo F. Effect of Glomus mosseae on chlorophyll content, chlorophyll fluorescence parameters, and chloroplast ultrastructure of beach plum (Prunus maritima) under NaCl stress. Photosynthetica. 2012;**50**:323-328. DOI: 10.1007/s11099-012-0035-5

[141] Talaat N, Shawky B. Protective effects of arbuscular mycorrhizal fungi on wheat (Triticum aestivum L.)

plants exposed to salinity. Environmental and Experimental Botany. 2014;**98**:20-31. DOI: 10.1016/j.envexpbot.2013.10.005

[142] Zhu X, Song F, Liu S, Liu T. Effects of arbuscular mycorrhizal fungus on photosynthesis and water status of maize under high temperature stress. Plant and Soil. 2011;**346**:189-199. DOI: 10.1007/s11104-011-0809-8

[143] Theocharis A, Clément C, Barka EA. Physiological and molecular changes in plants grown at low temperatures. Planta. 2012;**235**:1091-1105. DOI: 10.1007/s00425-012-1641-y

[144] Ruelland E, Zachowski A. How plants sense temperature. Environmental and Experimental Botany. 2010;**69**:225-232. DOI: 10.1016/j.envexpbot.2010.05.011

[145] Matteucci M, D'Angeli S, Errico S, Lamanna R, Perrotta G, Altamura MM. Cold affects the transcription of fatty acid desaturases and oil quality in the fruit of Olea europaea L. genotypes with different cold hardiness. Journal of Experimental Botany. 2011;**62**:3403-3420. DOI: 10.1093/jxb/err013

[146] Miransari M, Bahrami H, Rejali F, Malakouti M. Using arbuscular mycorrhiza to alleviate the stress of soil compaction on wheat (Triticum aestivum L.) growth. Soil Biology and Biochemistry. 2008;**40**:1197-1206. DOI: 10.1016/j.soilbio.2007.12.014

[147] Zhu J. Regulation of ion homeostasis under salt stress. Current Opinion in Plant Biology. 2003;**6**:441-445. DOI: 10.1016/S1369-5266(03)00085-2

[148] FAO. How to Feed the World 2050. High-Level Expert Forum. 2009

[149] Munns R. Genes and salt tolerance. The New Phytologist. 2005;**167**:645-663

[150] Rogers M, Craig A, Munns R, Colmer T, Nichols P, Malcolm C, et al. The potential for developing fodder plants for the salt-affected areas of southern and eastern Australia: An overview. Australian Journal of Experimental Agriculture. 2005;**45**:301-329. DOI: 10.1071/EA04020

[151] Akbarimoghaddam H, Galavi M, Ghanbari A, Panjehkeh N. Salinity effects on seed germination and seedling growth of bread wheat cultivars. Trakia Journal of Sciences. 2011;**9**:43-50

[152] Rajkumar M, Prasad MNV, Swaminathan S, Freitas H. Climate change driven plant-metal-microbe interactions. Environment International. 2013;**53**:74-86. DOI: 10.1016/j.envint.2012.12.009

[153] Farooq M, Hussain M, Wakeel A, Siddique KHM. Salt stress in maize : Effects , resistance mechanisms , and management. A review. Agronomy for Sustainable Development. 2015;**35**:461-481. DOI: 10.1007/s13593-015-0287-0

[154] Ma Y, Prasad MNV, Rajkumar M, Freitas H. Plant growth promoting rhizobacteria and endophytes accelerate phytoremediation of metalliferous soils. Biotechnology Advances. 2011;**29**:248-258. DOI: 10.1016/j.biotechadv.2010.12.001

[155] Nagajyoti PC, Lee KD, Sreekanth TVM. Heavy metals, occurrence and toxicity for plants: A review. Environmental Chemistry Letters. 2010;**8**:199-216. DOI: 10.1007/s10311-010-0297-8

[156] Schnoor T, Lekberg Y, Rosendahl S, Olsson P. Mechanical soil disturbance as a determinant of arbuscular mycorrhizal fungal communities in semi-natural grassland. Mycorrhiza. 2011;**21**:211-220. DOI: 10.1007/s00572-010-0325-3

[157] Asmelash F, Bekele T, Birhane E. The potential role of arbuscular mycorrhizal fungi in the restoration of degraded lands. Frontiers in Microbiology. 2016;**7**:1-15. DOI: 10.3389/fmicb.2016.01095

[158] Cardozo Júnior F, Carneiro R, Goto B, Bezerra A, Araújo A, Nunes L. Arbuscular mycorrhizal fungi in degraded lands in Northeast Brazil. African Journal of Microbiology Research. 2012;**6**:7198-7205

[159] Nannipieri P, Ascher J, Ceccherini MT, Landi L, Pietramellara G, Renella G. Microbial diversity and soil functions. European Journal of Soil Science. 2003;**54**:655-670. DOI: 10.1046/j.1351-0754.2003.0556.x

[160] Griffiths BS, Philippot L. Insights into the resistance and resilience of the soil microbial community. FEMS Microbiology Reviews. 2013;**37**:112-129. DOI: 10.1111/j.1574-6976.2012.00343.x

[161] Xiao D, Xiao S, Ye Y, Zhang W, He X, Wang K. Microbial biomass, metabolic functional diversity, and activity are affected differently by tillage disturbance and maize planting in a typical karst calcareous soil. Journal of Soils and Sediments. 2019;**19**:809-821. DOI: 10.1007/s11368-018-2101-5

[162] Allison VJ, Miller RM, Jastrow JD, Matamala R, Zak DR. Changes in soil microbial community structure in a tallgrass prairie Chronosequence. Soil Science Society of America Journal. 2005;**69**:1412-1421. DOI: 10.2136/sssaj2004.0252

[163] Drijber RA, Doran JW, Parkhurst AM, Lyon DJ. Changes in soil microbial community structure with tillage under long-term wheat-fallow management. Soil Biology and Biochemistry. 2000;**32**:1419-1430

[164] McGonigle T, Miller M. Winter survival of extraradical hyphae and spores of arbuscular mycorrhizal fungi in the field. Applied Soil Ecology. 1999;**12**:41-50. DOI: 10.1016/S0929-1393(98)00165-6

[165] Miller M, Mcgonigle T, Addy H. Functional ecology of vesicular arbuscular mycorrhizas as influenced by phosphate fertilization and tillage in an agricultural ecosystem. Critical Reviews in Biotechnology. 1995;**15**:241-255. DOI: 10.3109/07388559509147411

[166] Mozafar A, Anken T, Ruh R, Frossard E. Tillage intensity, mycorrhizal and nonmycorrhizal fungi, and nutrient concentrations in maize, wheat, and canola. Agronomy Journal. 2000;**92**:1117-1124. DOI: 10.2134/agronj2000.9261117x

[167] Castillo CG, Rubio R, Rouanet JL, Borie F. Early effects of tillage and crop rotation on arbuscular mycorrhizal fungal propagules in an Ultisol. Biology and Fertility of Soils. 2006;**43**:83-92. DOI: 10.1007/s00374-005-0067-0

[168] Galvez L, Douds D, Drinkwater L, Wagoner P. Effect of tillage and farming system upon VAM fungus populations and mycorrhizas and nutrient uptake of maize. Plant and Soil. 2001;**228**:299-308. DOI: 10.1023/A:1004810116854

[169] Antunes P, Koch A, Dunfield K, Hart M, Downing A, Rillig M, et al. Influence of commercial inoculation with Glomus intraradices on the structure and functioning of an AM fungal community from an agricultural site. Plant and Soil. 2009;**317**:257-266. DOI: 10.1007/s11104-008-9806-y

[170] Treseder K, Turner K. Glomalin in ecosystems. Soil Science Society of America Journal. 2007;**71**:1257-1266. DOI: 10.2136/sssaj2006.0377

[171] South S, Qi J, Lusch DP. Optimal classification methods for mapping agricultural tillage practices. Remote Sensing of Environment. 2004;**91**:90-97. DOI: 10.1016/j.rse.2004.03.001

[172] Al-Kaisi M, Yin X. Tillage and crop residue effects on soil carbon and carbon dioxide emission in corn-soybean rotations. Journal of Environmental Quality. 2005;**34**:437-445. DOI: 10.2134/jeq2005.0437

[173] Curaqueo G, Barea J, Acevedo E, Rubio R, Cornejo P, Borie F. Effects of different tillage system on arbuscular mycorrhizal fungal propagules and physical properties in a Mediterranean agroecosystem in Central Chile. Soil and Tillage Research. 2011;**113**:11-18. DOI: 10.1016/j.still.2011.02.004

[174] Liu X, Zhang J, Gu T, Zhang W, Shen Q, Yin S, et al. Microbial community diversities and taxa abundances in soils along a seven-year gradient of potato monoculture using high throughput pyrosequencing approach. PLoS One. 2014;**9**:1-10. DOI: 10.1371/journal.pone.0086610

[175] van Groenigen K, Bloem J, Bååth E, Boeckx P, Rousk J, Bodé S, et al. Abundance, production and stabilization of microbial biomass under conventional and reduced tillage. Soil Biology and Biochemistry. 2010;**42**:48-55. DOI: 10.1016/j.soilbio.2009.09.023

[176] Halloran I, Miller M, Arnold G. Absorption of P by corn (Zea mays L.) as influenced by soil disturbance. Canadian Journal of Soil Science. 1986;**66**:287-302

[177] McGonicle T, Evans D, Miller M. Effect of degree of soil disturbance on mycorrhizal colonization and phosphorus absorption by maize in growth chamber and field experiments. The New Phytologist. 1990;**116**:629-636. DOI: 10.1111/j.1469-8137.1990.tb00548.x

[178] Jansa J, Mozafar A, Kuhn G, Anken T, Ruh R, Sanders IR, et al. Soil tillage affects the community structure of mycorrhizal fungi in maize roots. Ecological Applications. 2003;**13**:1164-1176. DOI: 10.1890/1051-0761(2003)13[1164,STATCS]2.0.CO;2

[179] Fairchild M, Miller G. Vesicular-arbuscular mycorrhizas and the soil-disturbance-induced reduction of nutrient absorption in maize. III. Influence of P amendments to soil. The New Phytologist. 1990;**114**:641-650. DOI: 10.1111/j.1469-8137.1990.tb00435.x

[180] Wilson C. Environmental and human costs of commercial agricultural production in South Asia. International Journal of Social Economics. 2000;**27**:816-846. DOI: 10.1108/03068290010335244

[181] Fawole O, Alori E. Impact of chemical inputs on arbuscular mycorrhizal spores in soil :Response of AM spores to fertilizer and herbicides. Albanian Journal of Agricultural Sciences. 2017;**16**:10-13

[182] Balzergue C, Chabaud M, Barker D, Bécard G, Rochange S. High phosphate reduces host ability to develop arbuscular mycorrhizal symbiosis without affecting root calcium spiking responses to the fungus. Frontiers in Plant Science. 2013;**4**:1-15. DOI: 10.3389/fpls.2013.00426

[183] López-Ráez J, Shirasu K, Foo E. Strigolactones in plant interactions with beneficial and detrimental organisms: The Yin and Yang. Trends in Plant Science. 2017;**22**:527-537. DOI: 10.1016/j.tplants.2017.03.011

[184] García-Caparrós P, Lao M, Preciado-Rangel P, Sanchez E.

Phosphorus and carbohydrate metabolism in green bean plants subjected to increasing phosphorus concentration in the nutrient solution. Agronomy. 2021;**11**:1-16. DOI: 10.3390/agronomy11020245

[185] Kiers E, Duhamel M, Beesetty Y, Mensah J, Franken O, Verbruggen E, et al. Reciprocal rewards stabilize cooperation in the mycorrhizal symbiosis. Science (80-.). 2011;**333**:880-882. DOI: 10.1126/science.1208473

[186] Willmann M, Gerlach N, Buer B, Polatajko A, Nagy R, Koebke E, et al. Mycorrhizal phosphate uptake pathway in maize: Vital for growth and cob development on nutrient poor agricultural and greenhouse soils. Frontiers in Plant Science. 2013;**4**:1-15. DOI: 10.3389/fpls.2013.00533

[187] Kobae Y, Ohmori Y, Saito C, Yano K, Ohtomo R, Fujiwara T. Phosphate treatment strongly inhibits new arbuscule development but not the maintenance of arbuscule in mycorrhizal rice roots. Plant Physiology. 2016;**171**:566-579. DOI: 10.1104/pp.16.00127

[188] Jin H, Germida JJ, Walley FL. Suppressive effects of seed-applied fungicides on arbuscular mycorrhizal fungi (AMF) differ with fungicide mode of action and AMF species. Applied Soil Ecology. 2013;**72**:22-30. DOI: 10.1016/j.apsoil.2013.05.013

[189] Murillo-Williams A, Pedersen P. Arbuscular mycorrhizal colonization response to three seed-applied fungicides. Agronomy Journal. 2008;**100**:795-800. DOI: 10.2134/agronj2007.0142

[190] Buysens C, Dupré de Boulois H, Declerck S. Do fungicides used to control Rhizoctonia solani impact the non-target arbuscular mycorrhizal fungus Rhizophagus irregularis? Mycorrhiza. 2015;**25**:277-288. DOI: 10.1007/s00572-014-0610-7

[191] Zocco D, Fontaine J, Lozanova E, Renard L, Bivort C, Durand R, et al. Effects of two sterol biosynthesis inhibitor fungicides (fenpropimorph and fenhexamid) on the development of an arbuscular mycorrhizal fungus. Mycological Research. 2008;**112**:592-601. DOI: 10.1016/j.mycres.2007.11.010

[192] Marin M, Ybarra M, Fé A, García-Férriz L. Effect of arbuscular mycorrhizal fungi and pesticides on Cynara cardunculus growth. Agricultural and Food Science. 2002;**11**:245-251. DOI: 10.23986/afsci.5728

[193] Ipsilantis I, Samourelis C, Karpouzas DG. The impact of biological pesticides on arbuscular mycorrhizal fungi. Soil Biology and Biochemistry. 2012;**45**:147-155. DOI: 10.1016/j.soilbio.2011.08.007

[194] Karpouzas DG, Papadopoulou E, Ipsilantis I, Friedel I, Petric I, Udikovic-Kolic N, et al. Effects of nicosulfuron on the abundance and diversity of arbuscular mycorrhizal fungi used as indicators of pesticide soil microbial toxicity. Ecological Indicators. 2014;**39**:44-53. DOI: 10.1016/j.ecolind.2013.12.004

[195] Demir S, Celebi SZ, Kaya I, Korhan Sahar A. The applications of arbuscular mycorrhizal fungus (AMF) and atrazine on silage maize (Zea mays L.) at different irrigation regimes. Journal of Food, Agriculture and Environment. 2011;**9**:629-633

[196] Makarian H, Poozesh V, Asghari HR, Nazari M. Interaction effects of arbuscular mycorrhiza Fungi and soil applied herbicides on plant growth.

Communications in Soil Science and Plant Analysis. 2016;**47**:619-629. DOI: 10.1080/00103624.2016.1146744

[197] Power AG. Ecosystem services and agriculture: Tradeoffs and synergies. Philosophical Transactions of the Royal Society B. 2010;**365**:2959-2971. DOI: 10.1098/rstb.2010.0143

[198] Meyer A, Koifman S, Koifman R, Moreira J, de Chrisman JR, Abreu-Villaça Y. Mood disorders hospitalizations, suicide attempts, and suicide mortality among agricultural workers and residents in an area with intensive use of pesticides in Brazil. Journal of Toxicology and Environmental Health - Part A: Current Issues. 2010;**73**:866-877. DOI: 10.1080/15287391003744781

[199] Alavanja M, Bonner M. Occupational pesticide exposures and cancer risk: A review. The Journal of Toxicology and Environmental Health: Part B, Critical Reviews. 2012;**15**: 238-263. DOI: 10.1080/10937404.2012.632358

[200] Freire C, Koifman RJ, Koifman S. Hematological and hepatic alterations in brazilian population heavily exposed to organochlorine pesticides. Journal of Toxicology and Environmental Health - Part A: Current Issues. 2015;**78**:534-548. DOI: 10.1080/15287394.2014.999396

[201] Carvalho FP. Agriculture, pesticides, food security and food safety. Environmental Science & Policy. 2006;**9**:685-692. DOI: 10.1016/j.envsci.2006.08.002

[202] Skevas T, Oude Lansink AGJM, Stefanou SE. Designing the emerging EU pesticide policy: A literature review. NJAS - Wageningen. Journal of Life Science. 2013;**64-65**:95-103. DOI: 10.1016/j.njas.2012.09.001

[203] Lim WM, Yong JLS, Suryadi K. Consumers' perceived value and willingness to purchase organic food. Journal of Global Marketing. 2014;**27**:298-307. DOI: 10.1080/08911762.2014.931501

[204] Oliveira R, Ma Y, Rocha I, Carvalho M, Vosátka M, Freitas H. Arbuscular mycorrhizal fungi are an alternative to the application of chemical fertilizer in the production of the medicinal and aromatic plant Coriandrum sativum L. Journal of Toxicology and Environmental Health - Part A: Current Issues. 2016;**79**:320-328. DOI: 10.1080/15287394.2016.1153447

[205] Rocha I, Ma Y, Carvalho M, Magalhães C, Janoušková M, Vosátka M, et al. Seed coating with inocula of arbuscular mycorrhizal fungi and plant growth promoting rhizobacteria for nutritional enhancement of maize under different fertilisation regimes. Archives of Agronomy and Soil Science. 2019;**65**:31-43. DOI: 10.1080/03650340.2018.1479061

[206] Baslam M, Garmendia I, Goicoechea N. Enhanced accumulation of vitamins, nutraceuticals and minerals in lettuces associated with arbuscular mycorrhizal fungi (Amf): A question of interest for both vegetables and humans. Agriculture. 2013;**3**:188-209. DOI: 10.3390/agriculture3010188

[207] Bouzeriba T, Mehjin A, Mikaeel Y, Dennis S, Ahmed A, Peter D. Effects of arbuscular mycorrhizal fungi on growth and phosphorus uptake of maize (Zea mays L.) at different levels of soil phosphorus and soil moisture. Journal of Dryland Agriculture. 2021;7:22-33. DOI: 10.5897/joda2020.0063

[208] Selvakumar G, Shagol CC, Kim K, Han S, Sa T. Spore associated bacteria regulates maize root K+/Na+

ion homeostasis to promote salinity tolerance during arbuscular mycorrhizal symbiosis. BMC Plant Biology. 2018;**18**:1-13. DOI: 10.1186/s12870-018-1317-2

[209] Zhang T, Yang X, Guo R, Guo J. Response of AM fungi spore population to elevated temperature and nitrogen addition and their influence on the plant community composition and productivity. Scientific Reports. 2016;**6**:1-12. DOI: 10.1038/srep24749

[210] Abdelmoneim T, Moussa T, Almaghrabi O, Alzahrani H, Abdelbagi I. Increasing plant tolerance to drought stress by inoculation with arbuscular mycorrhizal fungi. Life Science Journal. 2013;**10**:3273-3280

[211] Othira JO. Effectiveness of arbuscular mycorrhizal fungi in protection of maize (Zea mays L.) against witchweed (Striga hermonthica Del Benth) infestation. Journal of Agricultural Biotechnology and Sustainable Development. 2012;**4**:37-44. DOI: 10.5897/jabsd12.007

[212] Yang Y, Shuzhen Z, Honglin H, Lei L, Bei W. Arsenic accumulation and speciation in maize as affected by inoculation with arbuscular mycorrhizal fungus glomus mosseae. Journal of Agricultural and Food Chemistry. 2009;**57**:3695-3701. DOI: 10.1021/jf900107y

[213] Bai J, Lin X, Yin R, Zhang H, Junhua W, Xueming C, et al. The influence of arbuscular mycorrhizal fungi on As and P uptake by maize (Zea mays L.) from As-contaminated soils. Applied Soil Ecology. 2008;**38**:137-145. DOI: 10.1016/j.apsoil.2007.10.002

[214] Wang F, Lin X, Yin R. Inoculation with arbuscular mycorrhizal fungus Acaulospora mellea decreases Cu phytoextraction by maize from Cu-contaminated soil. Pedobiologia (Jena). 2007;**51**:99-109. DOI: 10.1016/j.pedobi.2007.02.003

[215] Wu S, Cao Z, Li Z, Cheung K, Wong M. Effects of biofertilizer containing N-fixer, P and K solubilizers and AM fungi on maize growth: A greenhouse trial. Geoderma. 2005;**125**:155-166. DOI: 10.1016/j.geoderma.2004.07.003

[216] Karasawa T, Takebe M, Kasahara Y. Arbuscular mycorrhizal (AM) effects on maize growth and AM colonization of roots under various soil moisture conditions. Soil Science & Plant Nutrition. 2000;**46**:61-67. DOI: 10.1080/00380768.2000.10408762

[217] Liu W, Zhang Y, Jiang S, Deng Y, Christie P, Murray P, et al. Arbuscular mycorrhizal fungi in soil and roots respond differently to phosphorus inputs in an intensively managed calcareous agricultural soil. Scientific Reports. 2016;**6**:1-11. DOI: 10.1038/srep24902

[218] Hussain H, Qingwen Z, Hussain S, Hongbo L, Waqqas A, Li Z. Effects of arbuscular mycorrhizal Fungi on maize growth, root colonization, and root exudates varied with inoculum and application method. Journal of Soil Science and Plant Nutrition. 2021;**21**:1577-1590. DOI: 10.1007/s42729-021-00463-7

[219] Qin Z, Zhang H, Feng G, Christie P, Zhang J, Li X, et al. Soil phosphorus availability modifies the relationship between AM fungal diversity and mycorrhizal benefits to maize in an agricultural soil. Soil Biology and Biochemistry. 2020;**144**:107790. DOI: 10.1016/j.soilbio.2020.107790

[220] El-Samad HMA, El-Hakeem KNSA. Strategy role of mycorrhiza inoculation on osmotic pressure, chemical

constituents and growth yield of maize plant gown under drought stress. American Journal of Plant Sciences. 2019;**10**:1102-1120. DOI: 10.4236/ajps.2019.106080

[221] Hu J, Cui X, Wang J, Lin X. The non-simultaneous enhancement of phosphorus acquisition and mobilization respond to enhanced arbuscular mycorrhization on maize (Zea mays L.). Microorganisms. 2019;**7**:1-13. DOI: 10.3390/microorganisms7120651

[222] Polcyn W, Paluch-Lubawa E, Lehmann T, Mikuła R. Arbuscular mycorrhiza in highly fertilized maize cultures alleviates short-term drought effects but does not improve fodder yield and quality. Frontiers in Plant Science. 2019;**10**:1-16. DOI: 10.3389/fpls.2019.00496

[223] Hi A, Orluchukwu J. Response of three varieties of okra (Abelmoschus esculentus L. Moench) to arbuscular mycorrhizal Fungi (Gigaspora gigantea) in the humid tropics. International Journal of Agriculture and Earth Science. 2018;**4**:30-36

[224] Liu A, Hamel C, Elmi A, Costa C, Ma B, Smith D. Concentrations of K, Ca and Mg in maize colonized by arbuscular mycorrhizal fungi under field conditions. Canadian Journal of Soil Science. 2002;**82**:271-278. DOI: 10.4141/s01-022

[225] Qian L, Zhang W, Yan J, Han L, Gao W, Liu R, et al. Effective removal of heavy metal by biochar colloids under different pyrolysis temperatures. Bioresource Technology. 2016;**206**:217-224. DOI: 10.1016/j.biortech.2016.01.065

[226] Davaran Hagh E, Mirshekari B, Ardakani MR, Farahvash F, Rejali F. Optimizing phosphorus use in sustainable maize cropping via mycorrhizal inoculation. Journal of Plant Nutrition. 2016;**39**:1348-1356. DOI: 10.1080/01904167.2015.1086797

[227] Priyadharsini P, Muthukumar T. Arbuscular mycorrhizal fungus influence maize root growth and architecture in rock phosphate amended tropical soil. Anales de Biologia. 2017;**39**:211-222. DOI: 10.6018/analesbio.39.22

[228] Zhang W, Cao J, Zhang S, Wang C. Effect of earthworms and arbuscular mycorrhizal fungi on the microbial community and maize growth under salt stress. Applied Soil Ecology. 2016;**107**:214-223. DOI: 10.1016/j.apsoil.2016.06.005

[229] Al-Qarawi A, Al-Rowaily S, Al-Oud S. Interaction effects of arbuscular mycorrhizal Fungi and heavy metals on maize plants grown in cadmium, Lead and arsenic contaminated soil. Journal of Agricultural Chemistry and Biotechnology. 2009;**34**:8571-8586. DOI: 10.21608/jacb.2009.99282

[230] Dymytrov S, Sabluk V, Tanchyk S, Gumentyk M, Balagura O. Increasing maize productivity by presowing usage of biologies Mycofriend. In: Mikovital and Florobacillin. E3S Web Conf. Vol. 255. 2021. DOI: 10.1051/e3sconf/202125501006

[231] Lino IAN, Da Silva DKA, Martins JCR, Sampaio EVSB, Maia LC. Mycorrhizal inoculation and application of cattle manure in field-grown maize in semiarid conditions. Experimental Agriculture. 2019;**55**:866-874. DOI: 10.1017/S0014479718000443

[232] Brito ODC, Hernandes I, Ferreira JCA, Cardoso MR, Alberton O, Dias-Arieira CR. Association between arbuscular mycorrhizal fungi and Pratylenchus brachyurus in maize crop. Chilean Journal of Agricultural

Research. 2018;**78**:521-527. DOI: 10.4067/S0718-58392018000400521

[233] Berta G, Copetta A, Gamalero E, Bona E, Cesaro P, Scarafoni A, et al. Maize development and grain quality are differentially affected by mycorrhizal fungi and a growth-promoting pseudomonad in the field. Mycorrhiza. 2014;**24**:161-170. DOI: 10.1007/s00572-013-0523-x

[234] Celebi SZ, Demir S, Celebi R, Durak ED, Yilmaz IH. The effect of arbuscular mycorrhizal Fungi (AMF) applications on the silage maize (Zea mays L.) yield in different irrigation regimes. European Journal of Soil Biology. 2010;**46**:302-305. DOI: 10.1016/j.ejsobi.2010.06.002

[235] Usharani G, Sujitha D, Sivasakthi S, Saranraj P. Effect of arbuscular mycorrhizal (AM) fungi (Glomus fasciculatum L.) for the improvement of growth and yield of maize (Zea mays L.). Central European Journal of Experimental Biology. 2014;**3**:30-35

[236] Olawuyi OJ, Odebode AC, Olakojo SA, Popoola OO, Akanmu AO, Izenegu JO. Host-pathogen interaction of maize (Zea mays L.) and aspergillus Niger as influenced by arbuscular mycorrhizal fungi (Glomus deserticola). Archives of Agronomy and Soil Science. 2014;**60**:1577-1591. DOI: 10.1080/03650340.2014.902533

[237] Olawuyi O, Odebode A, Olakojo S, Adesoye A. Variation in maize tolerance to Striga lutea (lour) and influence of arbuscular mycorrhizal fungi. International Journal of Basic, Applied and Innovative Research. 2012;**1**:1-5

[238] Ramírez-Flores M, Perez-Limon S, Li M, Barrales-Gamez B, Albinsky D, Paszkowski U, et al. The genetic architecture of host response reveals the importance of arbuscular mycorrhizae to maize cultivation. Plant Biology. 2020;**9**:1-18. DOI: 10.7554/eLife.61701

[239] Sheng M, Tang M, Chen H, Yang B, Zhang F, Huang Y. Influence of arbuscular mycorrhizae on the root system of maize plants under salt stress. Canadian Journal of Microbiology. 2009;**55**:879-886. DOI: 10.1139/W09-031

[240] Zhang W, Wang C, Lu T, Zheng Y. Cooperation between arbuscular mycorrhizal fungi and earthworms promotes the physiological adaptation of maize under a high salt stress. Plant and Soil. 2018;**423**:125-140. DOI: 10.1007/s11104-017-3481-9

[241] Miller G, Spoolman S. In: Julet M, editor. Living in the Environment: Concepts, Connections, and Solutions. Sixteenth ed. California: Yolanda Cossio; 2009 ISBN 9780495556718

[242] Evans A. The Feeding of the Nine Billion: Global Food Securuty for the 21st Century. UK; 2009

[243] Hijri I, Sýkorová Z, Oehl F, Ineichen K, Mäder P, Wiemken A, et al. Communities of arbuscular mycorrhizal fungi in arable soils are not necessarily low in diversity. Molecular Ecology. 2006;**15**:2277-2289. DOI: 10.1111/j.1365-294X.2006.02921.x

[244] Bennett A, Bending G, Chandler D, Hilton S, Mills P. Meeting the demand for crop production : The challenge of yield decline in crops grown in short rotations. Biological Reviews. 2012;**87**:52-71. DOI: 10.1111/j.1469-185X.2011.00184.x

[245] Sangabriel-conde W, Maldonado-mendoza I, Mancera-lópez M, Cordero-ramírez J, Trejo-aguilar D, Negrete-yankelevich S. Glomeromycota associated with Mexican native maize

landraces in Los Tuxtlas, Mexico. Applied Soil Ecology. 2015;**87**:63-71. DOI: 10.1016/j.apsoil.2014.10.017

[246] Rabot E, Wiesmeier M, Schlüter S, Vogel HJ. Soil structure as an indicator of soil functions: A review. Geoderma. 2018;**314**:122-137. DOI: 10.1016/j.geoderma.2017.11.009

[247] Lehmann E, Leifheit A, Rillig M. Mycorrhizas and soil aggregation. In: Johnson N, Gehring C, Jansa J, editors. Mycorrhizal Mediation of Soil. Amsterdam: Elsevier; 2017. pp. 241-262 ISBN 9780128043127

[248] Cornejo P, Rubio R, Borie F. Mycorrhizal propagule persistence in a succession of cereals in a disturbed and undisturbed Andisol fertilized with two nitrogen sources. Chilean Journal of Agricultural Research. 2009;**69**:426-434. DOI: 10.4067/s0718-58392009000300016

[249] Rubin JGKR, Stürmer SL. Potencial de inóculo micorrízico e importância do comprimento do micélio para a agregação de solos de ambiente fluvial. Revista Brasileira de Ciencia do Solo - SCI Journal. 2015;**39**:59-68. DOI: 10.1590/01000683rbcs20150466

[250] Marinho F, Oehl F, da Silva I, Coyne D, Veras J, Maia L. High diversity of arbuscular mycorrhizal fungi in natural and anthropized sites of a Brazilian tropical dry forest (Caatinga). Fungal Ecology. 2019;**40**:82-91. DOI: 10.1016/j.funeco.2018.11.014

[251] Miller R, Jastrow J. The role of mycorrhizal Fungi in soil conservation. In: Bethlenfalvay G, Linderman R, editors. Mycorrhizae in Sustainable Agriculture. Madison: American Society of Agronomy; 1992. pp. 29-44

[252] Rillig M, Mummey D. Mycorrhizas and soil structure. The New Phytologist. 2006;**171**:41-53. DOI: 10.1111/j.1469-8137.2006.01750.x

[253] Wright S, Green V, Cavigelli M. Glomalin in aggregate size classes from three different farming systems. Soil and Tillage Research. 2007;**94**:546-549. DOI: 10.1016/j.still.2006.08.003

[254] Rillig M, Muller L, Lehmann A. Soil aggregates as massively concurrent evolutionary incubators. The ISME Journal. 2017;**11**:1943-1948. DOI: 10.1038/ismej.2017.56

[255] Olsson P, Thingstrup I, Jakobsen I, Bååth E. Estimation of the biomass of arbuscular mycorrhizal fungi in a linseed field. Soil Biology and Biochemistry. 1999;**31**:1879-1887. DOI: 10.1016/S0038-0717(99)00119-4

[256] Zhu Y, Miller R. Carbon cycling by arbuscular mycorrhizal fungi in soil-plant systems. Trends in Plant Science. 2003;**8**:407-409. DOI: 10.1016/S1360-1385(03)00192-4

[257] Mirás-Avalos JM, Antunes PM, Koch A, Khosla K, Klironomos JN, Dunfield KE. The influence of tillage on the structure of rhizosphere and root-associated arbuscular mycorrhizal fungal communities. Pedobiologia (Jena). 2011;**54**:235-241. DOI: 10.1016/j.pedobi.2011.03.005

[258] Bedini S, Turrini A, Rigo C, Argese E, Giovannetti M. Molecular characterization and glomalin production of arbuscular mycorrhizal fungi colonizing a heavy metal polluted ash disposal island, downtown Venice. Soil Biology and Biochemistry. 2010;**42**:758-765. DOI: 10.1016/j.soilbio.2010.01.010

[259] Wuest S, Caesar-Tonthat T, Wright S, Williams J. Organic matter addition, N, and residue burning effects

on infiltration, biological, and physical properties of an intensively tilled silt-loam soil. Soil and Tillage Research. 2005;**84**:154-167. DOI: 10.1016/j.still.2004.11.008

[260] Treseder K, Mack M, Cross A. Relationships among fires, fungi, and soil dynamics in Alaskan boreal forests. Ecological Applications. 2004;**14**:1826-1838

[261] Rillig M, Maestre F, Lamit L. Microsite differences in fungal hyphal length, glomalin, and soil aggregate stability in semiarid Mediterranean steppes. Soil Biology and Biochemistry. 2003;**35**:1257-1260. DOI: 10.1016/S0038-0717(03)00185-8

[262] Steinberg P, Rillig M. Differential decomposition of arbuscular mycorrhizal fungal hyphae and glomalin. Soil Biology and Biochemistry. 2003;**35**:191-194. DOI: 10.1016/S0038-0717(02)00249-3

[263] Nichols K, Wright S. Comparison of glomalin and humic acid in eight native U.S. soils. Soil Science. 2005;**170**:985-997. DOI: 10.1097/01.ss.0000198618.06975.3c

[264] Batten K, Six J, Scow K, Rillig M. Plant invasion of native grassland on serpentine soils has no major effects upon selected physical and biological properties. Soil Biology and Biochemistry. 2005;**37**:2277-2282. DOI: 10.1016/j.soilbio.2005.04.005

[265] Lovelock C, Wright S, Clark D, Ruess R. Soil stocks of glomalin produced by arbuscular mycorrhizal fungi across a tropical rain forest landscape. 2004;**92**:278-287

[266] Pohanka M, Vlcek V. Assay of Glomalin using a quartz crystal microbalance biosensor. Electroanalysis. 2018;**30**:453-458. DOI: 10.1002/elan.201700772

[267] Langley JA, Hungate BA. Mycorrhizal controls on belowground litter quality. Ecology. 2003;**84**:2302-2312. DOI: 10.1890/02-0282

[268] Leifheit E, Veresoglou S, Lehmann A, Morris E, Rillig M. Multiple factors influence the role of arbuscular mycorrhizal fungi in soil aggregation-a meta-analysis. Plant and Soil. 2014;**374**:523-537. DOI: 10.1007/s11104-013-1899-2

[269] Se S, Dj R. Mycorrhizal symbiosis. London: Second Ed; 1997

[270] Krishnamurthy K, Biswavas Misra B, Davis Bagyaraj J, Abd E, Hashem A, Alqarawi A, et al. The interaction between arbuscular mycorrhizal Fungi and endophytic Bacteria enhances plant growth of Acacia gerrardii under salt stress. Frontiers in Microbiology. 2016;7:1-15. DOI: 10.3389/fmicb.2016.01089

[271] Barthès B, Roose E. Aggregate stability as an indicator of soil susceptibility to runoff and erosion; validation at several levels. Catena. 2002;**47**:133-149. DOI: 10.1016/S0341-8162(01)00180-1

[272] Wilson G, Rice C, Rillig M, Springer A, Hartnett D. Soil aggregation and carbon sequestration are tightly correlated with the abundance of arbuscular mycorrhizal fungi: Results from long-term field experiments. Ecology Letters. 2009;**12**:452-461. DOI: 10.1111/j.1461-0248.2009.01303.x

[273] Gillespie AW, Farrell RE, Walley FL, Ross ARS, Leinweber P, Eckhardt KU, et al. Glomalin-related soil protein contains non-mycorrhizal-related heat-stable proteins, lipids and humic materials. Soil Biology and Biochemistry. 2011;**43**:766-777. DOI: 10.1016/j.soilbio.2010.12.010

[274] Gałązka A, Gawryjołek K, Perzyński A, Gałązka R, Księżak J. Changes in enzymatic activities and microbial communities in soil under long-term maize monoculture and crop rotation. Polish Journal of Environmental Studies. 2017;**26**:39-46. DOI: 10.15244/pjoes/64745

[275] Roldán A, Salinas-García JR, Alguacil MM, Caravaca F. Changes in soil enzyme activity, fertility, aggregation and C sequestration mediated by conservation tillage practices and water regime in a maize field. Applied Soil Ecology. 2005;**30**:11-20. DOI: 10.1016/j.apsoil.2005.01.004

[276] Roldán A, Salinas-García JR, Alguacil MM, Caravaca F. Soil sustainability indicators following conservation tillage practices under subtropical maize and bean crops. Soil and Tillage Research. 2007;**93**:273-282. DOI: 10.1016/j.still.2006.05.001

[277] Guo H, He X, Li Y. Spatial distribution of arbuscular mycorrhiza and glomalin in the rhizosphere of Caragana korshinskii Kom. in the Otindag sandy land, China. African Journal of Microbiology Research. 2012;**6**:5745-5753. DOI: 10.5897/ajmr11.1560

Chapter 5

Saffron Endomycorrhizae: Diversity and Effect on Plant Growth and Corm Formation

Samah Ourras, Ismail El Aymani, Najoua Mouden, Soumaya El Gabardi, Manal Adnani, Karima Selmaoui, Mariam Artib, Rachid Benkirane, Cherkaoui El Modafar, Amina Ouazzani Touhami and Allal Douira

Abstract

Saffron cultivation is an important alternative for marginalized areas. Due to low soil fertility and low water availability, arbuscular mycorrhizal fungi are an essential alternative for maintaining fertility and water economy, stimulating growth and protecting plants against soil diseases. Studies on the diversity of endomycorrhizal fungi in the rhizosphere of *Crocus sativus* in Taliouine (Tinfat), located in Morocco, revealed the impact of age saffron plantations. A greater endomycorrhizal fungi density was recorded in the rhizosphere of saffron plants from plots operated for 2 years (138.66/100 g of soil) over that occupied for 10 years. Seventeen morphotypes of collected spores belong to 5 genera: *Glomus* (seven species), *Acaulospora* (seven species), *Rhizophagus*, Densicitata, and *Funneliformis* (one species). The weak endomycorrhizal species richness can be explained by the accumulation of *C. sativus* residues over time and its allelopathic effect. The beneficial effect of composite endomycorrhizal inocula, originating from Moroccan saffron plantations, was obvious in the growth of saffron plants, mother bulb number, the leaves length, root, and vegetative masses. These inocula mycorrhized over time saffron plants' roots and can sporulate at the level of the rhizosphere of these plants. The use of composite inocula, as biofertilizers, can be one of the solutions for sustainable agriculture.

Keywords: Morocco, saffron, diversity, endomycorrhizal, composite endomycorrhizal inoculums, growth parameters, bulb multiplication

1. Introduction

Saffron (*Crocus sativus* L.) is an autumn flowering plant from the *Iridaceae* family, renowned worldwide for its red stigmas, which represent the most precious spice in the world [1], a kilogram of good quality saffron produced from *C. sativus* can cost over 2000 US dollars. Approximately 150,000 flowers are needed to produce 1 kilogram of dried saffron, and to growth is amount one would typically require some 2000 m^2 under cultivation per kg harvest [2, 3]. It has been cultivated in the Mediterranean area and Near East since the ancient period, used as a condiment for food, as a dye for textiles, and in traditional medicinal preparations [4]. Nowadays, saffron cultivation gaining interest due to its beneficial health effects, including antioxidant, anticancer, anti-inflammatory, and anti-depressive properties [5]. Iran, India, Greece, Morocco, and Spain are the main saffron-producing countries, with respective average annual productions of 180 tons, 9 tons, 5.5 tons, 3 tons, and 1 ton, recorded over the period 2004–2008 [6]. Saffron is also grown in small areas in Azerbaijan, Afghanistan, France, Switzerland, Italy, Turkey, Australia, and China [1, 3]. In recent years the productivity in Iran has increased enormously, Nevertheless, Spain remains the country most associated with this spice, notably thanks to its main production area, this position makes Spain currently control the world market of saffron and more particularly that of the United States, considered to be the main market of Spain [3], packs and re-exports about 40–50% of Iranian saffron. The United Arab Emirates is also an important importer of Iranian saffron [7].

Saffron is a hardy plant that thanks to its morphology and physiology can with stand very severe climatic conditions [8]. In Morocco, the saffron sector represents a major challenge for this local product both economically and socially. This sector is a promising way to reduce poverty and income inequalities in saffron-growing areas.

Indeed, it is one of the pillars of the economy of the Taliouine-Taznakht region, which is characterized by difficult soil and climate conditions [9], high rates of poverty and income inequality, and a high level of the rural exodus [10].

The annual production has increased according to ANDZOA from 3 tons of dry stigmas and an area of 600 ha in 2009 to 6.8 tons and about 1800 ha in 2018. The cultivation of saffron has been extended and intensified in recent years (Green Morocco Plan). The objective is to reach 9 tons in 2020 according to the contract program signed in 2016 between the Moroccan Interprofessional Federation of Saffron and the government. The cultivation area in Morocco is concentrated in the area of Taliouine–Tazenakht (**Figure 1**).

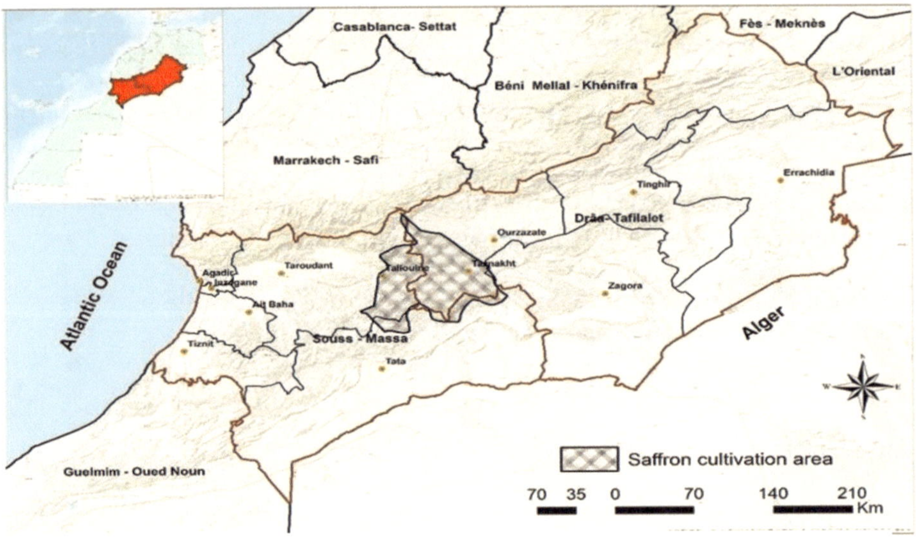

Figure 1.
Geographical distribution of saffron production area in Morocco.

According to the Regional Office of Agricultural Development (ORMVAO), the region of Souss-Massa ensures about 95% of the national production of saffron, which represents an annual turnover of 75 million DH. Around Two-thirds of the production is destined for the international market.

Moroccan saffron has a great reputation at national and international levels [11]. Data from the 'Office des Changes du Maroc' show that the main destination for "Moroccan" saffron is France. These data also show (mass imports registered with the customs services and which come essentially from the Islamic Republic of Iran or Greece. These quantities are redirected to Qatar or other lesser-known destinations. This intruding mass is cheaper, but affects the reputation of Moroccan saffron and competes with local productions.

Saffron cultivation management depends on the development of new technical production practices adapted to the pedoclimatic conditions of traditional areas for this crop. Among, the major constraints that limit its production and productivity is poor management of saffron cultivation, as it involves in adequate plant population, the incidence of corm rot disease [12], nutrient depletion, and lack of irrigation facilities. Moreover, environment and cultivation management affect strongly flower induction in C. sativus [13]. Accordingly, Increasing saffron yield and quality, reducing production costs, and flowering modulation may need new technology implementation as proposed in the Mediterranean environment viz., soilless cultivation systems were proposed but only limited and controversial reports are present in literature. Another option for saffron plant performance is through the use of biostimulants, or microorganisms applied to plants to enhance nutrition efficiency, abiotic stress tolerance, and/or crop quality traits, regardless of the irnutrient content. Soil microorganisms such as arbuscular mycorrhizal fungi (AMF) can form a mutualistic symbiosis with about 80% of land plant species, including several crops [14]. This association provides ecological stability to the environment [15, 16], and enhances water and nutrient uptakes such as phosphorus, nitrogen, and micronutrients, thus improving plant growth and resistance to biotic and abiotic stresses [17, 18].

This chapter presents an overview of saffron cultivation in Morocco and information on the diversity of arbuscular mycorrhizal fungi in the Taliouine region. Inocula, based on these fungi, originating from the saffron rhizosphere, were tested to promote plant growth and bulb production and multiplication.

2. Mycotrophic nature of Saffron plants

As saffron does not grow from seeds, reproduction is only possible by vegetative propagation, using the corms which can withstand a long dry dormant period before sprouting. Thus, corms are indispensable to saffron propagation reproducing vegetatively into corm lets that ultimately develop into new plants (**Figure 2**) [19].

The traditional area of saffron production in Morocco is characterized by very poor soils, however, the rhizosphere of cultivated saffron supports very interesting populations of microorganisms such as endomycorrhizae that are supposed to be essential for plant growth and ecosystem functioning.

Like most geophytes, saffron produces a coarse root system; these root systems have under gone extensive mycorrhization and benefit from this symbiosis [20]. Indeed, field-grown saffron is extensively mycorrhized [21, 22]. The Saffron plant

Figure 2.
A, B: Saffron fields in Talouine region; C: Saffron plant at flowering stage.

shows a high incidence of AMF colonization. It was observed that the corm, despite being the modified stem, shows AMF colonization [23]. Morphological examination of fine roots comprised in rhizospheric soil samples collected in October 2020 from 15 different plots of saffron located in the Taliouine area (Morocco) under light microscope revealed the presence of AMF structures inside the examined roots which showed different colonization levels (**Figure 3**).

Ourras et al. [24] pointed out a high mycorrhization rate of 95.02% which reflects the inoculum pressure or propagules rate infecting the surrounding substrate. Likewise, El Aymani et al. [12] have reported different root mycorrhizal frequencies of 93.33% and 96.67% associated with the rhizosphere of *C. sativus* grown in the Taliouine region. This mycorrhizal colonization demonstrates relevant interaction between *C. sativus* roots and native endomycorrhizal populations under field conditions through which mechanisms such as modulation of water transport [25], nutrient acquisition [26], and stimulating the exudation of root phosphatase [27, 28] under drought conditions, are active in mycorrhizal plants to alleviate environmental stressors.

With the identification of native arbuscular mycorrhizal fungi and their application, it could be possible to expand the saffron cultivated area and increase the performance of arable lands.

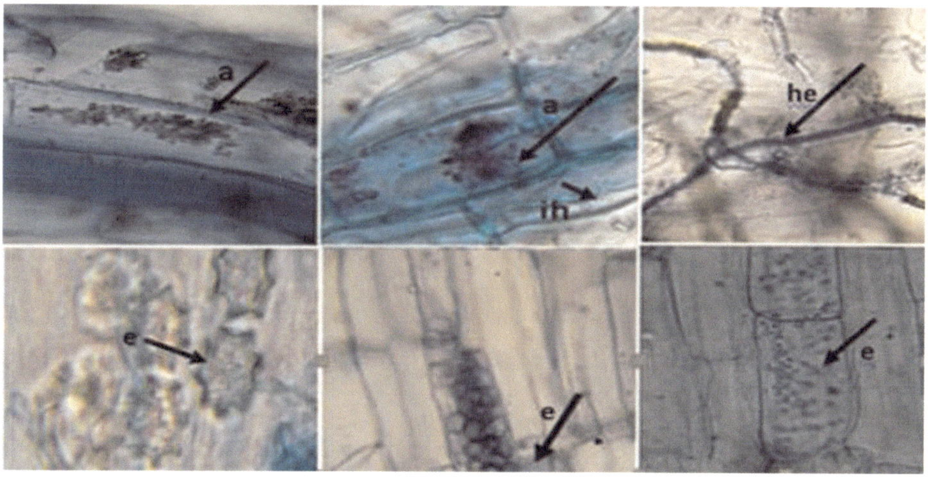

Figure 3.
AMF structures inside the examined roots of saffron plants grown in fields in Taliouine regions.

3. Diversity of arbuscular mycorrhizal fungi from moroccan saffron plantations areas

The knowledge of the current richness of Glomale an AM fungi (Glomerales; Glomeromycota) and biological diversity associated with plants grown in adverse environments could be necessary to expand the cultivated area and increase the performance of arable lands.

AMF diversity has been scarcely studied in the Taliouine-Taznakht region where saffron cultivation is exclusively concentrated [29].

This diversity was unveiled by direct spore isolation and morphological characterization of rhizospheric soil and roots samples from different sites in the main saffron production region. A variety of endomycorrhizal fungi has been noted in the rhizosphere of *C. sativus* grown in the Taliouine region (Tinfat) (**Figure 4**) [30]. These authors have detected 36 morphologically distinctive AMF species that were directly retrieved from field samples. All the encountered species in the different studied sites belong to six genera: *Glomus* (14 species), *Acaulospora* (10 species), *Scutellospora* (6 species), *Gigaspora* (2 species), *Pacispora* (2 species), and *Entrophospora* (1 species). The genus *Glomus* was the most dominant, it occurred in all studied sites with a proportion of distribution that varies between 8.33% and 30.56%.

A low number of AMF species have been reported by Ourras et al. [24]. Chamkhi et al. [31] detected 11 species of AMF spores among which *Funneliformis* and *Rhizoglomus* species were the most abundant (>35%). Yang et al. [32] reported that AMF soil community evolved in the function of environmental conditions. Another factor affects AMF survival and community composition in association with the saffron crop.

3.1 Age of plantation

Age of plantation seemed to be affected the richness of AM fungi, spore density and root colonization rates [12, 24, 31]. Indeed, the greatest richness of AM fungi was registered in the site at 4 years of successive exploitation by saffron (24 species),

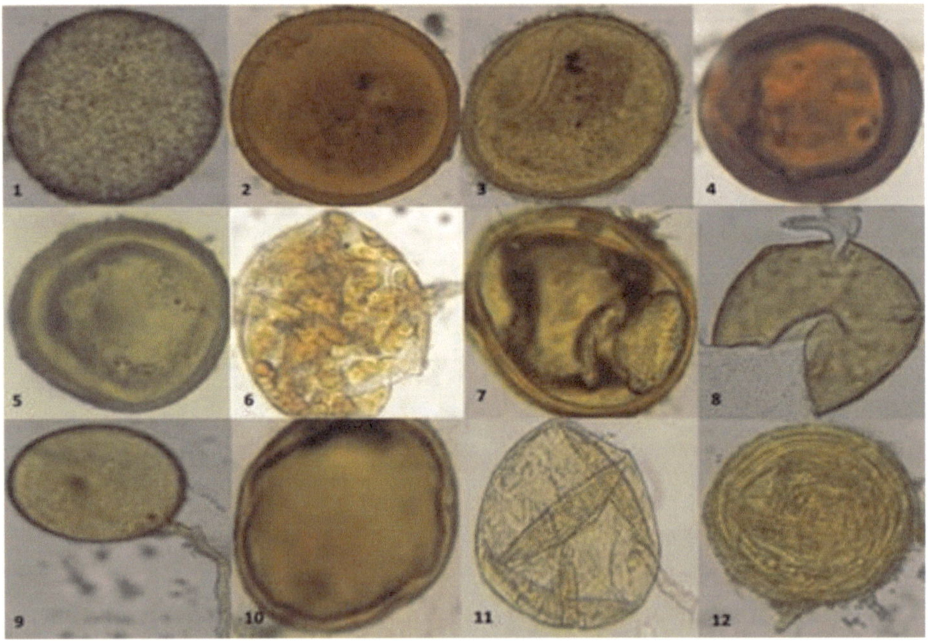

Figure 4.
Some indigenous endomycorrhizal fungals pecies isolated from the rhizosphere of Crocus sativus *grown in Taliouine areas.*

followed by the site at 6 years of occupation by saffron (21 species), while the lowest number of species was recorded at the sites of two, three and 10 years of soil use by saffron [33]. According to Yu et al. [34], stand age significantly changed the structure of the AM fungal community. The increase of spore density with field age has been cited in previous literature reviews [35, 36]. Rengifo-Del Aguila et al. [37] found that intra-radical colonization and AM fungal phylogenetic diversity increased with plantation age, while AM fungal richness was still constant across time but a significant compositional turnover was detected.

Many studies have demonstrated that the shift in the AM fungal community over ecological succession is associated with variations in soil properties induced by stand age, such as soil texture [38], pH [39], and nutrient availability [40]. Krüger et al. [41] pertain that P has become increasingly limited in old soils as the amount and availability of P declines over time. The changes in soil's chemical and physical properties can play an important role in decreasing saffron yields even after 6 years of cultivation [42].

4. AMF as biofertilizers on Saffron

Regarding the biological activity, as biofertilizers, AMF offers a way to replace, at least partially, the use of chemical fertilizers and pesticides seeing their fatal effects on the environment and health. In low soil fertility of saffron cultivation areas [43, 44], they can serve as an AMF inoculum which may be part of the solution for sustainable agriculture. Cases in point showed successful application on saffron production under controlled conditions.

4.1 Corm production

In addition to the spice yield, another economically important attribute of saffron is the number of replacement corms. Corms inoculation by composite AMF inoculum significantly initiated and stimulated the production of new corms (**Figure 5**) [30].

Such promoting effect of AMF on initiating the new underground stem propagules was observed previously too [18, 45–47]. Considering that tuber initiation in potato is hormonally mediated [48], it may be that in the present study too AMF affected hormone balance in the saffron plant, leading to earlier initiation and production of corms. Inoculation of corms with spores has enhanced the growth of corms and the number of spores per 10 g soil and results recorded as 13–52 with the highest occurrence in December–January [49]. Caser et al. [50] approved the fact that AMF symbiosis enhanced the production of replacement corms and reduced the occurrence of fungal diseases.

4.2 Growth response

Although AMF increased the number of corms produced, it increased also shoot fresh weight and root dry weight [51]. A significant increment of the overall growth parameters such as length of aerial and below ground of saffron plants occurred too after AMF colonization [30].

The use of arbuscular mycorrhizal symbionts as biostimulants positively affected saffron cultivation, improving the crop performances and the content of important nutraceutical compounds. In particular, the inoculum composed by *R. intraradices* and *F. mosseae* increased flower production and saffron yield [52]. Such compatibility between AMF fungi and host plant was previously observed in other plants also such as onion [47] and Potato [18] cultivars.

Van der Heijden et al. [53], found that in the grassland ecosystem, inoculation of 14 AM fungal species increased plant diversity by 105% and plant productivity by 42% compared to that in the ecosystem inoculated with only 1 AM fungal species. A study has also revealed that different AM fungi can absorb P at different distances from the

Figure 5.
New bulbs production; ***M:*** *Inoculated plant;* ***C:*** *Control.*

root of the plant, making the host plants inoculated with mixed AM fungal species grow better than those inoculated with AM fungal species.

Variations in the growth of saffron plants are possibly observed as a function of the age of saffron plantation as seen below (**Figure 6**).

Figure 6.
Performance of Crocus sativus *plants as affected by the presence of AMF and age of saffron plantations.*

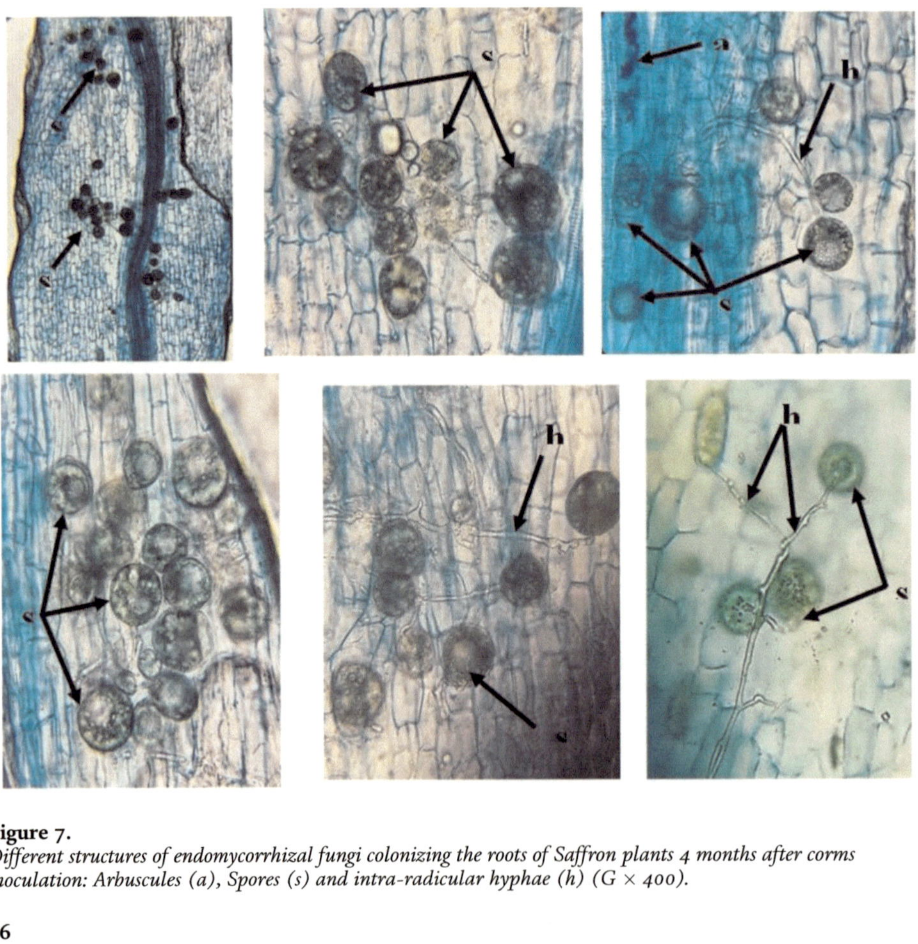

Figure 7.
Different structures of endomycorrhizal fungi colonizing the roots of Saffron plants 4 months after corms inoculation: Arbuscules (a), Spores (s) and intra-radicular hyphae (h) (G × 400).

Figure 8.
Quelques morphotypes de spores rencontrés au niveau de la rhizosphère des plants du safran provenant des cormes traitées avec un inoculum endomycorhiziens (S4), après quatre mois de culture. 1: Rhizophagus intraradices*, 2:* Glomus lamellosum, 3: Funneliformis geosporum, 4: Densicitata nigra, 5: Glomus microcarpum:, 6: Glomus deserticola.

4.3 Root colonization with AMF after corms inoculation

The positive influence of AMF on plant is otherwise viewed in roots, at this level, mycorrhized plants are shown if they have been able to establish a strong mutualistic relationship with composite AMF inoculum. Four months after inoculation in potted culture, the roots of the plants from the inoculated corms are richly mycorrhized and showed numerous arbuscules and vesicles with extended hyphal colonizing groots. Arbuscules and spores were the most dominant structures (**Figures** 7 and 8).

5. Conclusion

The growing interest in knowing and safeguarding the biological diversity associated with certain arid environments may contribute to improving ecologically the yields of agricultural crops. In the case of saffron, the rhizospheric soil harbored a diversified AMF community and its application as composite inoculum in substrate culture pertains a potential performance on growth and plant succession. As a perspective, future studies should be focused to know the influence of AMF on the growth of the saffron plant in open field conditions.

Acknowledgements

This study was conducted under the research project of the 2nd PMA 2020/11, entitled 'Effect of endomycorrhizal fungi inoculum on agro morphological behavior and productivity of saffron (*C. sativus*) under water and salinity stress' financed by 'Agence Nationale des Plantes Médicinales et Aromatique' et l'Université Ibn Tofail, Kénitra, Morocco'.

Author details

Samah Ourras[1], Ismail El Aymani[1], Najoua Mouden[2], Soumaya El Gabardi[1], Manal Adnani[1], Karima Selmaoui[1], Mariam Artib[1], Rachid Benkirane[1], Cherkaoui El Modafar[3], Amina Ouazzani Touhami[1] and Allal Douira[1*]

1 Laboratory of Plant, Animal and Agro-Industry Productions, Botany, Biotechnology and Plant Protection Team, Faculty of Sciences, Ibn Tofail University, Kenitra, Morocco

2 Laboratory of Molecular Chemistry and Environmental Molecules, Multidisciplinary Faculty of Nador-Mohammed 1st University Oujda, Morocco

3 Center of Agrobiotechnology, Research Unit Labeled CNRST (URL-CNRST05), Faculty of Sciences and Techniques Guéliz, Cadi Ayyad University, Marrakech, Morocco

*Address all correspondence to: douiraallal@hotmail.com

References

[1] Gresta F, Lombardo GM, Siracusa L, Ruberto G. Saffron, an alternative crop for sustainable agricultural systems. A review. Agronomy for Sustainable Development. Springer Verlag/EDP Sciences/INRA. 2008;**28**(1):95-112

[2] Aboudrare A, AW Hassan A, Lybbert TJ. Importance socio-économique du Safran pour les ménages des zones de montagne de la région de Taliouine-Taznakht au Maroc. Rev. Mar. Sci. Agron. Vét. 2014;**2**:5-14

[3] Wyeth P, Malik N. A Strategy for Promoting Afghan Saffron Exports. Report. Afghanistan: ICARDA and Washington State University; 2008

[4] Giupponi L, Ceciliani G, Leoni V, Panseri S, Pavlovic R, Guido L, et al. Quality traits of saffron produced in Italy: Geographical area effect and good practices. Journal of Applied Botany and Food Quality. 2019;**92**:336-342

[5] Mykhailenko O, Desenko V, Ivanauskas L, Georgiyants V. Standard operating procedure of Ukrainian saffron cultivation according to good agricultural and collection practices to assure quality and traceability. Industrial Crops and Products. 2020;**151**:112376

[6] Dubois, A., (2010). Analyse de la filière du safran au Maroc: Quelle perspective pour la mise en place d'une indication géographique? Thèse de Master of Science, N 107. Montpellier: CIHEAM-IAMM

[7] Ghorbani M. The efficiency of saffron's marketing channel in Iran. World Applied Sciences Journal. 2008;**4**(4):523-527

[8] Sepaskhah AR, Yarami N. Interaction effects of irrigation regime and salinity on flower yield and growth of saffron. The Journal of Horticultural Science and Biotechnology. 2009;**84**(2):216-222

[9] Garcin GD, Carral S. Le safran marocain entre tradition et marché. Etude de la filière du safran au Maroc, en particulier dans la région de Taliouine, province de Taroudant. Rapport de consultation. Maroc: Projet FAO/Association Migrations et Développement; 2007 http://www.mp-discussion.org/casablanca/doc/zaf.pdf

[10] Bouchelkha M. Etude Sociologique sur le Safran. *Rapport de consultation*. Maroc: Projet FAO/TCP/MOR/3201. Rome, Italie; 2009

[11] Lage M, Cantrell CL. Quantification of saffron (Crocus sativus L.) metabolites crocins, picrocrocin and safranal for quality determination of the spice grown under different environmental Moroccan conditions. Scientia Horticulturae. 2009;**121**(3):366-373

[12] El Aymani I, Qostal S, Mouden N, Selmaoui K, Touhami AO, Benkirane R, et al. Fungi associated with saffron (*Crocus sativus*) in Morocco. Plant Cell Biotechnology and Molecular Biology. 2019;**20**:1180-1188

[13] De Juan JA, Córcoles HL, Muñoz RM, Picornell MR. Yield and yield components of saffron under different cropping systems. Industrial Crops and Products. 2009;**30**(2):212-219

[14] Berruti A, Lumini E, Balestrini R, Bianciotto V. Arbuscular mycorrhizal fungi as natural biofertilizers: Let's benefit from past successes. Frontiers in Microbiology. 2016;**6**:1559

[15] Harley JL, Smith SE. Symbiose Mycorhizienne. Londres, Royaume-Uni: Academic Press; 1983

[16] Strullu DG. Les mycorhizes des arbres et des plantes cultivées. Lavoisier Paris. 3éme edit ed1991. p. 249

[17] Goussous SJ, Mohammad MJ. Comparative effect of two arbuscular mycorrhizae and N and P fertilizers on growth and nutrient uptake of onions. International Journal of Agriculture & Biology. 2009;**11**:463-467

[18] Lone R, Shuab R, Wani KA, Ganaie MA, Tiwari AK, Koul KK. Mycorrhizal influence on metabolites, indigestible oligosaccharides, mineral nutrition and phytochemical constituents in onion (*Alliumcepa L.*) plant. ScientiaHorti. 2015;**193**:55-61

[19] Hossein Zadeh H, Nassiri-Asl M. Avicenna's (Ibn Sina) the canon of medicine and saffron (Crocus sativus): A review. Phytotherapy Research. 2013;**27** (4):475-483

[20] Smith SE, Read DJ. Mycorrhizal Symbiosis. 2nd ed. London: Academic Press; 1997

[21] Kianmehr H. Vesicular arbuscular mycorrhizal spore population and infectivity of saffron (*Crocus sativus*) in Iran. New Phytologist. 1981;**88**(1):79-82

[22] Lone R, Razia SR, Koul KK. AMF association and their effect on metabolite mobilization, mineral nutrition and nitrogen assimilating enzymes in saffron (*Crocus sativus*) plant. Journal of Plant Nutrition. 2016;**39**:131852-131862

[23] Shuab R, Malla NA, Ahmad J, Lone R, Koul KK. Arbuscular mycorrhizal fungal symbiosis with saffron (Crocus sativus L.) Plant. J. New Biol. Rep. 2016; **5**(8)

[24] Ourras S, El Gabardi S, Mouden N, Chliyeh M, Selmaoui K, Benkirane R, et al. Diversity of arbuscular mycorrhizal fungi in the rhizosphere of saffron (Crocus sativus) plants along with age of plantation in Taliouine region in Morocco. Acta Biologica Szegediensis. 2021;**2**(65):199-209

[25] He JD, Zou YN, Wu QS, Kuča K. Mycorrhizas enhance drought tolerance of trifoliate orange by enhancing activities and gene expression of antioxidant enzymes. Scientia Horticulturae. 2020;**262**:108745

[26] Boutasknit A, Baslam M, Ait-El-Mokhtar M, Anli M, Ben-Laouane R, Douira A, et al. Arbuscular mycorrhizal fungi mediate drought tolerance and recovery in two contrasting carob (*Ceratonia siliqua L.*) ecotypes by regulating stomatal, water relations, and (in) organic adjustments. Plants. 2020;**9**(1):80

[27] Chen W, Meng P, Feng H, Wang C. Effects of arbuscular mycorrhizal fungi on growth and physiological performance of Catalpa bungei CA Mey. Under Drought Stress. Forests. 2020;**11**: 1117

[28] Cheng HQ, Zou YN, Wu QS, Kuča K. Arbuscular mycorrhizal fungi alleviate drought stress in trifoliate orange by regulating H+-ATPase activity and gene expression. Frontiers in Plant Science. 2021;**12**:659694

[29] Aziz L, Sadok W. Stratégies d'adaptation des producteurs du Safran de Taliouine (Maroc) face au changement climatique. Journal of Alpine Research| Revue de géographie alpine. 2015:103-102

[30] El Aymani I, Chliyeh M, Selmaoui K, Mouden M, El Gabardi S, Ouazzani Touhami A, et al. Effect of a composite endomycorrhizal inoculum on the growth of saffron plants (*Crocus sativus L.*) and the multiplication of

corms. Plant Cell Biotechnology and Molecular Biology. 2019c;**20**(23&24): 1122-1136

[31] Chamkhi I, Sbabou L, Aurag J. Endophytic fungi isolated from *Crocus sativus L.* (saffron) as a source of bioactive secondary metabolites. Pharmacognosy Journal. 2018;**10**(6)

[32] Yang C, Hamel C, Schellenberg MP, Perez JC, Berbara RL. Diversity and functionality of arbuscular mycorrhizal fungi in three plant communities in semi arid Grasslands National Park, Canada. Microbial Ecology. 2010;**59**(4):724-733

[33] El Aymani I, El Gabardi S, Artib M, Chliyeh M, Selmaoui K, Ouazzani TA, et al. Effect of the number of years of soil exploitation by saffron cultivation in Morocco on the diversity of endomycorrhizal fungi. Acta Phytopathologica et Entomologica Hungarica. 2019b;**54**(1):71-86

[34] Yu Z, Liang K, Wang X, Huang G, Lin M, Zhou Z, et al. Alterations in arbuscular mycorrhizal community along a chronosequence of teak (Tectona grandis) plantations in tropical forests of China. Frontiers in Microbiology. 2021;**12**

[35] Birhane E, Gebremedihin KM, Tadesse T, Hailemariam M, Solomon N. Exclosures restored the density and root colonization of arbuscular mycorrhizal fungi in Tigray, Northern Ethiopia. Ecol Process. 2017;**6**:33

[36] Burni T, Hussain F, Sharief M. Arbuscular mycorrhizal fungi (amf) associated with the rhizosphere of Mentha arvensis l., and M. longifolia huds. Pakistan Journal of Botany. 2011; **43**(6):3013-3019

[37] Aguila SRD, la Sota-Ricaldi D, Maria A, Corazon-Guivin MA, López-García Á. Phylogenetic diversity of arbuscular mycorrhizal fungal communities increases with crop age in coffea arabica plantations. Journal of Soil Science and Plant Nutrition. 2022:1-13

[38] Herrmann L, Lesueur D, Bräu L, Davison J, Jairus T, Robain H, et al. Diversity of root-associated arbuscular mycorrhizal fungal communities in a rubber tree plantation chronosequence in Northeast Thailand. Mycorrhiza. 2016;**26**(8):863-877

[39] Reyes HA, Ferreira PFA, Silva LC, da Costa MG, Nobre CP, Gehring C. Arbuscular mycorrhizal fungi along secondary forest succession at the eastern periphery of Amazonia: Seasonal variability and impacts of soil fertility. Applied Soil Ecology. 2019;**136**:1-10

[40] Wang Q, Ma M, Jiang X, Guan D, Wei D, Cao F, et al. Influence of 37 years of nitrogen and phosphorus fertilization on composition of rhizosphere arbuscular mycorrhizal fungi communities in black soil of Northeast China. Frontiers in Microbiology. 2020; **11**:539669

[41] Krüger M, Teste FP, Laliberté E, Lambers H, Coghlan M, Zemunik G, et al. The rise and fall of arbuscular mycorrhizal fungal diversity during ecosystem retrogression. Molecular Ecology. 2015;**24**(19):4912-4930

[42] Khozaei M, Kamgar Haghighi AA, Sepaskhah AR, Karimian NA. Effect of 10-year continuous saffron cultivation on physical and chemical properties of soil. Iran Agricultural Research. 2015; **34**(1):46-55

[43] Chamkhi I, Sbabou L, Aurag J. Application of Plant Growth Promoting Rhizobacteria (PGPR) as Biological Fertilizers for Improving the Growth and Yield of Saffron (*Crocus sativus L.*).

In: 1st Mediterranean Forum for Ph.D. Students and Young Researchers Designing Sustainable Agricultural and Food Production Systems under Global Changes in the Mediterranean 18-19 July 2016. Montpellier, France; 2016

[44] Chamkhi I, Abbas Y, Tarmoun K, Aurag J, Sbabou L. Morphological and molecular characterization of arbuscular mycorrhizal fungal communities inhabiting the roots and the soil of saffron (Crocus sativus L.) under different agricultural management practices. Archives of Agronomy and Soil Science. 2019;**65**(8):1035-1048

[45] Graham SO, Green NE, Hendrix JW. The influence of vesicular-arbuscularmycorrhizalfungi on growth and tuberization of potatoes. Mycologia. 1976;**68**(4):925-929

[46] Niemira BA, Safir GR, Hammerschmidt R, Bird GW. Production of prenuclear minitubers of potato with peat-based arbuscular mycorrhizal fungal inoculum. Agronomy Journal. 1995;**87**(5):942-946

[47] Shuab R, Rafiq LR, Jayanti NJ, Sharma V, Imtiyaz S, Koul KK. Benefits of inoculation of arbuscular mycorrhizal fungi on growth and development of onion (*Allium cepa*) plant. American-Eurasian J. Agric. & Environ. Sci. 2014; **14**(6):527-535

[48] Ewing EE. The role of hormones in potato (*Solanum tuberosum* L.) tuberization. In: Plant Hormones. Dordrecht: Springer; 1995. pp. 698-724

[49] Kianmehr H. Endotrophicmicorehyza of Saffron in Khorasan and its possible application. Gonabad, Iran: Proceeding of the Second National Symposium on Saffron and Medicinal plants; 1994

[50] Caser M, Victorino IMM, Demasi S, Berruti A, Lumini E, Bianciotto V, et al. In: XXX International 60 Horticultural Congress IHC2018: 61 International Symposium on Medicinal 62 and Aromatic Plants, Culinary Herbs and IV International Jujube Symposium and VI International Symposium on Saffron Biology and Technology. 2018b;**1287**: 441-446

[51] Mohebi-Anabat M, Riahi H, Zanganeh S, Sadeghnezhad E. Effects of arbuscular mycorrhizal inoculation on the growth, photosynthetic pigments and soluble sugar of Crocus sativus (saffron) in autoclaved soil. Int. J. Agron. Agric. Res. 2015;**6**:296-304

[52] Caser M, Victorino ÍMM, Demasi S, Berruti A, Donno D, Lumini E, et al. Saffron cultivation in marginal alpine environments: How AMF inoculation modulates yield and bioactive compounds. Agronomy. 2018a;**9**(1):12

[53] Van Der Heijden MG, Klironomos JN, Ursic M, Moutoglis P, Streitwolf-Engel R, Boller T, et al. Mycorrhizal fungal diversity determines plant biodiversity, ecosystem variability and productivity. Nature. 1998;**396**(6706): 69-72

Section 2

The Effect of Mycorrhizae on Plant Nutrition and Protection

Chapter 6

AM Fungi as a Potential Biofertilizer for Abiotic Stress Management

Malik A. Aziz, Shayesta Islam, Gousia Gani, Zaffar M. Dar, Amajad Masood and Syed H. Baligah

Abstract

Climate change and agricultural practices like unrestricted utilization of insecticides especially fertilizer and pesticides have amplified the effects of inanimate stress on the productivity of crops and degraded the environment. The need of the hour is to adopt eco-friendly crop management techniques, including the usage of arbuscular mycorrhizal fungi (AMF). AMFs are frequently referred to as bio-fertilizers. Mycorrhiza improves the movement and absorption of nutrients from soils, thereby limiting the demand for artificial fertilizers and avoiding the accretion of nutrients in soil. Reduced fertilizer use reduces the effects of fertilizer runoff and leaching on water quality and serves as a cost-effective method for farmers. Inanimate stressors (such as salt, drought, heat, cold, and mineral shortage) have emerged as the most serious dangers to global agricultural productivity. These stresses induce ion toxicity nutritional imbalance, hormonal inequalities which in turn influence plant growth and development, maturity, productivity etc. Some beneficial microorganisms, such as mycorrhizal fungi, live in mutualistic association with the roots of host plant in the rhizospheric region. Mycorrhiza significantly improves host plant resilience to a variety of animate and inanimate stresses. This chapter emphasizes the relevance of mycorrhizal fungi in stress reduction and their beneficial impacts on plants' production, growth and enlargement.

Keywords: arbuscular mycorrhizal fungi, plant growth, abiotic stresses, drought, salinity, temperature

1. Introduction

Agriculture production is declining globally, owing primarily to animate and inanimate pressures. Inanimate stressors inhibit plant enlargement, expansion of plants various levels like biochemical and molecular, resulting in massive drop in crop production [1]. Fungi and bacteria make up the large component of rhizospheric associated microflora and play an critical part in growth and development of plants. Mycorrhizal fungi live in mutualistic association with host plant roots, complementing and increasing plant growth, rate of production, and resistance; however, recent research indicates that mycorrhizal fungi also create induced systemic tolerance (IST) to animate and inanimate stimuli. Mycorrhizae improves absorption of nutrients and movement from soil, limiting the requirement for chemical fertilizers and preventing nitrate and phosphate increases in agricultural soils. Reduced fertilizer use decreases water contamination from fertilizer runoff and leaching, which benefits farmers' economies. Abiotic stressors have emerged as major risks to global agriculture output. These stresses, either alone or in conjunction, regulate critical metabolic activities of plants by producing physiological disturbances. The Fungi (mycorrhizae) in symbiotic connection with plant roots enhances contact area of roots, allowing efficient absorption of mineral and water from huge soil volumes. This obligatory mycorrhizal connection improves nutrition and water accessibility while also protecting the plant from a number of inanimate challenges [2]. Some activities employed by arbuscular mycorrhizal fungi (AMF), under particular circumstances including the synthesis of metabolic substances viz. amino acids, vitamins, photohormones, and solubilization and mineralization processes describe the growth promotion caused by this association [3]. AMF has also been shown to impact the manifestation of many reactive oxygen species (ROS) safeguarding enzymes produced due to various stresses [4]. Besides giving nutritional and structural advantages to plants, the other advantages provided include secondary metabolite production/accumulation for osmotic regulation, enhanced cycling of nitrogen, rate of photosynthesis, and tolerance to animate and inanimate stresses. Numerous investigations have found that AMF have the capacity to enhance tolerance against various animate as well as inanimate stresses like heavy metals, drought, and salinity, and pathogen attack [5].

2. Arbuscular mycorrhizal fungi

Arbuscular mycorrhizae fungi (AMF), are beneficial organisms and play an important part in plant nutrition and performance. The interaction of AMF symbiosis with positively charged bivalent ($Ca2+$, $Fe2+$) and trivalent cations ($Al3+$) improves immobile absorption and insoluble phosphate ions in soil [6]. The ability of AMF to evolve exterior hyphae networks that can increase external area approximately 40 fold and expose soil volume for absorption of nutrients by generating enzymes or releasing organic compounds is the primary function in this mutualism [7]. AMF can produce phosphatises to disintegrate phosphate from organic phosphorus containing compounds, increasing productivity in hard environments (deficiency of phosphorous). Extra radical hyphae are thought to be important in terms of ammonium absorption, mobilization of fixed microelements viz. Cu and Zn, and other cations from the soil like potassium, Magnesium, iron and calcium. Reports reveal that AMF promotes plant nutrition, when used as biofertilizer and hampers plant hormone balance, which regulates plant development (bioregulators) and lessens the influence of external pressures that is bio protector. This boosts biomass and yield while causing changes in several quality indicators [8]. AM fungi develop intimate association with host plants through intracellular structures called arbuscules in cortical cells of roots thus referring as symbiotic biotrophs. AMF are soil inhibiting fungi with the potential to boost nutrient absorption in plants and resilience to a variety of inanimate stressors [9]. Furthermore these depend on metabolic products of host produced during photosynthesis for completion of their life cycle, hence referred as obligate biotrophs. AMF provides growth benefits to plants through two ways viz., by improving water and mineral nutrient absorption from the surrounding soil and by protecting them from fungal infections [10]. As a result, AMF are beneficial endosymbionts that contribute to plant productivity and ecological function. They are critical for sustainable agricultural production [11].

3. Mechanism of mycorrhizal association

Arbuscular Mycorrhizae fungi completely rely on host plants for their nutrition hence referred to as obligate biotrophs. The various stages involved in symbiosis are

Stage 1: This is a very vital stage of the colonization process in which fungi scout for host plant.

Stage 2: The second stage involves the invasion of host roots by fungi for colonization and development.

Few chemical substances known as bioactive agents are produced by special cells or tissues that persuade different organisms to function. Similarly, substances such as strigolactones produced by the roots enable fungi to recognize their host as well as induce enlargement and expansion. The fungi respond to these stimuli by releasing a series of factors referred to as Myc (Mycorrhizal Factors) which also have a critical part in the interaction between AM fungi and nitrogen-fixing bacteria. This association is facilitated further through the production of seven genes (SYM genes) [12]. After spore germination, few hyphae branches approach the root of host and enter into the cortical cell wall and finally invade the internal cortical cells resulting in the formation of a greatly branched structure called arbuscular, which acts as an platform for nutritional exchange [13]. The Myc impulses are recognized by Myc Factor Receptor(s) of host which leads to the release of cytosolic calcium in root cells. Another membrane-based protein (SYMPK) is stimulated, which encodes for a receptor-like kinase having the ability to identify fungal signals. SYMPK has the potential to convert these impulses from the cytoplasm to the nucleus by phosphorylation of unrevealed substances through its kinase domain [14]. The localization of all downstream elements present in the cytoplasm stimulates the fast impulse transmission between the cells and nuclei, this continuous to and fro moment of Ca2+ concentration is possible due to alternate action of Ca2+ channels and transporters. The calcium to and fro moments are coded by a Calmodulin-Dependent Protein Kinase (CCaMK). CCaMK phosphorylates the product of one of the SYM genes (CYCLOPS). This ultimately results in the control of other genes and consequently root colonization [13]. The mutualistic interaction of Arbuscular Mycorrhizal (AM) fungi and the roots of higher plants is of broad nature. Various studies have revealed that AM symbiosis is a principle component to overcome various stresses and in enhancing tolerance against various stresses by bringing modifications in phonology of host.

4. Role of AMF as a bio-fertilizer

Biofertilizer is a combination of naturally existing substances utilized to enhance nutrient status of soil. These fertilizers are of remarkable significance to soil microflora and fauna as well as the efficient growth of plants [15]. Nearly from two decades, research investigations have focused on their multiple benefits to soil microflora and fauna and productivity. As a result, it is extensively assumed that AMF will be investigated as a substitute for inorganic fertilizers in the coming years, because mycorrhizal treatment can cause an efficient reduction in the quantity of chemical fertilizer input used, primarily phosphorus [16]. Constant use of inorganic fertilizers, herbicides, and fungicides has generated several difficulties for soil, plant, and human health, owing to their negative impact on food quality, soil micro biota, and air and water environment [17]. For excellent agricultural production, AMF is believed to have the tendency to decrease the need of chemical fertilizers by 50%, however it alters with the type of species and existing stress conditions. AM escalates the absorption efficiency of host plant roots 10 fold [18] and the effectiveness of immobile nutrients by 60 fold owing to penetration in to the nutrient depleted areas of soil. It has been reported that maize plants growing in loamy sand texture soil under water stress absorbed more phosphorus when inoculated with *Glomus etunicatum* compared to non-mycorrhizal plants [19]. Moreover Mycorrhiza has the potential alter the concentration of organic matter thereby plays a tremendous role in storage of carbon in soil [20] and changes the kinetic characteristics of roots resulting in nutrient uptake, thus it is clear indication the mycorrhiza has great role in enhancing productivity and cycling of nutrients [21, 22].

4.1 AMF and mineral nutrition

AMF colonization is widely thought to promote plant nutrient absorption. It is well known fact that inoculation with AMF can result in adequate escalation in accumulation of various macro- and micronutrients, leading to increase in production of photosynthates and thus enhanced biomass accumulation [23]. AMF have the capacity to escalate the absorption of inorganic substances, notably phosphate, in practically all plants [24]. AMF are also particularly productive at assisting plants in absorbing nutrients from mineral deficient soils. Apart from macronutrients, the association of AMF has been described to inflate the phyto-availability of microelements viz. Zinc and Copper [21, 22]. AMF increases the surface absorbency of the roots of host plant. Inflation in photosynthetic activity and other foliage activities are clearly related to an increase in the occurrence of AMF inoculation, which is clearly related to absorption of carbon, nitrogen and phosphorus, which approaches towards roots and increase tuber enlargement. AMF has been found to balance the absorption of nitrogen and phosphorus, thereby aiding plant enlargement at both high and low concentrations of P under diverse irrigation regimes.

The enhancement of growth in plants by fungi is more remarkable in tropical soils than temperate areas because of inherently low fertility in tropical areas. The phosphatic fertilizers added to the soil are fixed and became unavailable for plants, under low pH phosphorus gets fixed with Fe and Al and under high pH it gets fixed with Ca [25]. AM fungi has the ability to improve P- nutrition in plants through the enlargement of hyphae far off the root system, which permits investigation of fixed minerals [26]. The most significant function of AM is to improve Uptake of P in plants from low P concentrated areas owing to the greater surface area of hyphae and uptake mechanism. The mycorrhizal associated plants show rapid movement of P from roots to above ground parts because of steep gradient between the two. The steepness is due to rapid conversion of inorganic P in the above-ground parts compared to roots. AM also acts on phytate minerals (source of organic P) through the production of acid phosphatize to liberate the H2PO4 ions [27].

4.2 AMF and plant yield

Useful root zone microflora besides improving the nutrient content of crop, improve crop standard as well. For example, reports have revealed that colonization

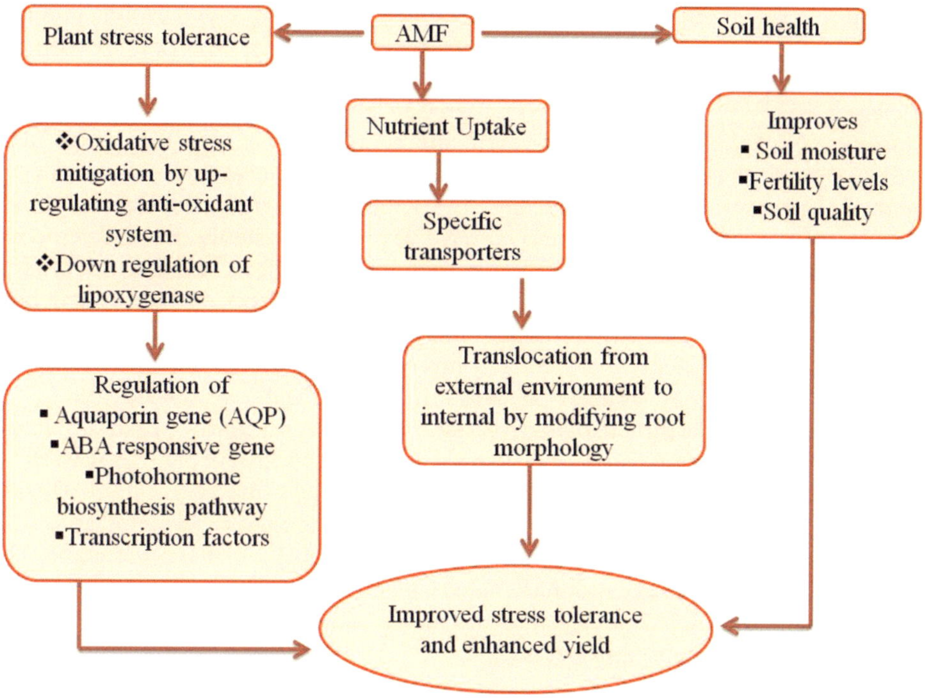

Figure 1.
A diagrammatic illustration of mycorrhizal roles in regulating numerous ecosystem processes and promoting plant development under abiotic stress conditions.

of strawberry with AMF escalated the concentration of secondary metabolites, leading to better antioxidant activity [28]. AMF has the potential to enhance nutritional standard by influencing the synthesis of carotenoids and other volatile chemicals. Further research [29] found that *Glomus versiforme* had higher levels of sugars, organic acids, vitamin C, flavonoids, and minerals, resulting in higher standard of citrus fruits. Mycorrhizal symbiosis increases anthocyanins, chlorophyll, cartenoids, total soluble phenolics, tocopherols, and a variety of mineral nutrients. Rouphael et al. [30] revealed that AMF might mitigate inanimate pressue by regulating soil pH and so preserving its horticultural value (**Figure 1**).

5. AMF and abiotic stresses

5.1 Drought

Drought is the dearth of sufficient levels of water in the root zone for normal functioning of plant. It is also referred as water deprivation or water stress. Drought stress effects in plants emerge due to deficiency of water in the rhizospheric region, a high transpiration rate, or the rapid formation of reactive oxygen species (ROS), and the subsequent onset of oxidative stress [31]. Drought stresses have a negative impact on expansion on plants owing to alteration in enzymatic activity, Ion and mineral absorption [32]. Symbiotic associations are thought to govern a number of physio-biochemical processes in plants, including greater osmotic adjustment, stomatal regulation through modulating ABA metabolism, increased proline buildup, and higher glutathione levels. Under drought conditions, the symbiotic connection of numerous plants with AMF may eventually boost biomass, LAI, length of roots and efficiency [33]. Li et al. [34], reported that in C3 plants viz. *Leymus chinensis* and C4 plants viz. *Hemarthria altissima* the growth and photosynthesis was enhance by AMF mediation through up-regulation of antioxidant system. Abiotic stressors like salt and drought generate significant decline in agricultural return. Furthermore, mineral depletion, water stress, salt stress and increase in pH, the existence of trace elements, and elevated temperatures are major issues in numerous regions of the world, especially in dry regions [35]. This mutualistic interaction has been reported to participate in a variety of biochemical and physiological activities, including (1) direct absorption and transfer of water and minerals by mycorrhizal fungi, (2) enhanced osmotic regulation, (3) improved gas exchange and Efficiency of water utilization, and (4) strong defense in opposition to oxidative destruction [36]. In contrast to non-mycorrhizal plants, mycorrhizal fungus can also change water control in plant growth by altering hormonal equilibrium signaling or by stimulating osmolytes in mycorrhizal plants (increased vigor or volume of products of photosynthesis and dissolvable sugars in the foliage symplasm). In drought conditions, AM inoculation of plants improved size and density of root hairs. These plants also have higher concentrations of methyl jasmonate, IAA, calmodulin, and nitric oxide in their roots, resulting in enhanced resistance to drought stress [37].

5.2 Salinity

Soil salinization is a well-known environmental phenomenon that poses a serious danger to international food safety. Salinity stress is common for suppressing plant growth by altering vegetative development and overall absorption rate, leading to lower quantity of yield. It also stimulates the extravagant production of reactive oxygen species. The resistance to salinity involves Na + and Cl- storage in cell vacuoles, which prevents Na + entry into the cell and its removal by transpiration. Efforts are being undertaken to investigate potential methods of increasing agricultural output on salt-affected soils. One such option is to apply AMF sparingly to reduce the negative effects of salinity on plants. Some research investigations have established the effectiveness of AMF in initiating growth and increasing production in plants subjected to salt stress [38]. EL-Nashar [39], demonstrated that MF increased enlargement rate. Foliage water potential, water utilization potential of *Antirrhinum majus* plant. Under saline conditions, mycorrhizal inoculation significantly increased rate of photosynthesis and other gas exchange properties, chlorophyll concentration, and water utilization potential in Ocimum basilicum L. [40]. Furthermore, Plants with AMF produce more jasmonic acid, salicylic acid, and other vital inorganic nutrients. Under salt stress conditions, for instance, total N, P, K+, Ca2+, and Mg2+ concentrations were more in AMF-treated Cucumis sativus plants than in untreated plants [41]. AMF inoculation can efficiently modulate critical growth regulator levels. Furthermore, AMF inoculation increased the build up of different organic acids, leading to enhanced osmoregulation mechanism in plant growth under salt stress. Salinity has an impact on crop morphology, physiological function, and yield, as well as a large amount of arable land. It has been discovered that tomato plant production is maximum at salinity concentration of 5 dSm^{-1}, however Salinity has been shown to reduce foliage area and dry matter concentration. Furthermore, the foliage was shown to be more vulnerable to salinity stress than the fruits because they contained more proline and Na [42]. Salinity influences all stages of plant growth, including germination, seedling, vegetative phase, and maturity. Salinity disrupts plant ionic adjustment and osmotic pressure, as well as cell membrane selectivity [43]. Salinity disrupts plant ionic homeostasis by amplifying ROS (reactive oxygen species), that negatively influences nutrient absorption, cell membranes, and different ultra structures, resulting in ionic and osmotic stress [44].

5.3 Heavy metals

AMF is commonly thought to reinforce plant development in heavy metal-contaminated soil owing to their ability to build up defense system of host plants and growth and development promotion. These trace elements accrete in all crops viz., (food crops, fruits, vegetables) and soils, giving rise to a variety of health risks. Under aluminum stress, the interaction of AMF with whet improved nutrition absorption [45]. Plants planted in soil augmented with Cd and Zn showed remarkable inhibition of under and above ground growth, chlorosis, and even mortality [46]. The significant effect of AMF on plant development and growth under acute stressful circumstances is usually due to the tendency of fungi to augment morphological and physiological processes that enhance plant biomass and, as a result, absorption of critical immobile elements like Cu, Zn, and P, resulting in decrease in harmful effects of metals on the host plants [47]. Decrease in concentration of metals by dilution in plant tissues and chelation in the root zone is believed to cause enhanced growth. AMF blend with Cd and Zn in the cell walls of mantle hyphae and cortical cells, limiting their absorption and leading in increased growth, yield, and nutritional status. AMF were highly effective in decreasing Cd detoxifying levels in rice [48]. Numerous processes occur as a result of the AMF, including metal compound immobilization/restriction, polyphosphate granule formation in the soil, adsorption to fungal cell wall chitin, and heavy metal chelation within the fungus.

5.4 Temperature

Plant community reactions to rising soil temperatures may be based on AMF interactions for long-term yield and production [49]. Heat stress has a remarkable affect on plant growth and development through different processes viz. (1) drop in plant vigor and seed germination inhibition, (2) retarded growth potential, (3) diminishing biomass (4) wilting and burning of foliage and reproductive parts, (5) abscission and senescence of foliage, (6) fruit destruction and change in color, (7) yield loss and cell death, and (8) increased oxidative stress. AMF-inoculated plants typically develop faster under heat stress than non-AMF-inoculated plants. Maya and Matsubara [50] demonstrated the symbiosis of AMF Glomus fasciculatum with plant expansion and enlargement, resulting in favorable alterations in growth under high temperature situations.

AMF can help plants tolerate cold stress. Furthermore, according to the majority of reports, diverse plants inoculated with AMF at low temperatures show good growth than non-AMF-inoculated plants [51]. AMF helps crops fight cold stress and subsequently improves plant development. Furthermore, AMF can regulate moisture content in the host plant, leading to escalation in plant secondary metabolites, so strengthening the plant defense system, and enhanced protein concentration, thereby assisting plants in combating cold stress situations [52]. During cold stress,

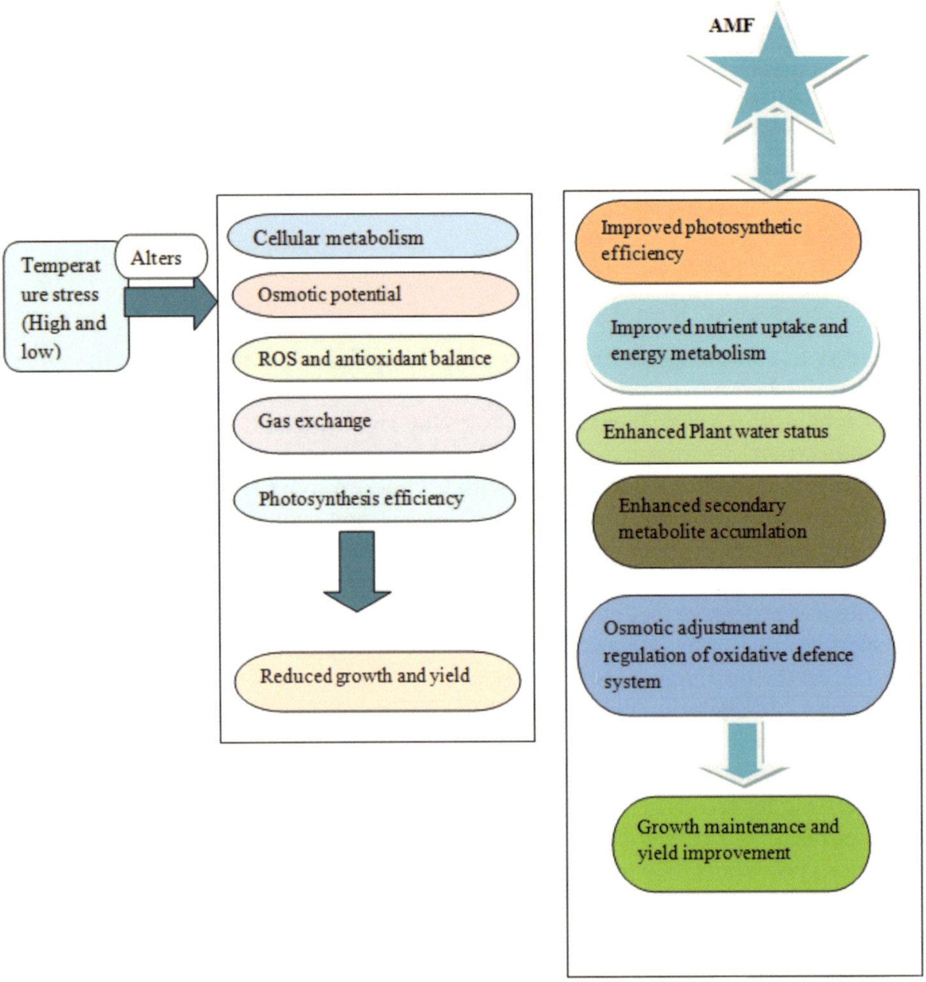

Figure 2.
Temperature stress in plants is alleviated by AMF inoculation.

Fungal species and recorded response	Host species involved
Pavithra and Yapa [56] reported that under Drought conditions AMF has enhanced proline content in foliage, leaf area index, rate of growth, photosynthesis, weight of seeds	*Glycine max L.*
Zhang et al. [57] revealed that *Funneliformis mosseae, Paraglomus occultum* under drought stress conditions resulted in increase in size of hyphae, water potential of foliage	*Poncirus trifoliate*
Sara et al. [58] reported that AMF under drought stress situations provided relief to plants and enhanced turgor pressure and absorption of minerals	*Olea europaea*
Pal and Pandey [59] reported various fungal species viz., *Glomus mosseae, Glomus fasciculatum, Gigaspora decipiens* led to escalation of plant growth, pigments and total chlorophyll under drought stress	*Triticum aestivum L.*
Pedranzani et al. [60] reported that *Rhizophagus irregularis* improved conductance of stomata, lipid peroxidation and liberation of hydrogen peroxide in above and underground parts under drought stress	*Digitaria eriantha*

Fungal species and recorded response	Host species involved
Rani [61] reported that *Glomus mosseae* under drought stress conditions escalated osmotic potential, activity of enzymes and ascorbic acid	*Triticum aestivum*
Goicoechea et al. [62] reported under drought stress conditions *Rhizophagus intraradices* increased micronutrient content in grains and biomass	*Triticum durum*
Yooyongwech et al. [63] reported that *Glomus* spp. has the tendency to adjust the osmotic potential of plants under drought by soluble carbohydrates and proline	*Ipomoea Batatas*
Mirshad and Pathur [64] reported that *Glomus* spp. under drought promoted absorption of nutrients, phenolics, biomass in plants	*Saccharm arundinaceum Retz.*
Zhao et al. [65] revealed that *Rhizophagus intraradices, strain BGCBJ09* improved absorption of essential elements in above ground parts of plant, enhanced dry weight in plants and water use efficiency	*Zea mays*
Ruiz-Lozano et al. [66] reported that *Rhizophagus irregularis, Glomus intraradices* improved biomass and efficiency of PS2 under drought	*Lettuce and tomato*
Amiri et al. [67] showed enhanced nutrient level, biomass and proteins associated with glomalin under drought by *Rhizophagus intraradices, Funneliformis mosseae*	*Pelargonium graveolens*
Boyer et al. [68] reported increase in survival rate under drought stress	*Fragaria ananassa*
Yang et al. [69] reported enhanced photosynthetic action, dry biomass under drought stress by *Funneliformis mosseae* and *Rhizophagus intraradices*	*Robinia pseudoacacia L.*
Grümberg et al. [70] reported that *Glomus* spp. escalated macronutrient content and water content in plants under drought	*G. max*
Asrar et al. [71] reported that *Glomus deserticola* has improved dimensions of plants like diameter, foliage per plant, chlorophyll and proline under drought	*Antirrhinum majus*
Tsoata et al. [72] reported that *Glomus intraradices, Gigaspora gregaria* enhanced the sugar and mineral concentration and decrease in proline content under drought	*Vigna subterranean*
Bayani et al. [73] reported that *Glomus intraradices* increased Root volume, P content, and activity of phosphatase enzyme under drought stress	*Hordeum vulgare*
Cabral et al. [74] reported under heat stress *Rhizophagus irregularis, Funneliformis mosseae, Claroideoglomus claroideum* resulted in increased root grain number, resource allocation, and nutrient composition	*Triticum aestivum* L.
Mathur et al. [75] reported that *Rhizophagus intraradices, Funneliformis mosseae, F. geosporum* increased plant height, chlorophyll a, stomatal conductance, transpiration rate, and photosynthetic rate under high temperature	*Zea mays*
Calvo-Polanco et al. [76] reported improvement in water conduction as well as aquaporin quantity and phosphorylation status under high temperature by *Rhizophagus irregularis*	*Solanum lycopersicum*
Lin et al. [77] reported nodule formation in roots, increased concentration of N and P under metal contamination by *Glomus mosseae*	*Sesbania rostrata*
Abdelhameed and Rabab [78] revealed that *Glomus clarum, G. monosporum, Gigaspora nigra* have enhanced antioxidant activity and malondialdehyde content in Cd contaminated soil	*Trigonella foenum –graecum* L.
Garg and Singh [79] reported increase in under -ground biomass, nutrient content, proline content in Cd and Zn contaminated soil by *Rhizophagus irregularis*	*Cajanus cajan* L.

Fungal species and recorded response	Host species involved
Hajiboland et al. [80] reported improvement in ion absorption, dry matter, chlorophyll concentration and other growth paramaters under saline stress by *Glomus intraradices.*	*Solanum lycopersicum* L.
Hashem et al. [41] reported that *Glomus etunicatum, Glomus intraradices, Glomus mosseae* has escalated biomass, pigment production in photosynthetic, and antioxidant enzyme	Cucumis sativus L.
Khalloufi et al. [81] reported increment in foliage area, number of leaves and concentration of growth hormones by *Rhizophagus irregularis* under salt stress	*Solanum lycopersicum* L.
Porcel et al. [82] has revealed that *Claroideoglomus etunicatum* resulted in improvement in PSII photochemistry quantum yield, net photosynthetic rate, and stomatal conductance under salt stress.	*Oryza sativa* L.
Hajiboland et al. [83] reported increment in above and under-ground dry mass, stomatal conductance, soluble sugars, free amino acids, and Na + and K+ absorption by *Claroideoglomus etunicatum* under salt stress	*Aeluropus littoralis*
Giri et al. [2] reported under salt stress *Glomus fasciculate* has escalated root and shoot biomass, as well as P, Zn, and Cu concentrations.	Acacia nilotica
Jixiang et al. [84] reported under saline stress *Glomus mosseae* increased colonization rate, seedling weight, water content, and P and N concentration.	Leymus chinensis

Table 1.
Brief overview of various fungal species and their recorded responses.

for example, AMF injected plants demonstrated higher water saving ability as well as usage efficiency [53]. The symbiotic AMF connection boosts the water-plant relationship while increasing gas exchange potential and osmotic balance. AMF improves chlorophyll production, resulting in a remarkable increase in the concentrations of different metabolites in plants exposed to cold stress conditions [54]. Low temperatures have a deleterious impact on plant metabolism. It can cause substantial harm to plant tissue, including yellowing of leaves, membrane damage, cell death, and changes in enzyme action and cytoplasm viscosity in vegetable plants. Chilling stress reduces photosynthetic efficiency and increases electrolyte leakage in watermelon seedlings [55] (**Figure 2**) (**Table 1**).

6. Conclusion

Arbuscular mycorrhizal fungi have cosmopolitan distribution in soil environment and live in mutualistic association with the roots of angiosperms and other plants. The mutualistic association between fungi and plant roots assist in absorption of macronutrients viz., P, N as well as trace elements viz. Zn, Fe, Cu. The various ways to achieve these functions effectively are increment in absorption area of plants, liberation of biochemical substances. These have significant role in cycling of nutrients by mobilizing the fixed elements as well as serving as sink for various elements. Furthermore, they have ability to decrease various biotic and abiotic stresses like drought, saline, water, temperature and resistance to diseases. These increase the availability of less accessible elements to plants. The significance of fungi in agricultural production and forestry is due to its contribution in growth and nutrition of plants. The utilization of AM as fertilizers is not deleterious to plants as chemical fertilizers. These microbe-mediated supplements are utilized to escalate the nutritional status of plants. The merit is they are eco-friendly and do not cause any detrimental effect to the environment besides increasing yield and providing protection against diseases.

Author details

Malik A. Aziz[1]*, Shayesta Islam[2], Gousia Gani[1], Zaffar M. Dar[3], Amajad Masood[4] and Syed H. Baligah[5]

1 Division of Basic Sciences and Humanities, FOA-Wadura, SKUAST-K, Srinagar, India

2 Division of Environmental Sciences, FOH-Shalimar, SKUAST-K, Srinagar, India

3 Division of Basic Sciences, FOH- Shalimar, SKUAST-K, Srinagar, India

4 Division of Agronomy, FOA-Wadura, SKUAST-K, Srinagar, India

5 Division of Soil Sciences, FOA-Wadura, SKUAST-K, Srinagar, India

*Address all correspondence to: drasif_skuast@yahoo.com

References

[1] Shanker A. Abiotic Stress Response in Plants - Physiological, Biochemical and Genetic Perspectives. London, UK: InTech; 2011

[2] Giri B, Kapoor R, Mukerji KG. Improved tolerance of acacia nilotica, to salt stress by arbuscular mycorrhiza, *Glomus fasciculatum*, may be partly related to elevated K/Na ratios in root and shoot tissues. Microbiology Ecology. 2007a;**54**:753-760

[3] Varma A, Prasad R, Tuteja N. Mycorrhiza—Nutrient Uptake, Biocontrol, Ecorestoration. Berlin, Germany: Springer; 2018

[4] Latef AAHA, Hashem A, Rasool S. Arbuscular mycorrhizal symbiosis and abiotic stress in plants: A review. Journal of Plant Biology. 2016;**59**(5):407-426

[5] Wu QS, Xia RX. Arbuscular mycorrhizal fungi influence growth, osmotic adjustment and photosynthesis of citrus under well-watered and water stress conditions. Journal of Plant Physiology. 2006;**163**(4):417-425

[6] Kloepper JW, Schroth MN. Plant growth-promoting rhizobacteria on radishes. In: Station de Pathologie Vegetale et Phyto-Bacteriologie, Editor. Proceedings of the 4th International Conference on Plant Pathogenic Bacteria. Vol. II. Tours: Gilbert-Clarey. 1978. pp. 879-882

[7] Okon Y. Azospirillum/Plant Associations. Boca Raton, Florida: CRC Press; 1994

[8] Stewart LI, Hamel C, Hogue R, Moutoglis P. Response of strawberry to inoculation with arbuscular myccorrhizal fungi under very high soil phosphorus conditions. Mycorrhiza. 2005;**15**:612-619

[9] Sun Z, Song J, Xin X, Xie X, Zhao B. Arbuscular mycorrhizal fungal proteins 14-3-3- are involved in arbuscule formation and responses to abiotic stresses during AM symbiosis. Frontiers Microbiology. 2018;**5**:9-19

[10] Jung SC, Martinez-Medina A, Lopez-Raez JA, Pozo MJ. Mycorrhiza-induced resistance and priming of plant defenses. Journal of Chemistry and Ecology. 2012;**38**:651-664

[11] Gianinazzi S, Golotte A, Binet MN, Van Tuinen D, Redecker D, Wipf D. Agroecology: The key role of arbuscular mycorrhizas in ecosystem services. Mycorrhiza. 2010;**20**:519-530

[12] Bonfante P, Genre A. Mechanism underlying beneficial plant fungus interactions in mycorrhizal symbiosis. Nature Communications. 2010;**1**(8)

[13] Parniske M. Arbuscular Mycorrhiza: The Mother of Plant Root Endosymbiosis. Nature Reviews Microbiology. 2008;**6**:763-775

[14] Gutjahr C, Parniske M. Cell and developmental biology of arbuscular mycorrhiza symbiosis. Annual Review of Cell and Developmental Biology. 2013;**29**:593-617

[15] Sadhana B. Arbuscular mycorrhizal fungi (AMF) as a biofertilizers—a review. International Journal of Current Microbiology and Applied Science. 2014;**3**(4):384-400

[16] Ortas I. The effect of mycorrhizal fungal inoculation on plant yield, nutrient uptake and inoculation effectiveness under long-term field conditions. Field Crops Research. 2012;**125**:35-48

[17] Yang S, Li F, Malhi SS, Wang P, Dongrang S, Wang J. Long term fertilization effects on crop yield and nitrate nitrogen accumulation in soil in Northwestern China. Agronomy Journal. 2004;**96**:1039-1049

[18] Bisleski RL. Phosphate transport and phosphate availability. Annual Review of Plant Physiology. 1973;**24**:225-252

[19] Muller I, Hofner W. Influence of arbuscular mycorrhiza on phosphorus uptake and recovery potential of maize (Zea mays L.) under water-stressed conditions. Mycological Research. 1991;**98**:593-603

[20] Ryglewicz PT, Anderson CP. Mycorrhizae alter quality and quantity of carbon below ground. Nature. 1994;**369**:58-60

[21] Smith SE, Read DJ. Mycorrhizal Symbiosis. San Diego: Academic Press; 1997a. p. 607

[22] Smith SA, Read DJ. Mycorrhizal Symbiosis. 2nd edn. Cambridge: Academic press; 1997b

[23] Mitra D, Navendra U, Panneerselvam U, Ansuman S, Ganeshamurthy AN, Divya J. Role of mycorrhiza and its associated bacteria on plant growth promotion and nutrient management in sustainable agriculture. International Journal of Life Science and Applied Sciences. 2019;**1**:1-10

[24] Nell M, Wawrosch C, Steinkellner S, Vierheilig H, Kopp B, Lössl A. Root colonization by symbiotic arbuscular mycorrhizal fungi increases sesquiterpenic acid concentrations in *Valeriana officinalis* L. Planta Medica. 2010;**76**:393-398

[25] Ozanne PG. Phosphate nutrition of plants–A general treatise. In: Kasawneh FE, Sample EC, Kamprath EJ, editors. The Role of Phosphorus in Agriculture. Crop Science Society America and Soil Science Society America, Madison: American Society of Agronomy; 1980. pp. 559-589

[26] Smith FA, Jacobsen I, Smith SE. Spatial differences in acquisition of soil phosphate between two arbuscular mycorrhizal fungi in symbiosis with Medicago truncatula. The New Phytologist. 2000;**147**(2):357-366

[27] Joner EJ, Briones R, Leyval C. Metal binding capacity of arbuscular mycorrhizal mycelium. Plant and Soil. 2000;**226**(2):227-234

[28] Castellanos-Morales V, Villegas J, Wendelin S, Vierheiling H, Eder R, Cardenas-Navarro R. Root colonization by the arbuscular mycorrhizal fungus *Glomus intraradices* alters the quality of strawberry fruit (*Fragaria ananassa* Duch.) at different nitrogen levels. Journal of the Science of Food and Agriculture. 2010;**90**:1774-1782

[29] Zeng L, Fu LJ, Fu LJ, Ming Yuan W. Effects of arbuscular mycorrhizal (AM) fungi on citrus quality under natural conditions. Southwest China of Journal of Agricultural Sciences. 2014;**27**:2101-2105

[30] Rouphael Y, Franken P, Schneider C, Schwarz D, Giovannetti M, Agnolucci M. Arbuscular mycorrhizal fungi act as bio-stimulants in horticultural crops. Scientia Horticulturae. 2015;**196**:91-108

[31] Barzana G, Aroca R, Paz JA, Chaumont F, Martinez-Ballesta MC, Carvajal M, et al. Arbuscular mycorrhizal symbiosis increases relative apoplastic water flow inroots of the host plant under both well-watered and drought stress conditions. Annals of Botany. 2012;**109**:1009-1017

[32] Ahanger MA, Agarwal RM. Potassium up-regulates antioxidant metabolism and alleviates growth inhibition under water and osmotic stress in wheat (*Triticum aestivum* L.). Protoplasma. 2017;**254**(4):1471-1486

[33] Gholamhoseini M, Ghalavand A, Dolatabadian A, Jamshidi E, Khodaei-Joghan A. Effects of arbuscular mycorrhizal inoculation on growth, yield, nutrient uptake and irrigation water productivity of sunflowers grown under drought stress. Agricultural Water Management. 2013;**117**:106-114

[34] Li J, Meng B, Chai H, Yang X, Song W, Li S. Arbuscular mycorrhizal fungi alleviate drought stress in C3 (*Leymus chinensis*) and C4 (*Hemarthria altissima*) grasses *via* altering antioxidant enzyme activities and photosynthesis. Frontiers in Plant Science. 2019;**10**:499

[35] Evelin H, Kapoor R, Giri B. Arbuscular mycorrhizal fungi in alleviation of salt stress: A review. Annals of Botany. 2009;**104**:1263-1280

[36] Marulanda A, Porcel R, Barea JM, Azcon R. Drought tolerance and antioxidant activities in lavender plants colonized by native drought-tolerant or drought sensitive Glomus species. Microbial Ecology. 2007;**54**:543-552

[37] Zou Y-N, Wang P, Liu C-Y, Ni Q-D, Zhang D-J, Wu Q-S. Mycorrhizal trifoliate orange has greater root adaptation of morphology and phytohormones in response to drought stress. Scientific Reports. 2017;7:41134

[38] Talaat NB, Shawky BT. Protective effects of arbuscular mycorrhizal fungi on wheat (*Triticum aestivum* L.) plants exposed to salinity. Environmental and Experimental Botany. 2014;**98**:20-31

[39] EL-Nashar YI. Response of snapdragon *Antirrhinum majus* L. to blended water irrigation and arbuscular mycorrhizal fungi inoculation: Uptake of minerals and leaf water relations. Photosynthetica. 2017;**55**(2):201-209

[40] Elhindi KM, El-Din SA, Elgorban AM. The impact of arbuscular mycorrhizal fungi in mitigating salt-induced adverse effects in sweet basil (*Ocimum basilicum* L.). Saudi Journal of Biological Science. 2017;**24**:170-179

[41] Hashem A, Alqarawi AA, Radhakrishnan R, Al-Arjani AF, Aldehaish HA, Egamberdieva D. Arbuscular mycorrhizal fungi regulate the oxidative system, hormones and ionic equilibrium to trigger salt stress tolerance in Cucumis sativus L. Saudi Journal of Biological Science. 2018a;**25**(6):1102-1114

[42] Babu M, Singh D, Gothandam K. The effect of salinity on growth, hormones and mineral elements in leaf and fruit of tomato cultivar PKM1. Journal of Animal and Plant Sciences. 2012;**22**:159-164

[43] Nawaz K, Hussain K, Majeed A, Khan F, Afghan S, Ali K. Fatality of salt. 2010

[44] Arif Y, Singh P, Siddiqui H, Bajguz A, Hayat S. Salinity induced physiological and biochemical changes in plants: An omic approach towards salt stress tolerance. Plant Physiology and Biochemistry. 2020;**156**:64-77

[45] Aguilera P, Pablo C, Fernando B, Fritz O. Diversity of arbuscular mycorrhizal fungi associated with *Triticum aestivum* L. plants growing in an andosol with high aluminum level. Agricultural Ecological Environment. 2014;**186**:178-184

[46] Moghadam HRT. Application of super absorbent polymer and ascorbic acid to mitigate deleterious effects of cadmium in wheat. Pesquisa Agropecuaria Tropical. 2016;**6**(1):9-18

[47] Miransari M. Arbuscular mycorrhizal fungi and heavy metal tolerance in plants. In: Wu QS, editor. Arbuscular Mycorrhizas and Stress Tolerance of Plants. Singapore: Springer Nature; 2017. pp. 174-161

[48] Li H, Luo N, Zhang LJ, Zhao HM, Li YW, Cai QY. Do arbuscular mycorrhizal fungi affect cadmium uptake kinetics, subcellular distribution and chemical forms in rice? Science of the Total Environment. 2016;**571**:1183-1190

[49] Bunn R, Lekberg Y, Zabinski C. Arbuscular mycorrhizal fungi ameliorate temperature stress in thermophilic plants. Ecology. 2009;**90**(5):1378-1388

[50] Maya MA, Matsubara Y. Influence of arbuscular mycorrhiza on the growth and antioxidative activity in *Cyclamen* under heat stress. Mycorrhiza. 2013;**23**(5):381-390

[51] Chen S, Jin W, Liu A, Zhang S, Liu D, Wang F. Arbuscular mycorrhizal fungi (AMF) increase growth and secondary metabolism in cucumber subjected to low temperature stress. Scientia Horticulturae. 2013;**160**:222-229

[52] Abdel Latef AA, Chaoxing H. Arbuscular mycorrhizal influence on growth, photosynthetic pigments, osmotic adjustment and oxidative stress in tomato plants subjected to low temperature stress. Acta Physiologiae Plantarum. 2011;**33**:1217-1225

[53] Zhu XC, Song FB, Xu HW. Effects of arbuscular mycorrhizal fungi on photosynthetic characteristics of maize under low-temperature stress. Acta Ecologica Sinica. 2010b;**21**:470-475

[54] Zhu XC, Song FB, Xu HW. Arbuscular mycorrhizae improve low-temperature stress in maize via alterations in host water status and photosynthesis. Plant and Soil. 2010a;**331**:129-137

[55] Shirani Bidabadi S, Mehralian M. Arbuscular mycorrhizal fungi inoculation to enhance chilling stress tolerance of watermelon. Gesunde Pflanzen. 2020;**72**:171-179

[56] Pavithra D, Yapa N. Arbuscular mycorrhizal fungi inoculation enhances drought stress tolerance of plants. Ground Water Sustainable Development. 2018;**7**:490-494

[57] Zhang F, Jia-Dong HE, Qiu-Dan NI, Qiang-Sheng WU, Zou YN. Enhancement of drought tolerance in trifoliate orange by mycorrhiza: changes in root sucrose and proline metabolisms. Notulae Botanicae Horti. Agrobotanici Cluj-Napoca. 2018;**46**:270

[58] Sara O, Ennajeh M, Zrig A, Gianinazzi S, Khemira H. Estimating the contribution of arbuscular mycorrhizal fungi to drought tolerance of potted olive trees (*Olea europaea*). Acta Physiology Plant. 2018;**40**:1-81

[59] Pal A, Pandey S. Role of arbuscular mycorrhizal fungi on plant growth and reclamation of barren soil with wheat (*Triticum aestivum* L.) crop. International Journal of Soil Sciences. 2016;**12**:25-31

[60] Pedranzani H, RodrãGuez-Rivera M, GutiaRrez M, Porcel R, Hause B, Ruiz-Lozano JM. Arbuscular mycorrhizal symbiosis regulates physiology and performance of *Digitaria eriantha* plants subjected to abiotic stresses by modulating antioxidant and jasmonate levels. Mycorrhiza. 2016;**26**:141-152

[61] Rani B. Effect of arbuscular mycorrhiza fungi on biochemical parameters in wheat *Triticum aestivum*

L. under drought conditions. [Doctoral dissertation], CCSHAU, Hisar. 2016

[62] Goicoechea N, Bettoni M, Fuertes-Mendiza'bal T, Gonzalez-Murua C, Aranjuelo I. Durum wheat quality traits affected by mycorrhizal inoculation, water availability and atmospheric CO2 concentration. Crop and Pasture Science. 2016;**67**:147-155

[63] Yooyongwech S, Samphumphuang T, Tisarum R, Theerawitaya C, Chaum S. Arbuscular mycorrhizal fungi (AMF) improved water deficit tolerance in two different sweet potato genotypes involving osmotic adjustments *via* soluble sugar and free proline. Scientia Horticulturae. 2016;**198**:107-117

[64] Mirshad PP, Pathur JT. Arbuscular mycorrhizal association enhances drought tolerance potential of promising bioenergy grass *Saccharum arundinaceum*, Retz. Environmental Monitoring and Assessment 2016;**188**:425

[65] Zhao R, Guo W, Bi N, Guo J, Wang L, Zhao J, et al. Arbuscular mycorrhizal fungi affect the growth, nutrient uptake, and water status of maize (*Zea mays*, L.) grown in two types of coal mine spoil under drought stress. Applied Soil Ecology. 2015;**88**:41-49

[66] Ruiz-Lozano JM, Aroca R, Zamarreño ÁM, Molina S, Andreo-Jiménez B, Porcel R. Arbuscular mycorrhizal symbiosis induces *strigolactone* biosynthesis under drought and improves drought tolerance in lettuce and tomato. Plant, Cell & Environment. 2015;**39**(2):441-452

[67] Amiri R, Nikbakht A, Etemadi N. Alleviation of drought stress on rose geranium *Pelargonium graveolen* L Herit. in terms of antioxidant activity and secondary metabolites by mycorrhizal inoculation. Scientia Horticulture. 2015;**197**:373-380

[68] Boyer LR, Brain P, Xu XM, Jeffries P. Inoculation of drought-stressed strawberry with a mixed inoculum of two arbuscular mycorrhizal fungi: Effects on population dynamics of fungal species in roots and consequential plant tolerance to water. Mycorrhiza. 2014;**25**(3):215-227

[69] Yang Y, Tang M, Sulpice R, Chen H, Tian S, Ban Y. Arbuscular mycorrhizal fungi alter fractal dimension characteristics of *Robinia pseudoacacia*, L. seedlings through regulating plant growth, leaf water status, photosynthesis, and nutrient concentration under drought stress. Journal of Plant Growth Regulation. 2014;**33**:612-625

[70] Grümberg BC, María UC, Shroeder A, Vargas-Gil S, Luna CM. The role of inoculum identity in drought stress mitigation by arbuscular mycorrhizal fungi in soybean. Biology and Fertility of Soils. 2015;**51**:1-10

[71] Asrar AA, Abdel-Fattah GM, Elhindi KM. Improving growth, flower yield, and water relations of snapdragon *Antirhinum majus* L. plants grown under well-watered and water-stress conditions using arbuscular mycorrhizal fungi. Photosynthetica. 2012;**50**:305-316

[72] Tsoata E, Njock SR, Youmbi E, Nwaga D. Early effects of water stress on some biochemical and mineral parameters of mycorrhizal *Vigna subterranea* (L.) Verdc. (Fabaceae) cultivated in Cameroon. International Journal of Agrononmy and Agricultural Research. 2015;7:21-35

[73] Bayani R, Saateyi A, Faghani E. Influence of arbuscular mycorrhiza in phosphorus acquisition efficiency and drought-tolerance mechanisms in barley *Hordeum vulgare* L. International Journal Biosciences. 2015;7:86-94

[74] Cabral C, Sabine R, Ivanka T, Bernd W. Arbuscular mycorrhizal fungi modify nutrient allocation and composition in wheat (Triticum aestivum L.) subjected to heat-stress. Plant and Soil. 2016;**408**(1-2):385-399

[75] Mathur S, Sharma MP, Jajoo A. Improved photosynthetic efficacy of maize *Zea mays* plants with arbuscular mycorrhizal fungi (AMF) under high temperature stress. Journal of Photochemistry Photobiology B. 2016;**180**:149-154

[76] Calvo-Polanco M, Sanchez-Romera B, Aroca R, Asins MJ, Declerck S, Dodd IC. Exploring the use of recombinant inbred lines in combination with beneficial microbial inoculants (AM fungus and PGPR) to improve drought stress tolerance in tomato. Environmental and Experimental Botany. 2016;**131**:47-57

[77] Lin AJ, Zhang XH, Wong MH, Ye ZH, Lou LQ, Wang YS. Increase of multi-metal tolerance of three leguminous plants by arbuscular mycorrhizal fungi colonization. Environmental Geochemistry and Health. 2007;**29**:473-481

[78] Abdelhameed RE, Rabab AM. Alleviation of cadmium stress by arbuscular mycorrhizal symbiosis. International Journal Phytoremediation. 2019

[79] Garg N, Singh S. Arbuscular mycorrhiza *Rhizophagus irregularis*, and silicon modulate growth, proline biosynthesis and yield in *Cajanus cajan*, L. Millsp. (pigeon pea) genotypes under cadmium and zinc stress. Journal of Plant Growth Regulation. 2017;**37**:1-18

[80] Hajiboland R, Aliasgharzadeh N, Laiegh SF, Poschenrieder C. Colonization with arbuscular mycorrhizal fungi improves salinity tolerance of tomato *Solanum lycopersicum* L. plants. Plant and Soil. 2010;**331**:313-327

[81] Khalloufi M, Martínez-Andújar C, Lachaâl M, Karray-Bouraoui N, Pérez-Alfocea F, Albacete A. The interaction between foliar GA3 application and arbuscular mycorrhizal fungi inoculation improves growth in salinized tomato *Solanum lycopersicum* L. plants by modifying the hormonal balance. Journal of Plant Physiology. 2017;**214**:134-144

[82] Porcel R, Redondogómez S, Mateosnaranjo E, Aroca R, Garcia R, Ruizlozano JM. Arbuscular mycorrhizal symbiosis ameliorates the optimum quantum yield of photosystem II and reduces non-photochemical quenching in rice plants subjected to salt stress. Journal of Plant Physiology. 2015;**185**:75-83

[83] Hajiboland R, Dashtebani F, Aliasgharzad N. Physiological responses of halophytic C4 grass, *Aeluropus littoralis* to salinity and arbuscular mycorrhizal fungi colonization. Photosynthetica. 2015;**53**(4):572-584

[84] Jixiang L, Yingnan W, Shengnan S, Chunsheng M, Xiufeng Y. Effects of arbuscular mycorrhizal fungi on the growth, photosynthesis and photosynthetic pigments of *Leymus chinensis* seedlings under salt-alkali stress and nitrogen deposition. Science of the Total Environment. 2017;**576**:234-241

Chapter 7

Mycorrhizal Association and Plant Disease Protection: New Perspectives

Julio Alves Cardoso Filho

Abstract

Soil fungi of the phylum Glomeromycota and plants form arbuscular mycorrhizal (AM) symbiosis. The AM fungi, during the symbiosis, establish a sink for plant photosynthate by utilizing it for biomass and metabolic energy, while the AM plants obtain nutrients and water through the AMF hyphae. The benefits of AM symbiosis on plant fitness include better mineral nutrition, especially those that are immobile in soil solution (e.g., phosphorus, copper, and zinc), and higher tolerance of mycorrhizal plants to abiotic stresses, such as drought, salinity, high soil temperature, presence of heavy metals, and others abiotic factors. Recent studies have revealed that AMF can suppress pests and plant diseases by the activation of defense regulatory genes. The knowledge of the mechanisms behind the induction of resistance by mycorrhizal symbiosis (mycorrhizal-induced resistance [MIR]) remains unknown. This chapter describes the current advanced status of the role of MIR in plant disease protection.

Keywords: plant defense, arbuscular mycorrhizas, mycorrhizal-induced resistance

1. Introduction

Mycorrhizas are complex symbioses formed by several components that determine the rate of colonization, such as the incidence of propagules, and the effects and functions of symbiosis on plants and ecosystems [1]. Mycorrhizal fungi improve the host plant's resistance to environmental stresses, while the host plant provides carbon in the photosynthates form, for fungal growth and reproduction [2–4]. Arbuscular mycorrhizas (AMs) are beneficial interactions formed by plant roots and soil fungi of the phylum Glomeromycota [1]. The regulation of mycotrophism determines the response of the host plant, and biotrophism controls the degree of colonization and production of propagules, guaranteeing the survival and evolution of this group of fungi [5, 6]. AM is the most widespread form of mycorrhizal association and is of great ecological and economic importance [7–9]. AMs are compartmentalized biological systems. Thus, they suffer from the influence of the environment and numerous edaphic factors of each component that directly or indirectly control the establishment, the functioning, and the occurrence of AM [10–15]. AM symbiosis is generally inhibited under high fertility and favored by low fertility, where colonization and sporulation are generally maximal. Phosphorus (P) fertilization in the soil optimizes plant growth and reduces colonization [16].

The protection against soil-borne pathogens has been reported in mycorrhizal plants [17–24]. The differential expression of several genes involved in plant defense against pathogen attack has been observed and may play a fundamental role in colonization [25–27]. This modulation seems to occur not only locally but also systemically [28]. This phenomenon, called "priming," results in an induction of basal resistance mechanisms upon subsequent pathogen attack [28–31]. In symbiotic systems, microbial-associated molecular patterns (MAMPs) from AM fungi are perceived and elicit a transient defense response, which later undergoes suppressed induction at the early stages of AM [32]. The absence of a hypersensitive response (HR) after AM fungi recognition may reflect a non-activation, a low level of defense response, or suppression of the plant defense system [30, 31]. The phenylpropanoid biosynthetic pathway, widely known for the action of phytoalexins, involves a series of enzymes that can be regulated at the transcriptional level by biotic and abiotic stimuli and is influenced by mycorrhization [30, 31]. In signaling the localized induction of defense genes considered specific, two types of mechanisms may be involved: acid endochitinase PR4 and β1,3-endoglycanase EG488 and the systemic suppression of the expression of genes encoding these enzymes and those involved in the phenylpropanoid metabolism [32]. The expression of genes related to the plant defense system is systemically suppressed during the establishment of AM symbiosis [33]. However, the expression of other genes is locally induced, and the modulation of these responses depends on external factors, such as the phosphate level [34, 35].

In this chapter, we want to discuss new perspectives and updates on mycorrhiza-induced resistance (MIR).

2. Resistance of plants to microorganisms

The plant defense system is multi-component, acting in a dynamic and coordinated manner, at the appropriate time and place and with adequate magnitude [36–39]. Plant resistance would be its ability to delay or prevent the colonization of its tissues by a phytopathogen, characterized by its dynamic and coordinated nature in a sequence of logical events, immediately after the pathogen contacts its tissues [40]. The growing development of technologies aimed at agriculture and increases in the use of inputs, especially pesticides, have contributed to increased agricultural productivity and adverse effects on the environment and human health. New plant protection measures have been highlighted; such as the induction of resistance (IR) has been used to control pests and diseases [41–43].

2.1 Non-host resistance in plants

Land plants are sessile, and despite not having an adaptive immune system similar to animals, they are resistant to most microorganisms that attack them in an attempt to invade their tissues [44]. This phenomenon characterizes resistance as a rule and susceptibility to some microorganisms. This recognition of non-adapted pathogens by non-host plants at the cellular level is called non-host resistance (NHR) [45]. Non-host resistance (NHR) is, by definition, the resistance displayed by all genotypes of a plant species to all genotypes of a pathogen species, being considered the most common and effective form of resistance to diseases that occur in nature [46].

Several genes appear to control the NHR in a given plant species; in addition, the number of genes involved is related to the specific genotype of the analyzed plant [46]. NHR "resistance" refers to the inability of a pathogen to complete its asexual or sexual life cycle on that host plant species [47]. NHR is a tool for breeding against the induction factors of biotic and abiotic stresses, aiming at the development of durable genetic resistance [48]. NHR is an elaborate defense system that protects plants against invasion by non-pathogenic or pathogenic organisms and involves physical and biochemical mechanisms [49, 50]. These defenses include a combination of preformed and inducible mechanisms [51] that act as a physical barrier to penetration and as a toxic barrier [52]. The first line of plant defense is the structural and biochemical mechanism that is present even before the inoculum deposition, called constitutive or passive [53]. The barrier imposed by the cuticle and cell wall is considered an important factor of NHR. The cuticle is structurally variable between plant species; is composed of cutin, waxes, and hydrocarbons; and is closely associated with the cell wall of the epidermal cells [54]. Others mechanisms are activated from the recognition of the pathogen, by the host plant that is active, inducible, or post-formed, and can also be structural and biochemical [55]. Among the biochemical post-formed responses is the accumulation of reactive oxygen species (ROS), phytoalexins, pathogenesis-related (PR) proteins, and HR [56]. However, most defense mechanisms are activated in response to infection by the pathogen [57]. During initial contact with the pathogen, pathogen-associated molecular patterns (PAMPs) [58], such as chitin and glucans in fungi, flagellin, and elongation factor Tu in bacteria, are recognized by the plant through pattern recognition receptors (PRRs) [59]. Recognition of PAMPs triggers signaling events and basal resistance, called PAMP-activated immunity (PTI) [60]. Pathogens have evolved effectors that interfere with different signaling processes involved in plant defense, suppressing PTI, promoting virulence, and triggering effector-activated susceptibility (ETS) [61]. Plants, in turn, acquired resistance (R) genes that detect pathogen-specific effectors, resulting in effector-activated immunity (ETI) [62]. ETI is a version of PTI, typically involving HR, a form of programmed cell death that limits colonization by the pathogen [60]. HR is mediated by the accumulation of ROS, which, in addition to having a toxic effect on the pathogen, can act as secondary messengers for the activation of defense responses [63–67]. Mitogen-activated protein kinases (MAPKs) modulate the activity of transcriptional regulators and phytohormones [68]. Salicylic acid (SA), ethylene, and jasmonic acid are the main hormones involved in signaling. Other hormones, such as abscisic acid and cytokinins, have also emerged as participants in signaling pathways for defense [69, 70]. As a result of the recognition of the pathogen by the host and the consequent activation of signal transduction pathways, several defense mechanisms are formed to limit the infection [71–78]. All plant species are capable of constitutively biosynthesizing chemical compounds with potential defensive function, suggesting that this ability is an evolutionary trait [79]. A feature of antimicrobial compounds is that some are found on the plant surface or accumulated in cells close to the host surface, especially vacuoles or organelles in epidermal cells, and released by hydrolytic enzymes after attack by the microorganism [80].

2.2 Host resistance in plants

The microorganisms that manage to overcome the mechanisms of RNH become pathogens of the plant species and need to act against the host resistance that the plant has [81]. Plants also have a surveillance system that detects/monitors the presence or activity of effector molecules within their tissues and cells. This surveillance system has specific receptors (R proteins) encoded by the plant's R genes [82]. The typical R proteins with nucleotide-binding and leucine-rich domains (NLRs) have an ATPase-binding domain (NB—nucleotide-binding ATPase) and a TIR (toll interleukin-1 receptor) or CC (coiled-coil) domain forming the TIR-NB-LRR or CC-NB-LRR proteins, respectively [82]. The NB domain serves ATP binding and hydrolysis/signaling cascade, and the LRR domain is responsible for activation/autoinhibition. In the LRR domain, the N-terminal activates the modulation, while the C-terminal is related to recognition specificity [83–85]. Proteins from the NLR genes in their inactive form (absence of the effector cognate) are located in the host cell cytoplasm bound to the plasma membrane, endoplasmic reticulum, or tonoplast. In the presence of the effector, the cognate can move to the nucleus, interacting with transcription factors or with other cytoplasmic proteins to start the signaling chain for the expression of defenses [86–89]. Effector detection triggers a signaling cascade that culminates in the expression of defense and HR genes to contain the pathogen's advance, resulting in race-specific resistance, the ETI immune response, also known as vertical resistance. ETI is one of the main components of host resistance [90, 91]. ETI is activated by the direct or indirect interaction between one or more effectors and one or more NLR proteins. In the Decoy model, where the NLR detects changes in a protein that mimics the target, the effector apparently does not have a defined biological function [92, 93]. In the Guard model, during infection, a modification of the target protein (monitored) occurs by the action of the effector releasing the R protein (monitor), thus allowing it to initiate the signaling cascade that leads to the induction of ETI [94]. Resistance conferred by R genes depends on the effector for virulence or adaptability of the pathogen and its evolutionary potential [95]. From the point of view of resistance durability, for resistance conferred by R genes (qualitative resistance), the evolutionary pressure on the pathogen conferred by quantitative resistance is significantly reduced, thus constituting a source of durable resistance [96]. In summary, resistance mediated by R genes is specific to certain races of the pathogen and effective against biotrophic and hemibiotrophic parasites. On the other hand, quantitative resistance provides a means for controlling biotrophic, hemibiotrophic, and necrotrophic pathogens, being effective against several races of a pathogen, providing broad-spectrum resistance, or in some specific cases, effective against several pathogens [97–103].

3. Plant priming for enhanced defense

To compensate for their sessile life and face a broad range of biotic and abiotic stresses, plants have evolved survival and adaptation strategies, such as inducing some stress memory or stress imprinting [104]. In plants, the IR is frequently associated with the accumulation of antimicrobial pathogenesis-related (PR) proteins and with the so-called priming of cells [89]. Priming for enhanced defense is a cellular process in biological and chemical IR immunity, including systemic acquired resistance (SAR), induced systemic resistance (ISR), and herbivore-induced resistance [105–109]. Plants maintain immune memory through priming, in the absence of such specialized cells [110, 111].

3.1 Epigenetic molecular mechanism of priming

3.1.1 Memory of plant immunization

Advances in the knowledge of epigenetic regulation in the plant multiple generation stress memory have provided new procedures and approaches for breeding crops and sustainable germplasm banks for future climate challenges [112, 113]. Plant stress memory is described under two categories: mitotic stress memory, or somatic memory, and meiotic stress memory, or transgenerational memory [114–116]. Plant stress memory associated with the inheritance of SAR is likely to be epigenetic [117, 118]. In plants, defense priming and SAR are associated with epigenetic modifications of histones, DNA methylation, increased signaling enzymes, and an accumulation of pattern recognition receptors (PRRs) at cell membranes [119, 120]. In summary, strategies exploiting epigenetic variations appear promising for crop resistance breeding [121–124].

4. Mycorrhiza-induced resistance: new perspectives

The effects of AM symbiosis on plant interactions with other organisms, such as the induction of resistance against plant pathogens, seem to result from the combination of multiple mechanisms that may operate simultaneously [125]. A proposed hypothesis is that colonization of roots by AM fungi primes defense mechanisms, leading to mycorrhiza-induced resistance (MIR) [126] by the activation of MAMP-triggered immunity (MTI) [127]. MIR is a low-cost type of induced resistance that may be among the reasons to explain why root associations with AM fungi have been conserved during evolution and are widespread among species [128, 129]. MIR includes a priming of defense-related plant genes and shares more elements with the ISR induced by rhizobacteria [130, 131]. The plant can restrict AMF colonization once the plant is already mycorrhizal, a phenomenon known as autoregulation [132]. The mechanisms operating in such autoregulation may also affect plant interactions with phytopathogens [133]. However, the molecular mechanisms that regulate the formation and establishment of AM symbiosis and the modulation of plant defense responses during MIR are still not understood [134]. According to Fiorilli et al. [135], who elucidate the molecular mechanisms underlying the establishment of AM symbiosis, we need to investigate the changes in transcripts and proteins in roots and leaves during the double (plant-AM fungus) and tripartite (plant-AM fungus-pathogen) interactions [136, 137]. The recognition of friend versus foe is still incompletely understood in signaling between the host plant and the pathogen interaction [138–141]. Another critical challenge is to elucidate the biological roles of receptor-like kinases (RLK) mediated by endocytosis in the plant interaction with microbes, aiming to elucidate the molecular mechanisms by which pathogens and non-pathogen microorganisms can reprogram the RLK trafficking [142, 143]. Besides, the role of MIR against viruses and foliar phytopathogens stays in the speculative field. In some cases, the susceptibility to the foliar pathogen is related to mycorrhizal plants compared with non-mycorrhizal plants. However, it is linked with the higher amounts of plant phosphorus available to viruses for their multiplication in infected and colonized plants [144, 145]. Recent advances indicate that the beneficial effects of MIR may not be related to mycorrhizal plant nutrition [140]. Thus, knowing the mechanisms of AM symbiosis physiology regulation under different environmental conditions is required to understand the MIR context of AM fungi-host plant dependency.

5. Conclusions and future prospects

MIR is a low-cost type of IR and has a positive effect from an ecological point of view. However, still, open questions require particular attention, among which is why infection by AM fungi does not elicit a defense response in roots. Besides, the role of MIR against viruses and foliar phytopathogens stays in the speculative field. Thus, knowing the mechanisms of AM symbiosis physiology regulation during MIR is required for their applications in sustainable agriculture.

Conflicts of interest

The author declares no conflict of interest.

Author details

Julio Alves Cardoso Filho
Center for Engineering and Agrarian Sciences Federal University of Alagoas, Rio Largo, Brazil

*Address all correspondence to: julio.filho@ceca.ufal.br

References

[1] Schüßler A, Schwarzott D, Walker C. A new fungal phylum, the Glomeromycota: evolution and phylogeny. Mycological Research. 2001;**105**:1413-1421. DOI: 10.1017/S0953756201005196

[2] Smith SE, Gianinazzi-Pearson V. Physiological interactions between symbionts in vesicular-arbuscular mycorrhizal plants. Annual Review of Plant Physiology and Plant Molecular Biology. 1988;**39**:22124. DOI: 10.1146/annurev.pp.39.060188.001253

[3] Smith SE, Read DJ, Mycorrhizal Symbiosis. 3rd ed. London: Academic Press; 2008. p. 815. ISBN: 978-0-12-370526-6

[4] Harrison MJ. Cellular programs for arbuscular mycorrhizal symbiosis. Current Opinion in Plant Biology. 2012;**15**:691-698. DOI: 10.1016/j.pbi.2012.08.010

[5] Wang B, Qiu YL. Phylogenetic distribution and evolution of mycorrhizas in land plants. Mycorrhiza. 2006;**16**:299-363. DOI: 10.1007/s00572-005-0033-6

[6] Soudzilovskaia NA, van Bodegom PM, Terrer C, Zelfde MV, McCallum I, Luke McCormack M, et al. Global mycorrhizal plant distribution linked to terrestrial carbon stocks. Nature Communications. 2019;**10**(1):5077.7. DOI: 10.1038/s41467-019-13019-2

[7] Francis R, Read DJ. The contributions of mycorrhizal fungi to the determination of plant community structure. Plant and Soil. 1994;**159**:11-25. DOI: 10.1007/BF00000091

[8] Jeffries P, Gianinazzi S, Perotto S, Turnau K, Barea JM. The contribution of arbuscular mycorrhizal fungi in sustainable maintenance of plant health and soil fertility. Biology Fertility of Soils. 2003;**37**:1-16. DOI: 10.1007/s00374-002-0546-5

[9] Smith JL, De Moraes CM, Mescher MC. Jasmonate- and salicylate-mediated plant defense responses to insect herbivores, pathogens and parasitic plants. Pest Management Science. 2009;**65**:497-503. DOI: 10.1002/ps.1714

[10] Redecker D, Schüßler A, Stockinger H, Stürmer SL, Morton JB, Walker C. An evidence-based consensus for the classification of arbuscular mycorrhizal fungi (Glomeromycota). Mycorrhiza. 2013;**23**:515-531. DOI: 10.1007/s00572-013-0486-y

[11] Redecker D, Raab P. Phylogeny of the Glomeromycota (arbuscular mycorrhizal fungi): recent developments and new gene markers. Mycologia. 2006;**98**:885-895. DOI: 10.3852/mycologia.98.6.885

[12] Oehl F, Sieverding E, Palenzuela J, Ineichen K, da Silva GA. Advances in Glomeromycota taxonomy and classification. IMA Fungus. 2011;**2**:191-199. DOI: 10.5598/imafungus.2011.02.02.10

[13] Krüger M, Krüger C, Walker C, Stockinger H, Schüßler A. Phylogenetic reference data for systematics and phylotaxonomy of arbuscular mycorrhizal fungi from phylum to species level. The New Phytologist. 2012;**193**:970-984. DOI: 10.1111/j.1469-8137.2011.03962.x

[14] Fitter AH. Darkness visible: reflections on underground ecology.

Journal of Ecology. 2005;**93**:231-243. DOI: 10.1111/j.0022-0477.2005.00990.x

[15] Rosendahl S. Communities, populations and individuals of arbuscular mycorrhizal fungi. The New Phytologist. 2008;**178**:253-266. DOI: 10.1111/j.1469-8137.2008.02378.x

[16] Das D, Paries M, Hobecker K, et al. Phosphate starvation response transcription factors enable arbuscular mycorrhiza symbiosis. Nature Communications. 2022;**13**:477. DOI: 10.1038/s41467-022-27976-8

[17] Azcon-Aguilar C, Barea JM. Arbuscular mycorrhizas and biological control of soil borne plant pathogens – an overview of the mechanisms involved. Mycorrhiza. 1997;**6**:457-464. DOI: 10.1007/s005720050147

[18] Whipps JM. Prospects and limitations for mycorrhizas in biocontrol of root pathogens. Canadian Journal of Botany. 2004;**82**:1198-1227. DOI: 10.1139/b04-082

[19] Barea JM, Pozo MJ, Azcon R, Azcon-Aguilar C. Microbial co-operation in the rhizosphere. Journal of Experimental Botany. 2005;**56**:1761-1778. DOI: 10.1093/jxb/eri197

[20] Pozo MJ, Azcón-Aguilar C. Unraveling mycorrhiza-induced resistance. Current Opinion in Plant Biology. 2007;**10**:393-398. DOI: 10.1016/j.pbi.2007.05.004

[21] Saldajeno MG, Chandanie WA, Kubota M, Hyakumachi M. Effects of interactions of arbuscular mycorrhizal fungi and beneficial saprophytic mycoflora on plant growth and disease protection. In: Siddiqui ZA, Akhtar MS, Futai K, editors. Mycorrhizae: Sustainable Agriculture and Forestry. Netherlands, Dordrecht: Springer; 2008. pp. 211-226. DOI: 10.1007/978-1-4020-8770-7_9

[22] Pozo MJ, Verhage A, García-Andrade J, García JM, Azcón-Aguilar C. Priming plant defenses against pathogens by arbuscular mycorrhizal fungi. In: Azcón-Aguilar C, Barea JM, Gianinazzi S, Gianinazzi-Pearson V, editors. Mycorrhizas: Functional Processes and Ecological Impact. Heidelberg: Springer-Verlag; 2009. pp. 137-149. DOI: 10.1007/978-3-540-87978-7_9

[23] Pozo MJ, Jung SC, López-Ráez JA, Azcón-Aguilar C. Impact of arbuscular mycorrhizal symbiosis on plant response to biotic stress: The role of plant defense mechanisms. In: Koltai H, Kapulnik Y, editors. Arbuscular Mycorrhizas: Physiology and Function. 2nd ed. Dordrecht: Springer; 2010. pp. 193-207. DOI: 10.1007/978-90-481-9489-6_9

[24] Dreischhoff S, Das IS, Jakobi M, Kasper K, Polle A. Local responses and systemic induced resistance mediated by ectomycorrhizal fungi. Frontiers in Plant Science. 2020;**14**(11):590063. DOI: 10.3389/fpls.2020.590063

[25] Van Wees SC, Van der Ent S, Pieterse CMJ. Plant immune responses triggered by beneficial microbes. Current Opinion in Plant Biology. 2008;**11**:443-448. DOI: 10.1016/j.pbi.2008.05.005

[26] van der Heijden MGA, Bardgett RD, van Straalen NM. The unseen majority: Soil microbes as drivers of plant diversity and productivity in terrestrial ecosystems. Ecology Letters. 2008;**11**:296-310. DOI: 10.1111/j.1461-0248.2007.01139.x

[27] Jung SC, Medina-Matinez A, Lopez-Raez JA, Pozo MJ.

Mycorrhiza-induced resistance and priming of plant defenses. Journal of Chemical Ecology. 2012;**38**:651-664. DOI: 10.1007/s10886-012-0134-6

[28] Conrath U, Beckers GJ, Flors V, García-Agustín P, Jakab G, Mauch F, et al. Priming: getting ready for battle. Molecular Plant-Microbe Interactions. 2006;**19**:1062-1071. DOI: 10.1094/MPMI-19-1062

[29] Ahmad S, Gordon-Weeks R, Pickett J, Ton J. Natural variation in priming of basal resistance: from evolutionary origin to agricultural exploitation. Molecular Plant Pathology. 2010;**11**:817-827. DOI: 10.1111/j.1364-3703.2010.00645.x

[30] Conrath U. Molecular aspects of defense priming. Trends in Plant Science. 2011;**16**:524-531. DOI: 10.1016/j.tplants.2011.06.004

[31] Pastor V, Luna E, Mauch-Mani B, Ton J, Flors V. Primed plants do not forget. Environmental and Experimental Botany. 2013;**94**:46-56. DOI: 10.1016/j.envexpbot.2012.02.013

[32] Diao F, Dang Z, Xu J, Ding S, Hao B, Zhang Z, et al. Effect of arbuscular mycorrhizal symbiosis on ion homeostasis and salt tolerance-related gene expression in halophyte Suaeda salsa under salt treatments. Microbiological Research. 2021;**245**:126688. DOI: 10.1016/j.micres.2020.126688

[33] Wang H, An T, Huang D, Liu R, Xu B, Zhang S, et al. Arbuscular mycorrhizal symbioses alleviating salt stress in maize is associated with a decline in root-to-leaf gradient of Na^+/K^+ ratio. BMC Plant Biology. 2021;**21**(1):457.7. DOI: 10.1186/s12870-021-03237-6

[34] Bahadur A, Batool A, Nasir F, Jiang S, Mingsen Q, Zhang Q, et al. Mechanistic insights into arbuscular mycorrhizal fungi-mediated drought stress tolerance in plants. International Journal of Molecular Sciences. 2019;**20**(17):4199.27. DOI: 10.3390/ijms20174199

[35] Kang J-N, Lee W-H, Won SY, Chang S, Hong J-P, Oh T-J, et al. Systemic expression of genes involved in the plant defense response induced by wounding in Senna tora. International Journal of Molecular Sciences. 2021;**22**:10073. DOI: 10.3390/ijms221810073

[36] Jones JDG, Dangl JL. The plant immune system. Nature. 2006;**444**:323-329. DOI: 10.1038/nature05286

[37] Andolfo G, Ercolano MR. plant innate immunity multicomponent model. Frontiers in Plant Science. 2015;**13**(6):987. DOI: 10.3389/fpls.2015.00987

[38] Mermigka G, Amprazi M, Mentzelopoulou A, Amartolou A, Sarris PF. Plant and animal innate immunity complexes: fighting different enemies with similar weapons. Trends in Plant Science. 2020;**25**(1):80-91. DOI: 10.1016/j.tplants.2019.09.008

[39] Duxbury Z, Wu CH, Ding P. A comparative overview of the intracellular guardians of plants and animals: NLRs in innate immunity and beyond. Annual Review of Plant Biology. 2021;**17**(72):155-184. DOI: 10.1146/annurev-arplant-080620-104948

[40] Devanna BN, Jaswal R, Singh PK, Kapoor R, Jain P, Kumar G, et al. Role of transporters in plant disease resistance. Physiologia Plantarum. 2021;**171**(4):849-867. DOI: 10.1111/ppl.13377

[41] Yao T, Gai XT, Pu ZJ, Gao Y, Xuan YH. From functional characterization to the application of SWEET sugar transporters in plant resistance breeding. Journal of

Agricultural and Food Chemistry. 2022; **70**(17):5273-5283.4. DOI: 10.1021/acs.jafc.2c00582

[42] Gautam T, Dutta M, Jaiswal V, Zinta G, Gahlaut V, Kumar S. Emerging roles of SWEET sugar transporters in plant development and abiotic stress responses. Cell. 2022;**11**(8):1303.12. DOI: 10.3390/cells11081303

[43] Breia R, Conde A, Badim H, Fortes AM, Gerós H, Granell A. Plant SWEETs: from sugar transport to plant-pathogen interaction and more unexpected physiological roles. Plant Physiology. 2021;**186**(2):836-852.11. DOI: 10.1093/plphys/kiab127

[44] Rai A, Sivalingam PN, Senthil-Kumar M. A spotlight on non-host resistance to plant viruses. Peer J. 2022;**31**(10):e12996. DOI: 10.7717/peerj.12996

[45] Lee S, Whitaker VM, Hutton SF. Mini review: potential applications of non-host resistance for crop improvement. Frontiers in Plant Science. 2016;**11**(7):997. DOI: 10.3389/fpls.2016.00997

[46] Oh S, Choi D. Receptor-mediated nonhost resistance in plants. Essays in Biochemistry. 2022;**66**(5):435-445.30. DOI: 10.1042/EBC20210080

[47] Panstruga R, Moscou MJ. What is the molecular basis of nonhost resistance? Molecular Plant-Microbe Interactions. 2020;**33**(11):1253-1264. DOI: 10.1094/MPMI-06-20-0161-CR

[48] Sahu PK, Sao R, Choudhary DK, Thada A, Kumar V, Mondal S, et al. Advancement in the breeding, biotechnological and genomic tools towards development of durable genetic resistance against the rice blast disease. Plants (Basel). 2022;**11**(18):2386.13. DOI: 10.3390/plants11182386

[49] Gupta N, Reddy K, Bhattacharyya D, et al. Plant responses to geminivirus infection: guardians of the plant immunity. Virology Journal. 2021;**18**:143. DOI: 10.1186/s12985-021-01612-1

[50] Sun Y, Gao M, Kang S, Yang C, Meng H, Yang Y, et al. Molecular mechanism underlying mechanical wounding-induced flavonoid accumulation in *Dalbergia odorifera* T. Chen, an endangered tree that produces chinese rosewood. Genes (Basel). 2020;**11**(5):478.28. DOI: 10.3390/genes11050478

[51] Herrera-González JA, Bautista-Baños S, Serrano M, Romanazzi G, Gutiérrez-Martínez P. Non-chemical treatments for the pre- and post-harvest elicitation of defense mechanisms in the fungi-avocado pathosystem. Molecules. 2021;**26**(22):6819.11. DOI: 10.3390/molecules26226819

[52] Vega-Muñoz I, Duran-Flores D, Fernández-Fernández ÁD, Heyman J, Ritter A, Stael S. Breaking bad news: dynamic molecular mechanisms of wound response in plants. Frontiers in Plant Science. 2020;**8**(11):610445. DOI: 10.3389/fpls.2020.610445

[53] Lee K, Seo PJ. Wound-induced systemic responses and their coordination by electrical signals. Frontiers in Plant Science. 2022;**18**(13):880680. DOI: 10.3389/fpls.2022.880680

[54] Wang X, Kong L, Zhi P, Chang C. Update on cuticular wax biosynthesis and its roles in plant disease resistance. International Journal of Molecular Sciences. 2020;**21**(15):5514.1. DOI: 10.3390/ijms21155514

[55] García T, Gutiérrez J, Veloso J, Gago-Fuentes R, Díaz J. Wounding induces local resistance but systemic susceptibility to Botrytis cinerea in pepper plants. Journal of Plant Physiology. 2015;**15**(176):202-209. DOI: 10.1016/j.jplph.2014.12.013

[56] Tavanti TR, Melo AAR, Moreira LDK, Sanchez DEJ, Silva RDS, Silva RMD, et al. Micronutrient fertilization enhances ROS scavenging system for alleviation of abiotic stresses in plants. Plant Physiology and Biochemistry. 2021;**160**:386-396. DOI: 10.1016/j.plaphy.2021.01.040

[57] Shinya T, Yasuda S, Hyodo K, Tani R, Hojo Y, Fujiwara Y, et al. Integration of danger peptide signals with herbivore-associated molecular pattern signaling amplifies anti-herbivore defense responses in rice. The Plant Journal. 2018;**94**(4):626-637. DOI: 10.1111/tpj.13883

[58] Taghavi M, Khosravi A, Mortaz E, Nikaein D, Athari SS. Role of pathogen-associated molecular patterns (PAMPS) in immune responses to fungal infections. European Journal of Pharmacology. 2017;**5**(808):8-13. DOI: 10.1016/j.ejphar.2016.11.013

[59] Saijo Y, Loo EP, Yasuda S. Pattern recognition receptors and signaling in plant-microbe interactions. The Plant Journal. 2018;**93**(4):592-613. DOI: 10.1111/tpj.13808

[60] Peng Y, van Wersch R, Zhang Y. Convergent and divergent signaling in PAMP-triggered immunity and effector-triggered immunity. Molecular Plant-Microbe Interactions. 2018;**31**(4):403-409. DOI: 10.1094/MPMI-06-17-0145-CR

[61] Yu TY, Sun MK, Liang LK. Receptors in the induction of the plant innate immunity. Molecular Plant-Microbe Interactions. 2021;**34**(6):587-601. DOI: 10.1094/MPMI-07-20-0173-CR

[62] Kothari IL, Patel M. Plant immunization. Indian Journal of Experimental Biology 2004;**42**(3):244-252. PMID: 15233292

[63] Yu X, Feng B, He P, Shan L. From chaos to harmony: responses and signaling upon microbial pattern recognition. Annual Review of Phytopathology. 2017;**55**:109-137. DOI: 10.1146/annurev-phyto-080516-035649

[64] Ye C, Zheng S, Jiang D, Lu J, Huang Z, Liu Z, et al. Initiation and execution of programmed cell death and regulation of reactive oxygen species in plants. International Journal of Molecular Sciences. 2021;**22**(23):12942.30. DOI: 10.3390/ijms222312942

[65] Locato V, De Gara L. programmed cell death in plants: an overview. Methods in Molecular Biology. 1743;**2018**:1-8. DOI: 10.1007/978-1-4939-7668-3_1

[66] Balint-Kurti P. The plant hypersensitive response: concepts, control and consequences. Molecular Plant Pathology. 2019;**20**(8):1163-1178. DOI: 10.1111/mpp.12821

[67] DeFalco TA, Zipfel C. Molecular mechanisms of early plant pattern-triggered immune signaling. Molecular Cell. 2021;**81**(17):3449-3467.2. DOI: 10.1016/j.molcel.2021.07.029

[68] Xiong L, Yang Y. Disease resistance and abiotic stress tolerance in rice are inversely modulated by an abscisic acid-inducible mitogen-activated protein kinase. The Plant Cell. 2003;**15**(3):745-759. DOI: 10.1105/tpc.008714

[69] Kachroo A, Kachroo P. Mobile signals in systemic acquired resistance. Current

Opinion in Plant Biology. 2020;**58**:41-47. DOI: 10.1016/j.pbi.2020.10.004

[70] Pieterse C, Van Wees S. Induced disease resistance. In: Lugtenberg B, editor. Principles of Plant-Microbe Interactions. Cham: Springer; 2015. pp. 123-133. DOI: 10.1007/978-3-319-08575-3_14

[71] Walters DR, Ratsep J, Havis ND. Controlling crop diseases using induced resistance: challenges for the future. Journal of Experimental Botany. 2013;**64**:1263-1280. DOI: 10.1093/jxb/ert026

[72] Fu ZQ, Dong X. Systemic acquired resistance: turning local infection into global defense. Annual Review of Plant Biology. 2013;**64**:839-863. DOI: 10.1146/annurev-arplant-042811-105606

[73] Gao Q-M, Zhu S, Kachroo P, Kachroo A. Signal regulators of systemic acquired resistance. Frontiers in Plant Science. 2015;**6**:228. DOI: 10.3389/fpls.2015.00228

[74] Pieterse CMJ, Zamoudis C, Berendsen RL, Weller DM, Van Wees CMS, Bakker AHMP. Induced systemic resistance by beneficial microbes. Annual Review of Phytopathology. 2014;**52**:347-375. DOI: 10.1146/annurev-phyto-082712-102340

[75] Nguyen Q-M, Iswanto ABB, Son GH, Kim SH. Recent advances in effector-triggered immunity in plants: new pieces in the puzzle create a different paradigm. International Journal of Molecular Sciences. 2021;**22**:4709. DOI: 10.3390/ijms22094709

[76] Gupta B, Bhaumik U, Sengupta A, Chakraborty M, Saha J, et al. Plant immunity: an enigma during biotic and abiotic stress. Journal of Plant Biochemistry and Physiology. 2015;**3**:e128. DOI: 10.4172/2329-9029.1000e128

[77] Jing B, Xu S, Xu M, Li Y, Li S, Ding J, et al. Brush and spray: a high-throughput systemic acquired resistance assay suitable for large-scale genetic screening. Plant Physiology. 2011;**157**(3):973-980. DOI: 10.1104/pp.111.182089

[78] Zhou M, Wang W. Recent advances in synthetic chemical inducers of plant immunity. Frontiers in Plant Science. 2018;**6**(9):1613. DOI: 10.3389/fpls.2018.01613

[79] Delventhal R, Rajaraman J, Stefanato FL, et al. A comparative analysis of nonhost resistance across the two *Triticeae* crop species wheat and barley. BMC Plant Biology. 2017;**17**:232. DOI: 10.1186/s12870-017-1178-0

[80] Vivaldo G, Masi E, Taiti C, et al. The network of plants volatile organic compounds. Scientific Reports. 2017;**7**(1):11050. DOI: 10.1038/s41598-017-10975-x

[81] Gill US, Lee S, Mysore KS. Host versus nonhost resistance: distinct wars with similar arsenals. Phytopathology. 2015;**105**(5):580-587. DOI: 10.1094/PHYTO-11-14-0298-RVW

[82] Du D, Zhang C, Xing Y, Lu X, Cai L, Yun H, et al. The CC-NB-LRR OsRLR1 mediates rice disease resistance through interaction with OsWRKY19. Plant Biotechnology Journal. 2021;**19**(5):1052-1064. DOI: 10.1111/pbi.13530

[83] Andolfo G, Sánchez CS, Cañizares J, Pico MB, Ercolano MR. Large-scale gene gains and losses molded the NLR defense arsenal during the Cucurbita evolution. Planta. 2021;**254**(4):82.24. DOI: 10.1007/s00425-021-03717-x

[84] Liu Y, Zeng Z, Zhang YM, Li Q, Jiang XM, Jiang Z, et al. An angiosperm

NLR Atlas reveals that NLR gene reduction is associated with ecological specialization and signal transduction component deletion. Molecular Plant. 2021;**14**(12):2015-2031.6. DOI: 10.1016/j.molp.2021.08.001

[85] Yan J, Fang Y, Xue D. Advances in the genetic basis and molecular mechanism of lesion mimic formation in rice. Plants (Basel). 2022;**11**(16):2169.21. DOI: 10.3390/plants11162169

[86] Wu JY, Xue JY, Van de Peer Y. Evolution of *NLR* resistance genes in magnoliids: dramatic expansions of *CNLs* and multiple losses of *TNLs*. Frontiers in Plant Science. 2021;**21**(12):777157. DOI: 10.3389/fpls.2021.777157

[87] Lapin D, Johanndrees O, Wu Z, Li X, Parker JE. Molecular innovations in plant TIR-based immunity signaling. Plant Cell. 2022;**34**(5):1479-1496.26. DOI: 10.1093/plcell/koac035

[88] Parker JE, Hessler G, Cui H. A new biochemistry connecting pathogen detection to induced defense in plants. The New Phytologist. 2022;**234**(3):819-826. DOI: 10.1111/nph.17924

[89] Maruta N, Burdett H, Lim BYJ, Hu X, Desa S, Manik MK, et al. Structural basis of NLR activation and innate immune signaling in plants. Immunogenetics. 2022;**74**(1):5-26. DOI: 10.1007/s00251-021-01242-5

[90] Nguyen QM, Iswanto ABB, Son GH, Kim SH. Recent advances in effector-triggered immunity in plants: new pieces in the puzzle create a different paradigm. International Journal of Molecular Sciences. 2021;**22**(9):4709.29. DOI: 10.3390/ijms22094709

[91] Naveed ZA, Wei X, Chen J, Mubeen H, Ali GS. The PTI to ETI continuum in phytophthora-plant interactions. Frontiers in Plant Science. 2020;**11**:593905. DOI: 10.3389/fpls.2020.593905

[92] Villegas-Vázquez EY, Xoconostle-Cázares B, Ruiz-Medrano R. An ancestry perspective of the evolution of PBS1 proteins in plants. International Journal of Molecular Sciences. 2021;**22**(13):6819.25. DOI: 10.3390/ijms22136819

[93] Chakraborty J, Ghosh P, Das S. Autoimmunity in plants. Planta. 2018;**248**(4):751-767. DOI: 10.1007/s00425-018-2956-0

[94] Ray SK, Macoy DM, Kim WY, Lee SY, Kim MG. Role of RIN4 in regulating PAMP-triggered immunity and effector-triggered immunity: current status and future perspectives. Molecules and Cells. 2019;**42**(7):503-511.31. DOI: 10.14348/molcells.2019.2433

[95] Pilet-Nayel ML, Moury B, Caffier V, Montarry J, Kerlan MC, Fournet S, et al. Quantitative resistance to plant pathogens in pyramiding strategies for durable crop protection. Frontiers in Plant Science. 2017;**27**(8):1838. DOI: 10.3389/fpls.2017.01838

[96] Veneault-Fourrey RM. Quantitative resistance linked to late effectors. The New Phytologist. 2021;**231**:1301-1303. DOI: 10.1111/nph.17462

[97] Andersen EJ, Ali S, Byamukama E, Yen Y, Nepal MP. Disease resistance mechanisms in plants. Genes (Basel). 2018;**9**(7):339.4. DOI: 10.3390/genes9070339

[98] Dong OX, Ronald PC. Genetic engineering for disease resistance in plants: recent progress and future perspectives. Plant Physiology. 2019;**180**(1):26-38. DOI: 10.1104/pp.18.01224

[99] Ceulemans E, Ibrahim HMM, De Coninck B, Goossens A. Pathogen effectors: exploiting the promiscuity of plant signaling hubs. Trends in Plant Science. 2021;**26**(8):780-795. DOI: 10.1016/j.tplants.2021.01.005

[100] Dinglasan E, Periyannan S, Hickey LT. Harnessing adult-plant resistance genes to deploy durable disease resistance in crops. Essays in Biochemistry. 2022;**66**(5):571-580. DOI: 10.1042/EBC20210096

[101] Jaswal R, Kiran K, Rajarammohan S, Dubey H, Singh PK, Sharma Y, et al. Effector biology of biotrophic plant fungal pathogens: current advances and future prospects. Microbiological Research. 2020;**241**:126567. DOI: 10.1016/j.micres.2020.126567

[102] Shao D, Smith DL, Kabbage M, Roth MG. Effectors of plant necrotrophic fungi. Frontiers in Plant Science. 2021;**12**:687713. DOI: 10.3389/fpls.2021.687713

[103] Mapuranga J, Zhang N, Zhang L, Chang J, Yang W. Infection strategies and pathogenicity of biotrophic plant fungal pathogens. Frontiers in Microbiology. 2022;**13**:799396. DOI: 10.3389/fmicb.2022.799396

[104] Kamle M, Borah R, Bora H, Jaiswal AK, Singh RK, Kumar P. Systemic acquired resistance (SAR) and induced systemic resistance (ISR): role and mechanism of action against phytopathogens. In: Hesham AL, Upadhyay R, Sharma G, Manoharachary C, Gupta V, editors. Fungal Biotechnology and Bioengineering. Fungal Biology. Cham: Springer; 2020. pp. 457-470. DOI: 10.1007/978-3-030-41870-0_20

[105] Zhu L, Huang J, Lu X, Zhou C. Development of plant systemic resistance by beneficial rhizobacteria: Recognition, initiation, elicitation and regulation. Frontiers in Plant Science. 2022;**13**:952397. DOI: 10.3389/fpls.2022.952397

[106] Tripathi D, Raikhy G, Kumar D. Chemical elicitors of systemic acquired resistance –salicylic acid and its functional analogs. Curr. Plant Biol. 2019;**17**:48-59. DOI: 10.1016/j.cpb.2019.03.002

[107] Yu Y, Gui Y, Li Z, Jiang C, Guo J, Niu D. Induced systemic resistance for improving plant immunity by beneficial microbes. Plants. 2022;**11**(3):386.30. DOI: 10.3390/plants11030386

[108] Yang Z, Zhi P, Chang C. Priming seeds for the future: Plant immune memory and application in crop protection. Frontiers in Plant Science. 2022;**29**(13):961840. DOI: 10.3389/fpls.2022.961840

[109] Serteyn L, Quaghebeur C, Ongena M, Cabrera N, Barrera A, Molina-Montenegro MA, et al. Induced systemic resistance by a plant growth-promoting rhizobacterium impacts development and feeding behavior of aphids. Insects. 2020;**11**(4):234.8. DOI: 10.3390/insects11040234

[110] Liu X, Quan W, Bartels D. Stress memory responses and seed priming correlate with drought tolerance in plants: an overview. Planta. 2022; 23;**255**(2):45. DOI: 10.1007/s00425-022-03828-z

[111] Galviz Y, Souza GM, Lüttge U. The biological concept of stress revisited: relations of stress and memory of plants as a matter of space–time. Theory of Expirement Plant Physiology. 2022;**34**:239-264. DOI: 10.1007/s40626-022-00245-1

[112] Liu H, Able AJ, Able JA. Priming crops for the future: rewiring stress memory. Trends in Plant Science. 2022;**27**(7):699-716. DOI: 10.1016/j.tplants.2021.11.015

[113] Zhang H, Lang Z, Zhu JK. Dynamics and function of DNA methylation in plants. Nature Reviews. Molecular Cell Biology. 2018;**19**(8):489-506. DOI: 10.1038/s41580-018-0016-z

[114] Varotto S, Krugman T, Aiese Cigliano R, Kashkush K, Kondić-Špika A, Aravanopoulos FA, et al. Exploitation of epigenetic variation of crop wild relatives for crop improvement and agrobiodiversity preservation. Theoretical and Applied Genetics. 2022;**9**. DOI: 10.1007/s00122-022-04122-y

[115] Preite V, Oplaat C, Biere A, Kirschner J, van der Putten WH, Verhoeven KJF. Increased transgenerational epigenetic variation, but not predictable epigenetic variants, after environmental exposure in two apomictic dandelion lineages. Ecology and Evolution. 2018; 19;**8**(5):3047-3059. DOI: 10.1002/ece3.3871

[116] Liu J, He Z. Small DNA methylation big player in plant abiotic stress responses and memory. Frontiers in Plant Science. 2020;**11**:595603. DOI: 10.3389/fpls.2020.595603

[117] Zhi P, Chang C. Exploiting epigenetic variations for crop disease resistance improvement. Frontiers in Plant Science. 2021;**12**:692328. DOI: 10.3389/fpls.2021.692328

[118] Leonetti P, Molinari S. Epigenetic and metabolic changes in root-knot nematode-plant interactions. International Journal of Molecular Sciences. 2020;**21**(20):7759.20. DOI: 10.3390/ijms21207759

[119] Dalakouras A, Ganopoulos I. Induction of promoter DNA methylation upon high-pressure spraying of double-stranded RNA in plants. Agronomy. 2021;**11**:789. DOI: 10.3390/agronomy11040789

[120] He Y, Li Z. Epigenetic environmental memories in plants: establishment, maintenance, and reprogramming. Trends in Genetics. 2018;**34**:856-866. DOI: 10.1016/j.tig.2018.07.006

[121] Dalakouras A, Papadopoulou KK. Epigenetic modifications: an unexplored facet of exogenous RNA application in plants. Plants. 2020;**9**:673. DOI: 10.3390/plants9060673

[122] Cohen SP, Leach JE. High temperature-induced plant disease susceptibility: more than the sum of its parts. Current Opinion in Plant Biology. 2020;**56**:235-241. DOI: 10.1016/j.pbi.2020.02.008

[123] Alonso C, Ramos-Cruz D, Becker C. The role of plant epigenetics in biotic interactions. The New Phytologist. 2019;**221**:731-737. DOI: 10.1111/nph.1540

[124] Annacondia ML, Markovic D, Reig-Valiente JL, Scaltsoyiannes V, Pieterse CMJ, Ninkovic V, et al. Aphid feeding induces the relaxation of epigenetic control and the associated regulation of the defense response in Arabidopsis. The New Phytologist. 2021;**230**(3):1185-1200. DOI: 10.1111/nph.17226

[125] Schoenherr AP, Rizzo E, Jackson N, Manosalva P, Gomez SK. Mycorrhiza-induced resistance in potato involves priming of defense responses against cabbage looper (Noctuidae: Lepidoptera). Environmental Entomology. 2019;**48**(2):370-381.3. DOI: 10.1093/ee/nvy195

[126] Dutilloy E, Oni FE, Esmaeel Q, Clément C, Barka EA. Plant beneficial bacteria as bioprotectants against wheat and barley diseases. Journal of Fungi. 2022;;**8**(6):632.14. DOI: 10.3390/jof8060632

[127] McDowell JM. Focus on activation, regulation, and evolution of MTI and ETI. Molecular Plant-Microbe Interactions. 2019;**32**(1):5. DOI: 10.1094/MPMI-11-18-0307-FI

[128] Filho JAC, Pascholati SF, Sabrinho RR. Mycorrhizal association and their role in plant disease protection. In: Hakee K, Akhtar M, editors. Plant. Cham: Soil and Microbes. Springer; 2016. pp. 95-143. DOI: 10.1007/978-3-319-29573-2_6

[129] Sanmartín N, Pastor V, Pastor-Fernández J, Flors V, Pozo MJ, Sánchez-Bel P. Role and mechanisms of callose priming in mycorrhiza-induced resistance. Journal of Experimental Botany. 2020;**71**(9):2769-2781. DOI: 10.1093/jxb/eraa030

[130] Vishwakarma K, Kumar N, Shandilya C, Mohapatra S, Bhayana S, Varma A. Revisiting plant–microbe interactions and microbial consortia application for enhancing sustainable agriculture: a review. Frontiers in Microbiology. 2020;**11**:560406. DOI: 10.3389/fmicb.2020.560406

[131] Wei Z, Friman VP, Pommier T, Geisen S, Jousset A, Shen Q. Rhizosphere immunity: targeting the underground for sustainable plant health management. Frontiers of Agricultural Science and Engineering. 2020;7(3):317-328. DOI: 10.15302/J-FASE-2020346

[132] Pons S, Fournier S, Chervin C, Bécard G, Rochange S, Frei Dit Frey N, et al. Phytohormone production by the arbuscular mycorrhizal fungus Rhizophagus irregularis. PLoS One. 2020;**15**(10):e0240886. DOI: 10.1371/journal.pone.0240886

[133] López Sánchez A, Pascual-Pardo D, Furci L, Roberts MR, Ton J. Costs and Benefits of Transgenerational Induced Resistance in Arabidopsis. Frontiers in Plant Science. 2021;**26**(12):644999. DOI: 10.3389/fpls.2021.644999

[134] Nishad R, Ahmed T, Rahman VJ, Kareem A. Modulation of Plant Defense System in Response to Microbial Interactions. Frontiers in Microbiology. 2020;**11**:1298. DOI: 10.3389/fmicb.2020.01298

[135] Fiorilli V, Vannini C, Ortolani F, et al. Omics approaches revealed how arbuscular mycorrhizal symbiosis enhances yield and resistance to leaf pathogen in wheat. Scientific Reports. 2018;**8**:9625. DOI: 10.1038/s41598-018-27622-8

[136] Mac Lean AM, Bravo A, Harrison MJ. Plant signaling and metabolic pathways enabling arbuscular mycorrhizal symbiosis. The Plant Cell. 2017;**29**(10):2319-2335. DOI: 10.1105/tpc.17.00555

[137] Benjamin G, Pandharikar G, Frendo P. Salicylic acid in plant symbioses: beyond plant pathogen interactions. Biology (Basel). 2022;**11**(6):861. DOI: 10.3390/biology11060861

[138] Antolín-Llovera M, Petutsching EK, Ried MK, Lipka V, Nürnberger T, Robatzek S, et al. Knowing your friends and foes--plant receptor-like kinases as initiators of symbiosis or defense. The New Phytologist. 2014;**204**(4):791-802. DOI: 10.1111/nph.13117

[139] Balint-Kurti P. The plant hypersensitive response: concepts,

control and consequences. Molecular Plant Pathology. 2019;**20**(8):1163-1178. DOI: 10.1111/mpp.12821

[140] Pozo de la Hoz J, Rivero J, Azcón-Aguilar C, Urrestarazu M, Pozo MJ. Mycorrhiza-induced resistance against foliar pathogens is uncoupled of nutritional effects under different light intensities. Journal of Fungi. 2021;**7**:402. DOI: 10.3390/jof7060402

[141] Wang H, Hao Z, Zhang X, Xie W, Chen B. Arbuscular mycorrhizal fungi induced plant resistance against fusarium wilt in jasmonate biosynthesis defective mutant and wild type of tomato. Journal of Fungi. 2022;**8**(5):422.20. DOI: 10.3390/jof8050422

[142] Dora S, Terrett OM, Sánchez-Rodríguez C. Plant-microbe interactions in the apoplast: Communication at the plant cell wall. The Plant Cell. 2022;**34**(5):1532-1550. DOI: 10.1093/plcell/koac040

[143] Pruitt RN, Gust AA, Nürnberger T. Plant immunity unified. Nature Plants. 2021;**7**(4):382-383. DOI: 10.1038/s41477-021-00903-3

[144] Kadam SB, Pable AA, Barvkar VT. Mycorrhiza induced resistance (MIR): a defense developed through synergistic engagement of phytohormones, metabolites and rhizosphere. Functional Plant Biology. 2020;**47**(10):880-890. DOI: 10.1071/FP20035

[145] Aguilera P, Ortiz N, Becerra N, Turrini A, Gaínza-Cortés F, Silva-Flores P, et al. Application of arbuscular mycorrhizal fungi in vineyards: water and biotic stress under a climate change scenario: new challenge for chilean grapevine crop. Frontiers in Microbiology. 2022;**3**(13):826571. DOI: 10.3389/fmicb.2022.826571

Chapter 8

Role of Mycorrhizae in Crop Protection

Stephen Larbi-Koranteng, Frederick Kankam, Joseph Adomako and Muntala Abdulai

Abstract

Mychorrizae are indigenous soil fungi that are found associated symbiotically with plant root system. They promote growth of the root system by protecting the plant from pathogen attack, acting directly or indirectly as biocontrol agents and offering plant resistance. These group of rhizosphere fungi also benefit from various biosynthetic substances produced by the root of the plant (root exudates). In this chapter, attempt is being made to present a balanced account of the various roles these fungi play in plant protection. This will give our cherish readers the opportunity to appreciate the mycorrhizal fungi as potential biocontrol agents or bioprotectants of soilborne plant pathogens.

Keywords: mycorrhizae, mycorrhizal fungi, bioprotectant, biocontrol agents, rhizosphere fungi

1. Introduction

With ever increasing world population and its impact, there is a significant pressure to feed the world by agriculturists. Demand for growth of major agriculture commodities is imperative. One major component that militate against achieving this objective is the effort by pests and diseases to reduce yield of agriculture crop production. The impact of both biotic and abiotic stress on production has a greater effect on our aim of increasing agricultural productivity. Many management practices especially pesticides application to mitigate pests and diseases incidence have directly negative influence on the surrounding environment [1, 2].

Pesticides use is regarded as one of the major and common agricultural management practices with a growing evidence of negative impacts on the ecosystem for their application. Any form of synthetic pesticides used ends up creating environmental as well as health concerns. These products may find their way contaminating water bodies resulting in contamination to human and aquatic lives, residual in agricultural produce, causing metabolic disorders to humans when contaminated foods are consumed. They also pose high financial cost to farmers with serious financial burden on output due to their expensive nature.

Many natural resources conservatives have called for reduction in the application of these pesticides and resort to natural, environmentally friendlier and healthy alternatives/practices that require reduction, if not complete their elimination [3, 4]. Many biological, chemical and physical factors also influence soil quality, among these are the microbial communities in the rhizosphere that contribute to soil quality thereby enhancing plant growth and health [5, 6]. The extent of microbial interaction among members have great significance. Among these microorganisms are the Plant Growth Promoting Rhizobacteria (PGPR) and the Arbuscular Mycorrhizae (AM). The later lives symbiotically with the plant in the rhizosphere due to its extensively hyphal network development thereby protecting the crop from pathogen attack [7], decreases biotic and abiotic stress and reducing disease incidence [6, 8]. Therefore, if the potentials of AM are properly harnessed in agriculture, it should be able to reduce sustainably the cost and use of synthetic pesticides in agriculture systems. This is because many soils contain the indigenous AM fungi that colonize the root systems [9], even though not all plants are dependent on Mycorrhizae, most increase yield when AM fungi are applied [10]. This is sure way to attain agricultural sustainability with the reduction in the pesticides and at the same time protecting the crops from pathogen attack and ensuring high yield. Therefore, this chapter tries to detail the role Mycorrhizae fungi play not only in protecting the plant as a biocontrol agents/bioprotectants of soil-borne pathogens but also promoting plant growth thereby realizing its full potential and ensuring maximum yield.

2. Importance of mycorrhizal fungi as a biocontrol agent in suppressive soils

The management of plant disease by chemical approach has been one of the classical methods in agriculture that has sustained productivity for ages. As much as this has been helpful, it has also resulted in nearly an uncontrollable levels of pesticide resistance among many plant pathogens. Also, their direct and indirect impacts have led to the destruction of non-target and beneficial soil organisms as well as raising various health concerns among human and animal populations [4, 11]. To manage this, an appreciable number of studies in recent times have focused on identifying and engineering micro-organisms (i.e., mycorrhiza, bacteria, fungi, and nematodes) that are naturally antagonistic to various plant disease causing pathogens. This approach comes as a more environmentally friendly approach to the application of synthetic pesticides [12, 13]. Among all the organisms, the mycorrhizal fungi are the commonest, largest in biomass and the most important beneficial fungi group. They also combine this with a target specific inhibitory or antagonistic reaction on various soil-borne phytopathogens [3]. By means of changing both the anatomical and morphological structures of plant roots, mycorrhizal fungi improves both the chemical and physical properties on the root-zone environment, hence activating various defensive and disease resistance systems in the plant [14]. Furthermore, they have the abilities to minimize the damage infringed by bacteria, fungi, nematode, as well as other phytopathogens of crops such as *Musa nana*, *Fragaria ananassa*, *Medicago truncatula*, *Cucumis sativus*, *Lycopersicon esculentum*, *Cucum ismelo*, *Olea europaea*, *Zea mays*, *Citrus reticulata*, *Solanum tuberosum*, among other plants [3, 4, 15].

Most mycorrhizal fungi, usually being present as a biotrophic symbiotic microorganisms in the soil rhizosphere usually have a common invasion and ecological niche as most soil-borne pathogens. This could mean that under fair conditions, there must be a spatial competition between pathogens and most mycorrhizal fungi. In this, mycorrhizal fungi, eg., the Arbuscular mycorrhizal fungi (AMF) have been reported to have the potential of reducing the initial and reinfection rates of most pathogens that infects the root epidemics. For example, various studies have reported some competitive relationships between the AMF and an array of plant pathogens. i.e., bacteria, fungi, and nematodes [16–19].

Mycorrhizal fungi played some significant roles in regulating plant growth and development. For example, cucumber plants were found to have some higher levels of zeatin, GA, and IAA when it was inoculated with *G. terrestris*. This increment was further observed to have a bearing with an enhancement in the plant's resistance ability to *Rhizoctonia solanacearum*. In general, mycorrhizal fungi proves significant in the development of plants by inducing synthesis of various plant signaling substances, and improving and enhancing activities of enzymes.

3. Mechanism of suppression of mycorrhizae bioprotectants

The production of healthy and disease-free plant and plant products with corresponding higher yield can directly or indirectly be linked with microorganisms in the soil rhizosphere. Due to the number of environmental concerns regarding the use of different biological control agents, and increasing pathogens resistant to pesticides, more stable and environmentally friendly alternatives are now been considered. AM fungi are not only useful as biofertilizers, but also as bio-stimulants due to their antagonistic capabilities against plant pathogens [20]. They are known to established symbiotic relationship with more than 80% of the plant species [21, 22]. In plant disease management, AMF has been considered as one of the reliable and available options as it is found to serve as a bioprotectant and plant stimulant in sustainable food production and ensure reduction in plant pathogen population to acceptable level without harmful effect to the environment. AMF has been used as a biological control agent in the reduction of incidence and severity of bacteria such as *Pseudomonas syringae, Erwinia carotovora* [23] and fungi such as *Fusarium* spp. [24], *Pythium* sp. [25], *Verticillium* sp. [26], *Sclerotinia* sp., *Phytophthora* sp., *Macrophomina* sp. [27], and nematodes such as *Radophulus* sp. [28].

There are numerous pathogens in the soil that cause diseases to plants and result in substantial reduction in plant yields. These pathogens have to be controlled to ensure food security around the world. Among the new and sustainable control alternatives is biological control that involve the use of antagonistic organisms to suppress damage activities of other organisms that cause diseases to plants [29]. Among the most promising biological control agents is the rhizosphere-competent fungi called mycorrhizae, which is capable of suppressing the activity of disease-causing organisms both major and minor beside their role in stimulating plant growth response. The roots of most plants are in symbiotic association with certain soil fungi and this association is called mycorrhiza [30]. The mycorrhiza has number of functions that include enhancement of nutrient uptake, improvement of soil structure and plant establishment, protection of plants against environmental stresses and suppression of plant diseases [31].

Plant roots colonization by AMF usually results in the decrease of the incidence and severity of the diseases caused by pathogens. The reduction in damage by AMF maybe as a result of changes in the morphology and plant root growth, biochemical and physiological changes in the plant, histopathological changes in the plant root, mycorrhizosphere effects that results in the modification of microbial population density, activation of host defense mechanisms, parasitism of nematodes by AMF and competition for photosynthetic products and colonization sites [32]. Among the various proposed biocontrol mechanisms for the plant diseases, the most effective biocontrol scheme could either be the result of all the mechanisms working together or as a separate entity. The major limitation in the use of AMF as a biocontrol agent could culminate from its obligate nature, the role of environmental influence on the various mycorrhiza symbiotic associations and limited understanding of the mechanism involved in the interaction processes. The objective of this chapter is therefore to throw more light on the mechanism of suppression of mycorrhizae bioprotectants.

Protection of plants by mycorrhizal fungi against disease causing organisms involves multiple mechanisms that include: production and changes in the exudation pattern, formation of physical barrier (fungal mantle) around the roots and synthesis of anti-fungal compounds by the plant roots in response to mycorrhiza symbiotic association [33]. For example, *Paxillus involutus* (Ectomycorrhizal fungi) was reported to successfully controlled *Fusarium moniliforme* and *Fusarium oxysporum* causing rot disease in *Pinus resinosa* as well as *Pisolithus tinctorius* (Ectomycorrhizal fungi) in controlling *Phytophthora cinnamomi* causing disease in sand pine [34]. Specific form of disease suppression may result from the activity of one or few antagonistic microbes.

The symbiotic association of AMF has been reported to induce plant host defense response both at early and later stage of invasion by the pathogen [35]. There have been reports of quick response in terms of plant host defense to pathogens by the mycorrhizal associated plants compare to those devoid of this symbiotic relationship and for that matter, AMF colonization has been proposed to act as a priming scheme for the pathogen resistance process [36, 37].

In a related defensive mechanism against plant pathogens, AM fungi have been involved in the activation of the plant defense response against pathogens and this include the expression of number of genes with their matching proteins (e.g., phenolics, cellulose deposition, chitinases, hydroxyproline-rich glycoproteins, phytoalexins, peroxidases and proteins relating to pathogenicity) [38, 39]. Both localized and systemic resistance to *Phytophthora parasitica* has been reported in tomato root system [40], Pathogenesis-related proteins are involved in triggering of the Systemic acquired resistance (SAR) defense mechanism [41]. The pathogen *Aphanomyces euteiches* causing disease on garden pea was biologically controlled after pre gene activation of the host defense response by mycorrhiza-related chitinolytic enzymes [38].

Mycorrhization have been recorded to change plant root exudation pattern and these alterations could indirectly affect the pathogen through alteration of the pH of soil environment or through production of inhibitory products. In a study involving symbiotic association of strawberry with mycorrhiza fungi, exudates released by the roots of the strawberry had shown to suppress the growth and sporulation of *Phytophthora fragariae* [42], as well as affect the germination of microconidia produced by *Fusarium oxysporum* in a related experiment [43, 44]. There is also evidence of direct antagonistic action by AMF against pathogens in the soil rhizosphere [45, 46].

In terms of improvement in the nutritional status or reduction of plant root damage by the pathogen, the increase supply of nutrient by mycorrhiza fungi to plants have been suggested to enhanced their tolerance level to pathogen damage and carbon drain from plants to the pathogen. AMF absorb nutrient via the external network of fungi hypha by solubilizing both macro and micro elements like Mn, Ca, Zn, Cu, N and P [47–49]. This nutrient uptake ensures healthy growth of the plant due readily or available nutrients supply to the plant that enhances the tolerance or resistance level of the plant to the pathogens [50]. The mycorrhizal fungi increase the rate at which phosphorous is absorbed by increasing the surface area, number of roots, growth and development of plant root hairs. The increase in phosphorous uptake in plant-mycorrhizal symbiotic relationship constitute the major mechanism for the AMF-mediated biocontrol [51].

With regards to the morphological alteration of the plant roots, mycorrhization has been reported to cause some changes in the morphology of the roots in spatial, structural, temporal and quantitative way [52, 53]. The AM produce arbuscles and vesicles both inter and intracellularly within plant root. Any pathogen that encounters with ectomycorrhizal fungus has to first of all deal with the external and multilayer network of hyphae known as mantle and inner cortical cells which serve as physical barrier to invasive pathogens and play a critical role in enhancing the population of the useful microorganisms in the soil with corresponding production of growth promoting elements by PGPRs that increase the plant resistance to pathogens [54–56]. In aromatic plant (e.g., basil), the root length and toot tip numbers, level of branching and fresh weight of the plant have been reported to be altered independently based on the type of AMF involving in the colonization process [53].

Plant roots colonized by AM fungi have enlarged length and diameter with profuse branches [57, 58]. Plant roots were found to accumulate an increased deposition of lignin and chitinases content [59] as well increase the resistance of plant root system to pathogens when in association with AM fungi. Incidence and severity of diseases caused by *Phytophthora parasitica* were found to decrease in AMF association with plant as compare with non-mycorrhizal roots [22]. AMF associated plants produced a lot of arginine that were found to suppress Thielaviopsis spore formation and large amount of proteins, phytoalexins and peroxidases [58, 60, 61] that induce plant resistance to pathogens.

AMF is found to prevent infection of the root during root colonization by decreasing the access sites to the pathogen as well as stimulate plant host defense mechanism as it was reported in reducing the incidence and severity of root-knot nematodes [62]. Number of mechanisms have been reported to increase stress tolerance of plants by AM fungi and this include the formation of a complex network hypha by AM fungi around the plant roots that block intruding pathogens. In an apple seedling trial, an apple replant disease triggered by phytotoxic myxomycetes has been successfully suppressed by AM fungi such as *Glomus fasciculatum* and *G. macrocarpum* [63]. AMF are also known to provide protection to plants against pathogenic bacteria that affect roots in the soil. Disease caused by *P. syringae* on tomato plant have been drastically reduced in plant-mycorrhiza symbiotic association [23, 33]. The various protective and suppressive mechanisms involve in this include: indirect effects (chemical interactions; physical protection); and indirect mechanisms e.g., isoflavonoids, increase nutrition uptake by plants; changes in the morphology of the plant roots by increased lignification.

Competition by AMF with pathogens for infection site on the plant root is well documented. In the competition for the site, AMF usually inhabit the location on the plant root surface where the pathogen require to penetrate the root or it pre-establishes itself in the cells so that the site cannot be occupied by any new invasive pathogen [40, 64]. In other cases, Mycorrhizal fungi and pathogens causing plant diseases, more often than not live in the same niche that bring them into physical contact to compete for the limited resources (nutrient and space) in the rhizosphere [65]. AMF is also known to compete with the other pathogens for carbon. The AMF colonize the roots of the plants and make use of the carbohydrate from the plant, thereby leaving limited amount of carbon to be utilized by the competing pathogen and this explains the rationale behind the biocontrol strategy implore by AM fungi [36, 66, 67]. There are diverse AMF species that show different carbon sink strength in the roots of plants associated with mycorrhiza and thus have shown different inhibitory or antagonistic effect against plant pathogens [68, 69]. For example, in nematode trial, *Meloidogyne incognita* reproduction factor was found to be reduced when in association with AMF prior to inoculation [70]. Elucidation and protective capability of the mycorrhizal symbiotic association with variable expression of the traits in relation to their ability to protect plants have been well documented [16]. AM fungi in association with plants results in biochemical changes in host tissues, reduction in plant stress, uptake of phytonutrients, changes in plant root anatomy and morphology, trigger systemic resistance, and competition for the limited resources such as nutrient and space [40].

4. Action of AM fungi against plant pathogens

With increasing cost of pesticides and the negative effect of this on human health and the environment as well as pathogen resistance, AM fungi offers potential for more sustainable and environmentally friendlier alternative for sustainable agriculture. These fungi are nevertheless most important habitat of the rhizosphere and their activity has direct influence on disease incidence and severity especially on root diseases [71]. There are several reports of possible use of AM fungi in the biocontrol of plant diseases [72–74]. One communality among all these reported evidence are that, AM interactions with plant pathogens tends to reduce their damage to plants caused by fungi and nematodes; a symbiotic association with these plants enhances resistance or tolerance.

In the interaction between AM fungi and plant parasitic nematodes (PPM), for instance, the PPM are known to be very common agricultural soil inhabitants world over and cause extensive damages to many crop species. By their actions, they can be ectoparasites or endoparasites (semi-endoparasites and migratory endoparasites), sedentary endoparasites and causes about 50–60% yield losses and many often these damages are aggravated when other pathogens capitalize on them to cause severe diseases.

Both the nematodes and AM fungi tend to stablish relationship in the rhizosphere due to their common interest in nutrient provided by the host plant. The interaction between these two would have opposite effect on growth and yield that will tend to favor the host plant [75].

Also, plant pathogenic fungi are one of the common occupants of the soil matrix and causes wide range soil-borne diseases. The soil serves as host to these pathogens and cause severe damages to the roots of susceptible hosts. Soil-borne diseases caused by phytopathogenic fungi are also difficult to control due to their ability to develop over seasoning structures such as chlamydospore, sclerotia, rhizomorph, etc. The presence of AM fungi and their interaction with these plant pathogenic fungi in the rhizosphere gives the advantage to the AM fungi to exert its s opposite effect/influence on the plant pathogenic fungi thereby protecting the plant from their attacks, promoting plant growth and enhancing yield of the plant [76, 77].

Finally, there also reports on several other plant pathogens establishing opposite relationship with the AM fungi such as the bacteria, mycoplasma, plant viruses etc. thereby reducing disease incidence and severity in their interactions with AM fungi [78, 79].

5. Use of mycorrhizal fungi in plant growth and disease suppression

Plants are major source of energy for both human and animals providing about 80% of food consumed by humans and primary source of nutrition for livestock. Production of adequate food to feed the ever growing global is threatened by the high prevalence of diseases caused by biotic agents such as bacteria, fungi, nematodes, viruses and oomycetes. Plant diseases reduces quantity and quality of yield, thereby affecting food security and safety of produce for consumption. It is estimated that diseases account for yield loss ranging from 13 to 22% with billions in economic losses due to inputs purchase for their management [80]. Yield and storage losses attributed to diseases have significantly been linked to global starvation and malnutrition millions of people [80, 81]. Diseases reduce yield of plants by altering several physiological process such as the absorption and transportation of water and nutrients needed for plant use, photosynthesis, flower and fruit development [82].

Plant diseases results from positive interactions of host, pathogen and environment. To overcome the negative impact of diseases on plant growth, multiple strategies have been developed and successfully used to manipulate host-pathogen interactions to favor growth of the host whilst suppressing reproduction, establishment and transmission of the pathogen. Some of the approaches to controldiseases include the use of chemicals, physical, genetic and cultural means. Host resistant approach is economical, effective and environmentally friendly, however, rapid breakdown due to continuous pathogen evolution limits its use in commercial and modern crop production tilted towards intensification and mono-cropping which provides ideal environment for pathogen evolution. In situations where reliance on resistant varieties to suppress diseases has not been achieved, utilization of chemicals have become inevitable. Chemicals are highly effective but its harsh effect on human and animal health, non-target organisms and the environment resulting from excessive use has unfortunately defeated its mass promotion and utilization. An alternative to chemical pesticide is the use of biological control. According to [83, 84], biological control limits diseases causing pathogen, improves plant immunity, modifies environment through efficient cropping systems. Biological control agents offer advantages over chemical control agents by being antagonistic to specific pathogens with less risk compared to chemical pesticides. Contrary to its benefit, application of BCAs is heavily challenged by several biotic and abiotic factors as well as frequent pathogen evolution which makes field application frequently inconsistent [85]. Notwithstanding this, recent reports have shown that application of mycorrhizal fungi strains as biological control agents is an important option to reduce threats posed by diseases.

Mycorrhizal fungi exists closely with over 80% of plants species on land offering plethora of benefits to its host. These fungi may reside within the cortex of plant roots. Mycorrhiza fungi-host association contributes significantly to carbon, nitrogen and phosphorus cycling in the ecosystem. In addition to these, mycorrhiza fungal activities improve water uptake by increasing quantity of available soil water [86] thereby improving plant productivity, diversity and contributing significantly to plant growth and fitness. According to [87] these fungi alters root morphology by increasing root branching and growth to favor root vigor due to the high nutrient uptake [88] hence influencing plant growth and yield. Mycorrhizal fungi have successfully been used and reported to increase growth and yield of several crops such as carrot [89], yam [90], maize [91].

Biocontrol by mycorrhizal strains against multiple diseases is achieved by triggering defense mechanisms in the host to improve plant tolerance to pathogens. Earlier studies [92–94] have shown improved tolerance and suppressive ability of plants to vascular diseases caused by Fusarium, Verticillium and Bacteria. Other studies have demonstrated improved tolerance of cucumber, olive, date palm, and tomato seedlings to fusarium and bacteria wilt following application of mycorrhizal strains [95, 96]. Research findings by [21, 97] showed that multiple root branching resulting from root alteration due to interaction with host reduced infection of *Phytophthora fragariae* in strawberry. Root necrosis in cowpea caused by *Rhizoctonia solani* and *Pythium aphanidermatum* in pepper were reported to decrease in the presence of both *Glomus clarum* and *Gnypeta deserticola*. Although most success has been achieved in the use of Mycorrhizal fingi (MF) to manage soil borne fungal pathogens, other works have reported on their biocontrol potential against aerial pathogens like *Alternaria solani* in tomato [37] and other necrotrophic and biotrophic pathogens [98]. Apart from fungal pathogens, suppression of Plant Parasitic Nematodes (PPN) by MF has been reported in plants such as banana, coffee and tomato [99, 100]. Similarly, [101, 102] concluded that AMF can attack soybean cyst nematodes and reduce severity of nematode infection in crops such as soybeans, cotton, cucumbers, tomatoes and citrus. MF antagonizes activities of PPN by reducing infection, reproduction and enhances tolerance. Although many research outputs have concluded on the biocontrol potential of MF, mass application in the field is sporadic due to variability in performance, on host, pathogen isolate and environmental condition [103]. There is the need therefore to improve communication on the efficacy and safety associated with MF application biocontrol agents.

6. Mycorrhizal fungi for sustainable agricultural systems

Sustainable agricultural systems use available natural resources to achieve acceptable level of productivity, food quality, and quantity without compromising the environmental impacts [27]. As defined, sustainable agriculture is the use of ecological sound, economic viable and socially responsible practices to obtain higher productivity, numerous plant health practices contribute to sustainable agriculture through the control of soil-borne diseases by increasing soil microbial activity thereby enhancing symbiosis, competition, and parasitism within the rhizosphere [104, 105]. The current research focus has been the search for suitable alternatives to the use commercial synthetic pesticides. Many have been achieved though, but the efficient exploration of microorganisms to improve soil fertility and at the same time enhancing plant growth and protection is being pursued.

To improve crop protection, synthetic pesticides have been used extensively to mitigate effects of pest and diseases, over reliance has been a problem and therefore biological processes that will enhance plant health such as mycorrhizae, earthworm and other symbionts should be encouraged [106, 107]. Mycorrhizae association with plants are beneficial for sustainable agriculture as they reduce pests and diseases infestation [7].

7. Conclusion

This chapter has demonstrated that Mycorrhizae fungi especially AM fungi play critical role in plant protection. Apart from making nutrient available for the plant uptake, AM fungi provide protection to the plant thereby enhancing plant growth and yield. The AM fungi live symbiotically with the plant by benefiting from nutrient rich environment provided by the plant in the rhizosphere and the plant also benefit from the AM fungi through deprivation of other pathogen from getting direct contact with the plant thereby enhancing its ability to resist pathogens attack.

Conflict of interest

The authors declare no conflict of interest.

Author details

Stephen Larbi-Koranteng[1*], Frederick Kankam[2], Joseph Adomako[3] and Muntala Abdulai[4]

1 Department of Crop and Soil Sciences, College of Agriculture Education, Akenten Appiah-Menka University of Skills Training and Entrepreneurial Development, Mampong Ashanti, Ghana

2 Faculty of Agriculture, Department of Agricultural Biotechnology, Food and Consumer Sciences, University for Development Studies, Tamale, Ghana

3 CSIR-Crop Research Institute, Plant Health Division, Kumasi, Ghana

4 Department of Horticulture and Crop Production, School of Agriculture and Technology, University of Energy and Natural Resources, Dormaa Ahenkro, Ghana

*Address all correspondence to: larbikoranteng@yahoo.com

References

[1] Mozafar A, Anken T, Ruh R, Frossard E. Tillage intensity, mycorrhizal and non mycorrhizal fungi, and nutrient concentrations in maize, wheat, and canola. Agronomy Journal. 2000;**92**:1117-1124

[2] Carter MR, Campbell AJ. Influence of tillage and liquid swine manure on arbuscular mycorrhiza to reduce the stressful effects of soil compaction on corn (*Zea mays* L.) growth. Soil Biology and Biochemistry. 2006;**39**:2014-2026

[3] Allsup CM, Lankau RA, Paige KN. Herbivory and soil water availability induce changes in arbuscular mycorrhizal fungal abundance and composition. Microbial Ecology. 2022;**84**(1):141-152. DOI: 10.1007/s00248-021-01835-3

[4] Razak NA, Gange AC. Multitrophic interactions between arbuscular mycorrhizal fungi, foliar endophytic fungi and aphids. Microbial Ecology (Springer). 2021:1-11. DOI: /10.1007/s00248-01937-y

[5] Miransari M, Bahrami HA, Rejali F, Malakouti MJ, Torabi H. Using arbuscular mycorrhiza to reduce the stressful effects of soil compaction on corn (Zea mays L.) growth. Soil Biology and Biochem. 2007;**39**:2014-2026

[6] St-Arnaud M, Vujanovic V. Effects of the arbuscular mycorrhizal symbiosis on plant diseases and pests. In: Hamel C, Plenchette C, editors. Mycorrhizae in Crop Production. New York: Haworth; 2007. pp. 67-122

[7] Smith SE, Read DJ. Mycorrhizal symbiosis. London: Academic; 1997

[8] Aliasgarzad N, Neyshabouri MR, Salimi G. Effects of arbuscular mycorrhizal fungi and *Bradyrhizobium japonicum* on drought stress of soybean. Biologia. 2006;**61**:324-328

[9] Covacevich F, Echeverría HE, Andreoli YE. Micorrización vesículo arbuscular espontánea entrigoen función de la disponibilidad de fósforo. Ciencia del Suelo. 1995;**13**:47-51

[10] Al-Karaki GN, Al-Raddad A, Clark RB. Water stress and mycorrhizal isolate effects on growth and nutrient acquisition of wheat. Journal of Plant Nutrition. 1998;**21**:891-902

[11] Huang NX, Enkegaard A, Osborne LS, Ramakers PMJ, Messelink GJ, Pijnakker J, et al. The banker plant method in biological control. Critical Reviews in Plant Sciences. 2011;**30**:259-278

[12] Larkin RP, Fravel DR. Mechanisms of action and dose response relationships governing biological control of Fusarium wilt of tomato by non-pathogenous Fusarium spp. Phytopathology. 1999;**89**:1152-1161

[13] Meyer SLF, Roberts DP. Combinations of biocontrol agents for management of plant-parasitic nematodes and soil-borne plant pathogenic fungi. Journal of Nematology. 2002;**34**:1-8

[14] Aseel DG, Rashad YM, Hammad SM. Arbuscular mycorrhizal fungi trigger transcriptional expression of flavonoid and chlorogenic acid biosynthetic pathways genes in tomato against tomato mosaic virus. Scientific Reports. 2019;**9**:9692. DOI: 10.1038/s41598-019-46281-x

[15] Begum N, Qin C, Ahanger MA, Raza S, Khan MI, Ashraf M, et al. Role

of arbuscular mycorrhizal fungi in plant growth regulation: Implications in abiotic stress tolerance. Frontiers in Plant Science. 2019;**10**:1068

[16] Whipps JM. Prospects and limitations for mycorrhizas in biocontrol of root pathogens. Canadian Journal of Botany. 2004;**1227**:1198-1227

[17] Castillo P, Nico AI, Azcón-Aguilar C, Del Río Rincón C, Calvet C, Jiménez-Díaz RM. Protection of olive planting stocks against parasitism of root-knot nematodes by arbuscular mycorrhizal fungi. Plant Pathology. 2006;**55**:705-713

[18] Tabin T, Arunachalam A, Shrivastava K, Arunachalam K. Effect of arbuscular mycorrhizal fungi on damping-off disease in Aquilaria agallocha Roxb. Seedlings. Tropical Ecology. 2009;**50**:243-248

[19] Zhou J, Chai X, Zhang L, George TS, Wang F, Feng G. Different arbuscular mycorrhizal fungi co-colonizing on a single plant root system recruit distinct microbiomes. mSystems. 2020;**5**(6):e00929-e00920

[20] Whipps JM. Microbial interactions and biocontrol in the rhizosphere. Journal of Experimental Botany. 2001;**52**:487-511

[21] Norman JR, Atkinson D, Hooker JE. Arbuscular mycorrhizal fungal-induced 704 alteration to root architecture in strawberry and induced resistance to the root pathogen 705 *Phytophthora fragariae*. Plant and Soil. 1996;**185**:191-198

[22] Vigo C, Norman JR, Hooker JE. Biocontrol of the pathogen *Phytophthora parasitica* by arbuscular mycorrhizal fungi is a consequence of effects on infection loci. Plant Pathology. 2000;**49**:509-514

[23] García-Garrido JM, Ocampo JA. Effect of VA mycorrhizal infection of tomato on damage cuased by *Pseudomonas syringae*. Soil Biology and Biochemistry. 1989;**21**:165-167

[24] Sohrabi M, Mohammadi H, Mohammadi AH. Influence of AM fungi, *Glomus mosseae* and *Glomus intraradices* on Chickpea growth and root-rot disease caused by Fusarium solani f. sp. pisi under greenhouse conditions. Journal of Agricultural Science and Technology. 2015;**17**:1919-1929

[25] El-Mohamedy RSR. Biological control of Pythium root rot of broccoli plants under greenhouse conditions. International Journal of Agricultural Technology. 2012;**8**:1017-1028

[26] Mason TBS. Assessing direct and indirect effects of the fungicide Flutriafol on Arbuscular mycorrhizal fungi in controlling cotton root rot [Master thesis] Texas Tech University. 2015. pp. 88

[27] Harrier LA, Watson CA. The potential role of arbuscular mycorrhizal (AM) fungi in the bioprotection of plants against soil-borne pathogens in organic and/or other sustainable farming systems. Pest Management Science. 2004;**60**:149-157

[28] Jaizme-Vega MC, Tenoury P, Pinochet J, Jaumot M. Inter-actions between the root knot nematode *Meloidogyne incognita* and *Glomus mosseae* in banana. Plant and Soil. 1997;**196**:27-35

[29] Pal KK, McSpadden Gardener B. Biological control of plant pathogens. The Plant Health Instructor (APSnet). 2006: 1-29. DOI: 10.1094/PHI-A-2006-1117-02

[30] Marschner P, Rengel Z. Nutrient cycling in terrestrial ecosystems. In: Marschner P, Rengel Z, editors.

Nutrient Cycling in Terrestrial Ecosystems. Soil Biology. Vol. 10. New York: Springer; 2007. XX +397p. ISBN: 978-3-540-68026-0

[31] Barea JM. Mycorrhiza/bacteria interactions on plant growth promotion. In: Ogoshi A, Kobayashi L, Homma Y, Kodama F, Kondon N, Akino S, editors. Plant Growth-promoting Rhizobacteria, Present Status and Future Prospects. Paris: OECD; 1997. pp. 150-158

[32] Siddiqui ZA, Mahmood I. Role of plant symbionts in nematode management: A Review. Bioresource Technology. 1995;**54**:217-226

[33] Garcia-Garrido JM, Tribak M, Rejon-Palomares A, Ocampo JA, Garcia-Romera I. Hydrolytic enzymes and ability of arbuscular mycorrhizal fungi to colonize roots. Journal of Experimental Botany. 2000;**51**:1443-1448

[34] Ross EW, Marx DM. Susceptibility sand pine to *Phytophthora cinnamomi*. Phytopathology. 1972;**62**:1197-1200

[35] García-Garrido JM, Ocampo JA. Regulation of the plant defence response in arbuscular mycorrhizal symbiosis. Journal of Experimental Botany. 2002;**53**:1377-1386

[36] Azcón-Aguilar C, Jaizme-Vega MC, Calvet C. The contribution of arbuscular mycorrhizal fungi to the control of soil-borne plant pathogens. In: Gianinazzi S, Schüepp H, Barea JM, Haselwandter K, editors. Mycorrhizal technology in agriculture. Switzerland: Birkhäuser; 2002. pp. 187-197

[37] Pozo MJ, Azcón-Aguilar C. Unraveling mycorrhiza-induced resistance. Current Opinion in Plant Biology. 2007;**4**:393-398

[38] Slezack S, Negrel J, Bestel-Corre G, Dumas-Gaudot E, Gianinazzi S. Purification and partial amino acid sequencing of a mycorrhiza-related chitinase isoform from *Glomus mosseae* inoculated roots of *Pisum sativum* L. Planta. 2001;**213**:781-787

[39] Guillon C, St-Arnould M, Hamel C, Jabaji-Hare SH. Differential and systemic alteration of defence-related gene transcript levels in mycorrhizal bean plants with *Rhizoctonia solani*. Canadian Journal of Botany. 2002;**80**:305-315

[40] Cordier C, Pozo MJ, Barea MJ, Gianinazzi S, Gianinazzi-Pearson V. Cell defense responses associated with localized and systemic resistance to *Phytophthora parasitica* induced in tomato by an arbuscular mycorrhizal fungus. Molecular Plant-Microbe Interactions. 1998;**11**:1017-1028

[41] Ryals JA, Neuenschwander UM, Willts MG, Molina A, Steiner HY, Hunt MD. Systemic acquired resistance. Plant Cell. 1996;**8**:1809-1819

[42] Norman JR, Hooker JE. Sporulation of *Phytophthora fragariae* shows greater stimulation by exudates of non-mycorrhizal than by mycorrhizal strawberry roots. Mycological Research. 2000;**104**:1069-1073

[43] Scheffknecht S, Mammerler R, Steinkellner S, Vierheilig H. Root exudates of mycorrhizal tomato plants exhibit a different effect on microconidia germination of *Fusarium oxysporum* f. sp. lycopersici than root exudates from non-mycorrhizal tomato plants. Mycorrhiza. 2006;**16**:365-370

[44] Scheffknecht S, St-Arnaud M, Khaosaad T, Steinkellner S, Vierheilig H. An altered root exudation pattern through mycorrhization affecting microconidia germination of the highly

specialized tomato pathogen *Fusarium oxysporum* f. sp. lycopersici (Fol) is not tomato specific but also occurs in Fol non-host plants. Canadian Journal of Botany. 2007;**85**:347-351

[45] St-Arnaud M, Hamel C, Caron M, Fortin JA. Endomycorrhizes VA etsensibilité aux maladies: Synthèse de lalittérature et méc anismesd'interactionprobables. In: Charest C, Bernier R, editors. Mycorrhizal symbiosis. Frelighsburg, QC: Orbis; 1995. pp. 51-87

[46] Filion M, St-Arnaud M, Fortin JA. Direct interaction between the arbuscular mycorrhizal fungus *Glomus intraradices* and different rhizosphere microorganisms. The New Phytologist. 1999;**141**:525-533

[47] Parniske M. Arbuscular mycorrhiza: The mother of plant root endosymbiosis. Nature Reviews Microbiology. 2008;**6**:763-775

[48] Baum C, El-Tohamy W, Gruda N. Increasing the productivity and product quality of vegetable crops using arbuscular mycorrhizal fungi: A review. Scientia Horticulturae (Amsterdam). 2015;**187**:131-141

[49] Clark RB, Zeto SK. Mineral acquisition by arbuscular mycorrhizal plants. Journal of Plant Nutrition. 2000;**23**:867-902

[50] Singh N, Singh D, Singh N. Effect of *Glomus bagyaraji* inoculation and phosphorus amendments on Fusarium wilt of chickpea. Agricultural Research Journal. 2017;**54**:236-243

[51] Bodker L, Kjoller R, Rosendahl S. Effect of phosphate and the arbuscular mycorrhizal fungus *Glomus intraradices* on disease severity of root rot of peas (*Pisum sativum*) caused by *Aphanomyces euteiches*. Mycorrhiza. 1998;**8**:169-174

[52] Berta G, Trotta A, Fusconi A, Hooker JE, Munro M, Atkinson D, et al. Arbuscular mycorrhizal induced changes to plant growth and root system morphology in *Prunus cerasifera*. Tree Physiology. 1995;**15**:281-293

[53] Copetta A, Lingua G, Berta G. Effects of three AM fungi on growth, distribution of glandular hair, and of essential oil production in Ocimum basilicum L. var. Genovese. Mycorrhiza. 2006;**16**:485-494

[54] Hakeem KR, Akhtar MS, Abdullah SNA. Plant, Soil and Microbes.1, Implications in Crop Science. Vol. 1. Gewerbestrasse 11, 6330 Cham, Switzerland: Springer International Publishing AG; 2016. p. 366

[55] Linderman RG. Mycorrhizal interactions with the rhizosphere microflora: The mycorrhizosphere effect. Phytopathology. 1988;**78**:366-371

[56] Sylvia DM, Sinclair WA. Suppressive influence of *Laccaria laccata* on *Fusarium oxysporum* and on Douglas-fir seedlings. Phytopathology. 1983;**73**:384-389

[57] Schellenbaum L, Gianinazzi S, Gianinazzi-Pearson V. Comparison of acid so1uble protein synthesis in roots of endomycorrhizal wild type *Pisum sativum* and corresponding isogenic mutants. Journal of Plant Physiology. 1992;**141**:2-6

[58] Hooker JE, Black KE, Perry RL, Atkinson D. Arbuscular mycorrhizal induced alteration to root longevity of poplar. Plant and Soil. 1995;**172**:327-329

[59] Ziedan EH, Elewa IS, Mostafa MH, Sahab AF. Application of mycorrhizae for controlling root diseases of sesame. First International Congress MCOMED. Moracoo, 11-13 October 2010. 2010. p. 97

[60] Baltruschat H, Schănbeck F. Studies on the influence of endotrophic mycorrhizae on the injection of tobacco by *Thielaviopsis basicala*. Phytopathology Z. 1975;**84**:172-188

[61] Morandi D, Bailey JA, Gianinazzi-Pearson V. Isoflavonoid accumulation in soybean roots infected with vesicular-arbuscular mycorrhizal fungi. Physiological and Molecular Plant Pathology. 1984;**24**:357-364

[62] Linderman RG. Role of VAM fungi in biocontrol. In: Pfleger FL, Linderman RG, editors. Mycorrhizae and Plant Health. St Paul: APS Press; 1994. pp. 1-26

[63] Catska V. Interrelationship between vesicular-arbuscular mycorrhiza and rhizosphere microflora in apple replant disease. Biologia Plantarum. 1994;**36**:99-104

[64] Cordier C, Gianinazzi S, Gianinazzi-Pearson V. Colonization pattern of root tissue by *Phytophthora nicotianae* var. parasitica related to reduced disease in mycorrhizal tomato. Plant Soil. 1998a;**185**:223-232

[65] Smith GS. The role of phosphorous nutrition in interactions of vesicular arbuscular mycorrhizal fungi with soil-borne nematodes and fungi. Phytopathology. 1988;**78**:371-374

[66] Singh R, Adholeya A, Mukerji KG. Mycorrhiza in control of soil-borne pathogens. In: Mukerji KG, Chamola BP, Singh J, editors. Mycorrhizal Biology. New York: Kluwer; 2000. pp. 173-196

[67] Xavier LJC, Boyetchko SM. Arbuscular mycorrhizal fungi in plant disease control. In: Arora DK, editor. Fungal Biotechnology in Agricultural, Food, and Environmental Applications. New York: Dekker; 2004. pp. 183-194

[68] Lerat S, Lapointe L, Gutjahr S, Piché Y, Vierheilig H. Carbon partitioning in a split-root system of arbuscular mycorrhizal plants is fungal and plant species dependent. The New Phytologist. 2003a;**157**:589-595

[69] Lerat S, Lapointe L, Piché Y, Vierheilig H. Variable carbon sink strength of different *Glomus mosseae* strains colonizing barley roots. Canadian Journal of Botany. 2003b;**81**:886-889

[70] Dos Anjos ÉCT, Cavalcante UMT, Gonçalves DMC, Pedrosa EMR, Santos V, Maia LC. Interactions between an arbuscular mycorrhizal fungus (*Scutellospora heterogama*) and the rootknot nematode (*Meloidogyne incognita*) on sweet passion fruit (Passiflora alata). Brazilian Archives of Biology and Technology. 2010;**53**:801-809

[71] Linderman RG. VA mycorrhizae and soil microbial interactions. In: Bethelenfalvay GJ, Linderman RG, editors. Mycorrhizae in Sustainable Agriculture. Vol. 54. Madison, WI: ASA Special Publication; 1992. pp. 45-70

[72] Mukerji KG. Mycorrhiza in control of plant pathogens: Molecular approaches. In: Mukerji KG, Chamola BP, Upadhyay RK, editors. Bio-technological Approaches in Biocontrol of Plant Pathogens. New York: Kluwer Academic/ Plenum; 1999. pp. 135-155

[73] Siddiqui ZA, Mahmood I, Khan MW. VAM fungi as prospective biocontrol agents for plant parasitic nematodes. In: Bagyaraj DJ, Verma A, Khanna KK, Kehri HK, editors. Modern Approaches and Innovations in Soil Management. Rastogi: Meerut, India; 1999. pp. 47-58

[74] Barea JM, Pozo MJ, Azcon R, Azcon-Aguilar C. Microbial co-operation

in the rhizosphere. Journal of Experimental Botany. 2005;**56**:1761-1778

[75] Siddiqui ZA, Akhtar MS. Synergistic effects of antagonistic fungi and a plant growth promoting rhizobacterium, an arbuscular mycorrhizal fungus, or composted cow manure on the populations of *Meloidogyne incognita* and growth of tomato. Biocontrol Science and Technology. 2008;**18**:279-290

[76] Akhtar MS, Siddiqui ZA. Biocontrol of a root-rot disease complex of chickpea by *Glomus intraradices*, *Rhizobium* sp. and *Pseudomonas straita*. Crop Protection. 2008a;**27**:410-417

[77] Akhtar MS, Siddiqui ZA. *Glomus intraradices*, *Pseudomonas alcaligenes*, *Bacillus pumilus* as effective biocontrol agents for the root-rot disease complex of chickpea (*Cicer arietinum* L.). Journal of General Plant Pathology. 2008b;**74**:53-60

[78] MingQin G, Yu C, FengZhen W. Resistance of the AM fungus Eucalyptus seedlings against *Pseudomonas solanacearum*. Forest Research. 2004;**17**:441-446

[79] Garcia-Chapa M, Batlle A, Lavina A, Camprubi A, Estaun V, Calvet C. Tolerance increase to pear decline Phytoplasma in mycorrhizal OHF-333 pear root stock. In: Llacer G, editor. XIX Int. Symp. on Virus and Virus Like Diseases of Temperate Fruit Crops-Fruit Tree Diseases. Valencia, Spain: International Society for Horticultural Science (Paper presented in July 21-25, 2003); 2004

[80] Savary S, Willocquet L, Pethybridge SJ, Esker P, McRoberts N, Nelson A. The global burden of pathogens and pests on major food crops. Nature Ecology & Evolution. 2019;**3**:430-439

[81] FAO, IFAD, and WFP. The state of food insecurity in the world. Meeting the 2015 international hunger targets: Taking stock of uneven progress. Rome: FAO. 2015

[82] Al-Sadi AM. Impact of plant diseases on human health. International Journal of Nutrition, Pharmacology, Neurological Diseases [serial online]. 2017;7:21-22 Available from: https://www.ijnpnd.com/text.asp?2017/7/2/21/205287

[83] Bragard C, Caciagli P, Lemaire O, Lopez-Moya JJ, MacFarlane S, Peters D, et al. Status and prospects of plant virus control through interference with vector transmission. Annual Review of Phytopathology. 2013, 2013;**51**:177-201

[84] Poveda J, Abril-Urias P, Escobar C. Biological control of plant-parasitic nematodes by filamentous fungi inducers of resistance: Trichoderma, mycorrhizal and endophytic fungi. Frontiers in Microbiology. 2020;**11**:992

[85] Muhsen TAA, Mahdi SYA-A, Batoal ZA. Effect of arbuscular mycorrhizal fungi as a biocontrol agent and organic matter against fusarium wilt in tomato. Journal of Genetic and Environmental Resources Conservation. 2015;**3**(3):237-224

[86] Boutaj H, Meddich A, Wahbi S, Moukhli A, El Alaoui-Talibi Z, Douira A, et al. Improvement of growth and development of olive tree by mycorrhizal autochthonous inoculum. Research Journal of Biotechnology. 2020;**15**:76-84

[87] Gutjahr C, Paszkowski U. Multiple control levels of root system remodeling in arbuscular mycorrhizal symbiosis. Frontiers in Plant Science. 2013;**4**:204. DOI: 10.3389/fpls.2013.00204

[88] Schouteden N, De Waele D, Panis B, Vos CM. Arbuscular mycorrhizal fungi for the biocontrol of plant-parasitic nematodes: A review of the mechanisms involved. Frontiers in Microbiology. 2015;**6**:1280 ISSN: 1664-302X

[89] Kim SJ, Eo J, Lee E, Park H, Eom A. Effects of arbuscular mycorrhizal Fungi and soil conditions and crop plant growth. Mycobiology. 2017;**45**(1):20-24

[90] Oyetunji OJ, Osonubi O. The roles of improved cropping systems and an arbuscular mycorrhizal fungus on yam productivity in degraded soil. Crop Research. 2008;**35**:245-254

[91] Olowe OM, Olawuyi OJ, Sobowale AA, Odebode AC. Role of arbuscular mycorrhizal fungi as biocontrol agents against *Fusarium verticillioides* causing ear rot of *Zea mays* L. (Maize). Current Plant Biology. 2018;**15**:30-37

[92] Kumari B, Mallick MA, Solanki MK, Solanki AC, Hora A, Guo W. Plant Growth Promoting Rhizobacteria (PGPR): Modern Prospects for Sustainable Agriculture. In: Ansari R, Mahmood I, editors. Plant Health Under Biotic Stress. Singapore: Springer; 2019. DOI: 10.1007/978-981-13-6040-4_6

[93] Bidellaoui B, Segarra G, Hakkou A, Trillas MI. Beneficial effects of *Rhizophagus irregularis* and *Trichoderma asperellum* strain T34 on growth and fusarium wilt in tomato plants. Journal of Plant Pathology. 2019;**101**:121-127. DOI: 10.1007/s42161-018-0159-y

[94] Wang Y, Bao X, Li S. Effects of arbuscular mycorrhizal fungi on rice growth under different flooding and shading regimes. Frontiers in Microbiology. 2020;**12**:752-756

[95] Meddich A, Ait El Mokhtar M, Bourzik W, et al. Optimizing Growth and Tolerance of Date Palm (Phoenix dactylifera L.) to Drought, Salinity, and Vascular Fusarium-Induced Wilt (*Fusarium oxysporum*) by Application of Arbuscular Mycorrhizal Fungi (AMF). In: Giri B, Prasad R, Varma A, editors. Root Biology. Soil Biology 52. Cham: Springer; 2018

[96] Ait Rahou Y, Ait El Mokhtar M, Anli M, Boutasknit A, Ben-Laouane R, Douira A, et al. Use of mycorrhizal fungi and compost for improving the growth and yield of tomato and its resistance to *Verticillium dahlia*. Archives of Phytopathology and Plant Protection. 2020;**54**(13-14):665-690

[97] Gamalero E, Pivato B, Bona E, Copetta A, Avidano L, Lingua G, et al. Interactions between a fluorescent pseudomonad, anarbuscular mycorrhizal fungus and a hypovirulent isolate of *Rhizoctonia solani* affect plant growth and root architecture of tomato plants. Plant Biosystems - An International Journal Dealing with all Aspects of Plant Biology. 2010;**144**:582-591. DOI: 10.1080/11263504.2010.489315

[98] Veresoglou SD, Rillig CM. Suppression of fungal and nematodes plant pathogens through arbuscular mycorrhizal fungi. Biology Letters. 2012;**8**(2):214-217

[99] Alban R, Guerrero R, Toro M. Interactions between a root-knot nematode (*Meloidogyne exigua*) and arbuscular mycorrhizae in coffee plant development (*Coffea arabica*). American Journal of Plant Sciences. 2013;**4**:19-23. DOI: 10.4236/ajps.2013.47A2003

[100] Koffi MC, Vos C, Draye X, Declerck S. Effects of *Rhizophagus irregularis* MUCL41833 on the reproduction of *Radopholus similis*

in banana plantlets grown under *in vitro* culture conditions. Mycorrhiza. 2013;**23**:279-288. DOI: 10.1007/s00572-012-0467-6

[101] Rodrigues E, Silva MT, Calandrelli A, Miamoto A, Rinaldi LK, Moreno BP. Pre-inoculation with arbuscular mycorrhizal fungi affects essential oil quality and the reproduction of root lesion nematode in *Cymbopogon citratus*. Mycorrhiza. 2021;**31**:613-623

[102] De Sá CSB, Campos MAS. Arbuscular mycorrhizal fungi decrease *Meloidogyne enterolobii* infection of Guava seedlings. Journal of Helminthology. 2020;**2020, 94**:e183

[103] Salvioli A, Bonfante P. Systems biology and "omics" tools: A cooperation for next-generation mycorrhizal studies. Plant Science. 2013;**203-204**:107-114. DOI: 10.1016/j.plantsci.2013.01.001

[104] Jawson MD, Franzluebbers AJ, Deborah K, Galusha DK, Aiken RM. Soil fumigation within monoculture and rotations: Response of corn and mycorrhizae. Agronomy Journal. 1993;**85**(6):1174-1180

[105] Knudsen KA, Soler AP, Johnson KR, Wheelock MJ. Interaction of alpha-actinin with the cadherin/catenin cell-cell adhesion complex via alpha-catenin. Journal of Cell Biology. 1995;**130**:67-77

[106] Woomer PL, Swift MJ. The Biological Management of Tropical Soil Fertility. Chichester, UK: Wiley/UK: TSBF and Sayce; 1994

[107] Swift MJ. Toward the second paradigm: Integrated biological management of soil. Paper presented for the FERTBIO Conference, Brazil 1998

Chapter 9

Arbuscular Mycorrhizal Fungi (AMF) in Optimizing Nutrient Bioavailability and Reducing Agrochemicals for Maintaining Sustainable Agroecosystems

Addisu Ebbisa

Abstract

Arbuscular mycorrhizal fungi (AMF) are one of the essential components of the soil microbiome playing a crucial role in nutrients cycling and mediation of plant responses to different environmental stresses. They also play pivotal role in controlling soil erosion, enhancing phytoremediation, and eliminating other harmful microorganisms and then sustaining agroecosystem. Several studies have investigated the positive effects of mycorrhizal symbiosis as biofertilizers those are capable of reducing use of chemical fertilizer by 25–90% particularly NPK depending on crop species, soil type, and management practices, while increasing productivity in the range of 16–78%. Similarly, AMF can also act as bio-controllers and decrease the application rate and frequency of pesticides. This is directly translated to the primary role of AMF in the sustaining agriculture services. Thus, understanding the interaction between AMF-soil, and plant plays a vital role in benefitting societies and agro-industries. In this regard, this review attempted to explore how can AMF symbiosis reduce agro-chemicals and maintain sustainable human welfare. It also addresses impact of agrochemicals on crop production and the main factor influencing the success of AMF symbioses. Generally, if this is applied wisely it keeps the food safe, the soil healthy, the water clean, the climate stable, and the ecosystem flourishing.

Keywords: AMF symbioses, argo-chemical, biofertilizer, bio-controllers, agro-ecosystem

1. Introduction

Rapidly growing human population and modern agricultural system resulted in increased demand for agro-chemicals such as pesticides, fertilizers, preservatives, and disinfectant [1, 2] in which their excessive and indiscriminate use severely affects biodiversity, air, water, and soil healthy [1, 3–5]. They might also exert deleterious effects on human health and exacerbate subsequent socioeconomic effects on communities' livelihoods by disturbing the ecological balance [2, 3, 5–7]. Furthermore, climate change, an ever-increasing human population, depletion of global rock phosphorus, and growing energy prices make current fertilizer production unsustainable and represent sizeable challenges to global food security [8]. These disastrous consequences promote new strategies that can reduce and/or substitute agrochemicals in sustainable way without jeopardizing human health and ecosystem services [2, 9].

Considering such alarming situations, beneficial microbial inoculants are proposed as a "clean and ecofriendly" option in agriculture sectors for their potential role in food safety and sustainable crop production [3, 10, 11]. They act as biofertilizers, bioherbicide, biopesticides, and biocontrol agents, which minimize the negative impact of chemical input and consequently increase the quantity and quality of produced food [2]. Among soil microbe, arbuscular mycorrhizal fungi (AMF) symbiosis is one of the most promising that partially or fully supplement agrochemicals and reduce their consecutive negative impacts [12, 13].

AMF promote plant growth by bringing morpho-physiological and biochemical changes in host plants by serving as "biofertilizers and bio-protectors" in sustainable way [14, 15] and providing water and mineral nutrients to the plant [16, 17]. This services can occur through the direct pathway (by roots) and by AMF pathway [13, 18–20]. Furthermore, it boosts the health of the subsequent crop by improving soil aggregation, providing nutrients, enhancing abiotic stress tolerance, protection against pathogens [9, 12], and altering the accumulation of contaminants in plants and is essential for the sustainable management of agricultural ecosystems and biodiversity [1, 16, 21–23]. It is the key for production of pesticide-free food crops and ensuring that high yields correspond [23, 24].

In general, it is more effective for improvement of crop yield, growth, and development in areas with low levels of agrochemical inputs (e.g., Africa and South America) [25–27]. Similarly, low soil fertility optimizes the expression of the multiple beneficial effects of AMF in agro-ecosystem and reduces nutrient seepage to the environment [10, 28]. This relationship is the best scenario and alternate technology for both farmers and society to increase the utilization efficiency of scarce nonrenewable fertilizers such as rock phosphate [18, 29], and use of agrochemicals is catastrophically hampered ecosystem [6, 30]. This is also a strategy to enhance the sustainability of agricultural systems through promoting internal regulatory ecosystem processes while reducing chemical fertilizer use without the concomitant loss of crop yield [26, 31]. However, the application of AMF has not been fully adopted by farmers so far [26]. Therefore, to optimize AMF effects on nutrient bioavailability and ecosystem service to achieving future food security in more sustainable agricultural systems, understanding the linkage of AMF with soil and plant nutrient dynamic plays a vital role for benefitting societies and agro-industries. In this regard, this chapter attempted to summarize published results on contribution of AMF in reducing agrochemical use in agriculture for sustainable maintenance of human welfare and ecological service. It also addresses main factor influencing the success of AMF symbioses and inoculation.

2. Characteristics of arbuscular mycorrhizal fungi (AMF)

Mycorrhizal (fungus-root) fungi are specialized members of the vast population of microorganisms, which are morphologically and physiologically diverse in nature that colonize the rhizosphere [16]. Among mycorrhizal symbiosis, AMF symbiosis is one of the most ancient and widespread than other types of mycorrhizal associations [16, 32]. It is belong to the Phylum *Glomeromycota*, *Glomus* spp., which are reproduced asexually and associate with approximately 80% of terrestrial plant species [33], including forest trees, wild grasses, and many crops [16, 34].

AMF is the name given to the endosymbiotic association of a plant root and a fungus from the *Glomeromycota*, which form two unique structures: a highly branched structure formed inside root cells called *arbuscules* that is considered to be the key element of the symbiotic nutrient exchanges and a balloon-like structure called *vesicles* for storage of nutrients within the plant root cortical cells of the host (**Figure 1**) [35]. AMF are classified into three (*Archaeosporomycetes*, *Glomeromycetes*, and *Paraglomeromycetes*) and the five orders: Archaeosporales (e.g., *Geosiphon pyriformes*, *Archaeospora trappei*), Diversisporales (e.g., *Scutellospora calospora*, *Acaulospora laevis*, *Entrophospora infrequens*), *Gigasporales* (e.g., *Gigaspora margarita*, *G. rosea*), Glomerales (e.g., *Glomus intraradices*, *G. mosseae*, *G. geosporum*), and Paraglomerales (e.g., *Paraglomus occultum*, *P. laccatum*) [17]. They are obligate biotrophs endophytes [34] that completely depend on a host plant for carbon and energy [32] due to lack of the ability to absorb carbohydrates from other source. This carbon cost is offset by the positive effects of AMF on plant growth and soil quality [10]. There are two types of root colonization in AMF symbiosis [36] (1) *Arum*-type, in which the symbiont spreads intercellularly between cortical root cells, forming terminal arbuscules on intracellular hyphal branches (2) while in the *Paris*-type, the fungus grows directly from cell to cell within the cortex and forms intracellular hyphal coils and intercalary arbuscules along the coils [17, 37].

In AM roots the fungus penetrates intercellularly and intracellularly into the root cortex, whereas in ectomycorrhizal (ECM) roots the fungus only penetrates intercellularly into the root cortex (**Figure 1**) [17]. An ECM root is characterized by the presence of three structural components: a sheath or mantle, *Hartig net* (complex intercellular network system), and extraradical mycelium [16]. The mycelium can take nitrate and ammonium from the soil and transfer to the plants [3]. However, AMF mainly uptake ammonium from the soil and transfer to the plants [38, 39]. The host plant root exudates of Strigolactone bring recursive spore germination and

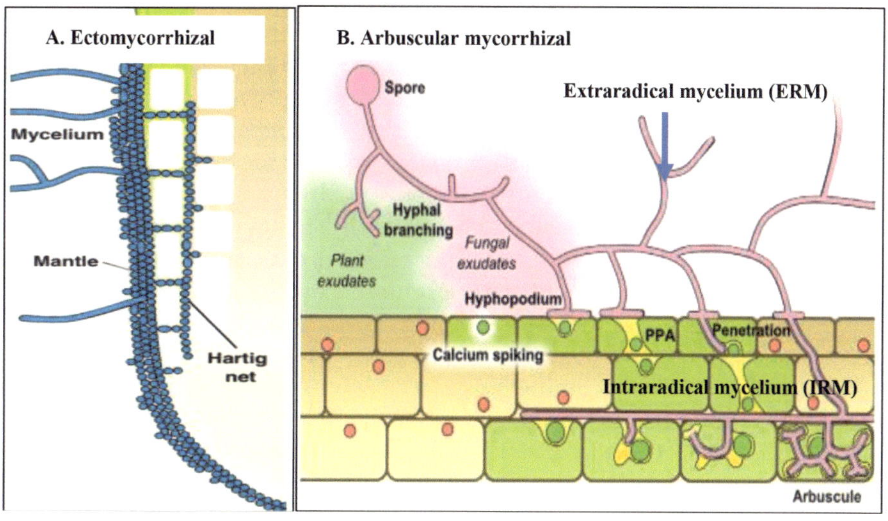

Figure 1.
Process of root colonization in ectomycorrhizal (A) and arbuscular mycorrhizal (B) (adopted from [33]).

hyphal branching, while *Myc factors* secreted by mycorrhiza perceived by host roots to trigger the signal transduction pathway or common symbiosis (SYM) pathway lead to adhesion of a hyphopodium to the root surface via Ca^{2+} spiking. Fungal hyphae emerging from this hyphopodium penetrate into the root through the prepenetration apparatus, which guides the fungal hyphae through the root cells toward the cortex. Eventually, a highly branched arbuscule occupies most of the cell volume, forming an extensive surface for nutrient exchange (**Figure 1**) [33].

AMF produce a high number of spores that grow faster even under different stress conditions [40] and produce thick-walled hyphae that penetrate the host root and extend from the roots out into the soil where they interface with soil particles [9]. Then they create a network of extraradical mycelium structure, which increases the fungal absorbing surface and facilitates the translocation of mineral nutrients from soil to host plants (**Figure 1**) [10, 16]. Despite its coenocytic nature, the mycelium that is formed within the root, the intraradical mycelium (IRM) differs morphologically and functionally from the ERM, the mycelium that grows into the soil. The ERM absorbs nutrients from the soil and transfers these nutrients to the host root. The IRM on the other hand releases nutrients into the interfacial apoplast and exchanges them against carbon from the host [17, 18]. Such a fungal structure represents one of the critical elements of the AMF symbiosis and provides increased surface area for nutrient uptake and bridges nutrient depletion zones [10, 37]. These extraradical hyphae acquire phosphate and initiate the colonization of other species [34, 41].

3. Contribution of AMF to plant productivity and agroecosystems

Sustainability of agricultural ecosystems can be restored by stimulating soil life and internally regulated ecosystem processes [31]. The real significance of mycorrhizal fungi is that they connect the primary producers of ecosystems and enable the flow of energy-rich compounds required for nutrient mobilization [32]. Mycorrhizal fungi are one of the commonly occurring living organisms in soil providing many ecosystem services, including N-fixation, soil carbon cycling, plant nutrition, soil erosion control by soil binding capacity [13], soil pollutants remediation, biodiversity, plant water economy [42], and enhanced C-sequestration [22]. AMF and rhizobia can act synergistically and stimulate plant productivity by supplying different limiting nutrients to the plant (e.g., N by rhizobia and P by AM fungi) [43].

Enhanced nutrient uptake and stress resistance are some of the mechanisms by which AM fungi can enhance plant productivity. AMF symbiosis is probably more favorable in conservative and sustainable agriculture to having the potentiality of major beneficial functions such as: (1) increased productivity in the range of 16–78% by gaining more N, P, and other less mobile nutrients increased [24, 43]; (2) increased water uptake and water holding capacity that initiate drought tolerance; (3) increased tolerance to other abiotic stresses such as soil salinity, heavy metal toxicity, etc.; (4) overcoming biotic stresses and offering bio-protection against pathogen; (5) improved soil quality; (6) enhanced plant vigor and yield, thus leading to the production of safe and high-quality foods, able to promote human health (**Figure 2**) [16, 44]. This is determined by fungal the strain, climate, soil type, and cultural practices (fertilization level) in which the soil type is the

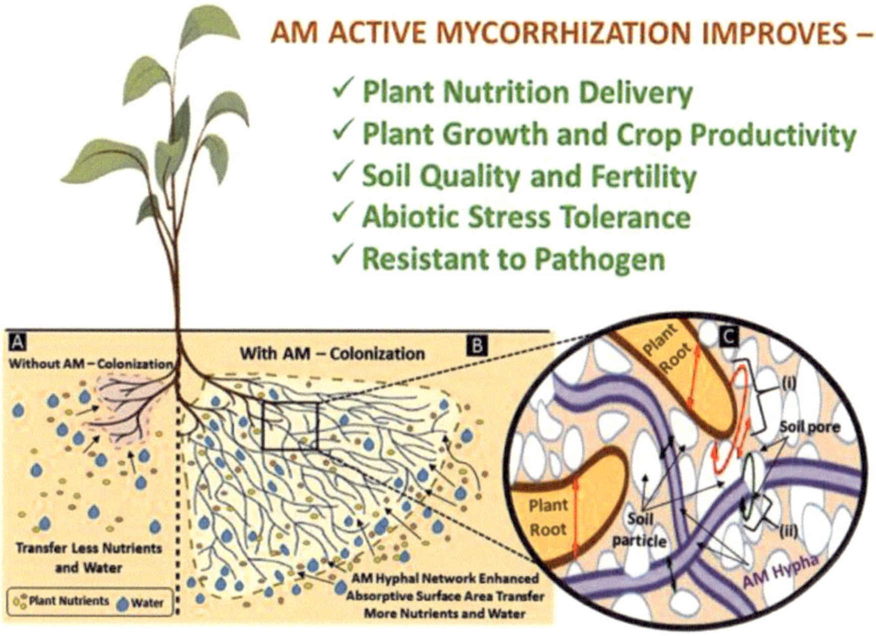

Figure 2.
Schematic representation of brief function of arbuscular mycorrhizal fungi (AMF) (source: [44]).

fundamental criterion to determine effective AMF symbiosis strains or species [24]. It is also to improve soil fertility, as they produce glomalin upon accumulation in soil that aids soil in soil stabilization [42, 45]. Furthermore, AMF prevent leaching losses, phosphorus (60%), and ammonium (7.5%) from grassland microcosms during periods of heavy rainfall that cause top environmental threats to ecosystems worldwide [46].

AMF fungi receive 100% of their carbon from the plant, and this increase in carbon flow to the roots, estimated at up to 20% of the plant's photosynthate, translates to a huge amount of carbon worldwide that plays a significant role in carbon cycling between the atmosphere and biosphere [34]. Bender et al. [47] have demonstrated that AMF contribute to reducing emissions of N_2O by increasing N immobilization into microbial or plant biomass, which results in the reduction of soluble N in the soil and, consequently, in a limitation of denitrification. Thus, AMF could have an indirect influence on potent greenhouse gas (GHG) emissions through change of the physical conditions of soil that influence the production and transport of GHG in soil [26].

The AMF symbiosis can reduce nutrient loss from ecosystems in three main ways: (1) by improving crop nutrient extraction capacity [43] allowing the production of good yield at lower levels of soil fertility; (2) by increasing soil aggregation via physical particle enmeshment and cementing with "sticky" exudates, which results in better soil nutrient storage and retention [48]; and (3) by promoting growth of host crops, thus increasing the size of this desirable nutrient sink [46]. On average, plants inoculated with *G. mosseae* yielded 13% more biomass relative to non-mycorrhizal plants [49]. Therefore, AMF are vital endosymbionts playing an effective role in plant productivity and the functioning of the ecosystem service.

4. Impact of agrochemicals on crop production and ecosystem service

Excess use of agrochemicals caused tremendous a reduction in soil macro-fauna diversity and promotes the accumulation of toxic compounds in soils that severely harm the environment [6, 30]. Pesticides significantly reduced the diseases and increased the grain yield, yet more resistant pesticides to degradation by abiotic (physical, chemical, and other factors) and biotic (living organisms in the soil food web) agencies, leach into the lower strata of the soil, then absorbed by plant roots, and accumulate in the food chain and are ultimately biomagnified in the food web that is hazardous to human health. They may also affect non-target crop and potentially non-target endangered species by transporting from the sprayed area to non-target areas [6].

Several synthetic fertilizers contain acid radicals, such as hydrochloride and sulfuric radicals, and hence increase the soil acidity and adversely affect soil and plant health [2]. Highly persistent and toxic agrochemicals are available in water, fish, vegetables, and human fluids, which are in turns hazardous to human and ecosystem [30]. These chemical inputs gain access into human body systems through three major means: (i) oral ingestion, (ii) infiltration through the skin, and (iii) breathing causing chronic disease [30] from respiratory disorders and musculoskeletal illnesses due to lack of knowledge of the caution code for hazardous agrochemicals.

5. The possible mechanism of AMF for improving nutrients bioavailability and reducing chemical fertilizer

5.1 How can AMF symbiosis reduce chemical fertilizer?

5.1.1 Increasing the surface area for nutrient absorption of plant roots

The uptake of mineral nutrients from the soil by plants is greatly aided by mutualistic associations with mycorrhizal fungi, which have two pathways for nutrients uptake such as direct pathway that involves the uptake of nutrients via high- or low-affinity uptake transporters in the plant root hairs and the mycorrhizal pathway that involves rapid uptake and translocation of nutrients by fungal transporters in the ERM along the intraradical mycelium (IRM) to the plant root cortex [18, 34, 50, 51]. AMF colonization enhances plant growth performance, root morphology, and leaf nutrient levels that could greatly increase the root-soil interface area [52, 53]. AMF extraradical hyphae can reach a soil volume beyond the nutrient and water depletion zone of roots and may extend up to 8 cm from the root surface, and 1 g of soil contains up to 200 m fungal hyphae [10, 15]. This length is a common parameter used to quantifying fungal hyphae, which greatly increases the surface area for the uptake of immobile nutrients particularly P, N, and Cu [15]. Extraradical hyphae explore soil volumes for nutrient extraction [27] by physically and enzymatically expanding the rhizosphere. Furthermore, fungal hyphae are much thinner than roots and are therefore able to penetrate smaller pores and extract water under dry conditions [54]. It is also improving the efficiency of nutrient cycling and reducing nutrient losses from the soil-plant system [46] that ensures adequate nutrient availability in infertile or less fertile soil.

In additionally water, fungi are able to extract and assimilate soil P from non-plant-available forms such as DNA or P bound to mineral particles [8]. The AMF hyphal network is also able to uptake K and other important micronutrients such as

Mg, Zn, Cu, Ca, S, Na, Mn, B, Mo, Fe, Al, and Si, essential for plant growth [16] and led to a mobilization of starch reserves in the apex of grapevine in winter, which was possibly responsible for enhancing root development [44, 55]. Mycorrhizal infection enhances plant growth by increasing nutrient uptake and significant delivery system for P (80%), Cu (60%), N (25%), Zn (25%), and K (10%) via increases in the absorbing surface area, by mobilizing sparingly available nutrient sources, or by excretion of chelating compounds or ectoenzyme [56].

5.1.2 Induces the expression of plant nutrient transporters

It is well known that AMF symbiosis specifically induces the expression of plant Pi transporters (PT4) (**Figure 3**) [3, 33] reducing the risk of wasteful P loss, thus preserving the quality of water and aquatic ecosystems [46]. In addition to the Pi transporters, mycorrhiza-inducible ammonium transporters (AMT) that not only deliver nutrients to the root cells but also trigger signaling that enables conditions for arbuscule maintenance [26]. Furthermore, the reviews [26] describe that the role AMF symbiosis improves the sulfur, K, and Zn nutritional status of the host plant, affecting the expression of plant transporters. AMF also produce other hydrolytic enzymes such as pectinase, cellulase, hemicellulases, xylanase, and chitinase, which may participate in key steps to mineralize organic chemicals [1].

5.1.3 Reduce biotic and abiotic stress

Mycorrhizal fungi have also a lot of "non-nutritional" effects on plant physiology often alleviating plant stress caused by biotic and abiotic factors such as promoting osmotic adjustment under drought and salinity stresses [16, 57]. AMF symbiosis can also stimulate the synthesis of plant secondary metabolites, which are important for increased plant tolerance to abiotic and biotic stresses. A plant that is associated with

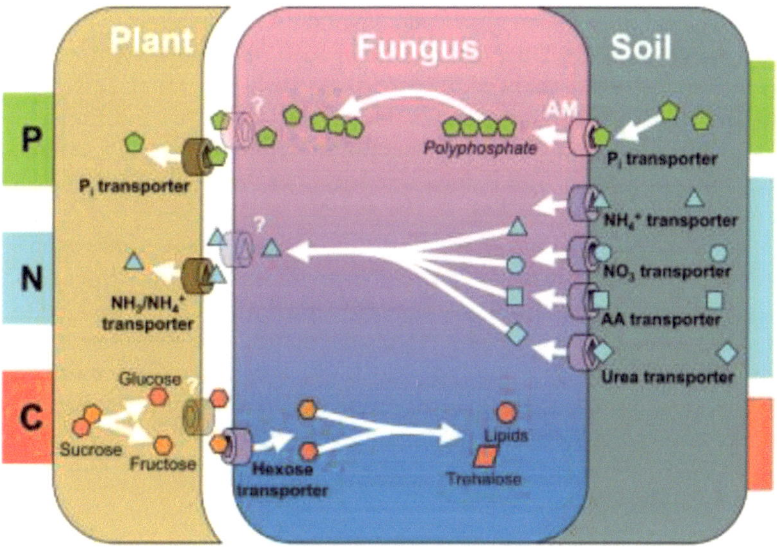

Figure 3.
Scheme summarizing the main nutrient exchange processes in AMF symbiosis (adapted from [33]).

diverse community of AMF may have a sort of "insurance" against fluctuating condition [12] through enhancing the activities of antioxidant enzymes (such as superoxide dismutase (SOD), catalase (CAT), peroxidase (POD), and glutathione reductase (GR) [58]) and reduction in H_2O_2 production, then protecting plant cells from deleterious impact of reactive oxygen species (ROS). In plant tissues, excessive ROS induce oxidative damage of lipids, proteins, and nucleic acids that influence normal structure and function. AMF alleviate the oxidative stress from contaminants via mediating the plant antioxidant system, allowing plants to grow better under contamination stress [1] and contribute to better growth under normal as well as stress conditions [59]. This healthy and better growth plant increases nutrient capture by AMF crops [60].

Several studies suggest that enhanced tolerance of AMF plants to water deficit may involve modulation of drought-induced genes by enhancing osmotic adjustment, gas exchange, and water use efficiency of the host plants [57, 61]. This may play a role in the increased tolerance of AMF plants to stress while increasing plant biomass and uptake efficiency of immobile nutrients such as P, Zn, and Cu and decreasing metal toxicity to plants. Therefore, application of AMF in agriculture is used to reclamation waste places to productive agriculture field, which is a promising alternative to conventional fertilization practices, with a view to sustainable agriculture [26].

5.1.4 Improve synergistic interaction of beneficial soil microorganism

AMF colonization can induce quantitative and qualitative changes in root exudates and subsequently modify microbial community structure of the rhizosphere [58]. Particularly, extraradical mycelium fungi provide a direct pathway for translocation of photosynthetically derived carbon to microsites in the soil and a large surface area for interaction with other microorganisms [32]. This enhances synergistic interaction of soil microorganism such as nitrogen fixation P solubilization and the production of phytohormones, siderophores, and antibiotics [37] and then reduces N and P chemical fertilizer application by 10–25% [3, 62], resulting in better root nodulation, nutrient uptake, and plant yield [9]. Rhizobia and AMF fungi often interact synergistically, and their exoenzymes play pivotal roles in accessing, mobilizing, and transferring nutrients [27]. Dual inoculation of such fungi with a Rhizobium and other bacterium on plant enhanced the growth and other beneficial effects, namely resistance to disease and tolerance to adverse soil and climatic conditions [63]. Similarly, co-inoculation of AMF with rhizobium stimulated the N fixation efficiency while improving N transfer and P uptake resulting in the yield advantages of legume/non-legume intercropping [39, 64].

Use of selected native bacterial consortiums reduces the use of nitrogen fertilizer by up to 25%, increasing the productivity of rice cultivation [65]. Other researchers reported that only 25–50% of the recommended N, P, and K rates were required by inoculated crops compared with non-inoculated plants. While the rate of 75% N, P, and K was required by other crops (potato, tomato, pepper, and plantain). Thus, the use of AMF inoculation could lead to reduction of agrochemicals application and reduces their potentially negative impact agro-ecosystem.

5.1.5 Improve soil rhizosphere and plant absorption efficacy

AMF fungal mycelia can facilitate the formation of water-stable soil aggregates via biological, biophysical, and biochemical-based mechanisms such as entanglement and enmeshment of soil particles by AMF extraradical mycelia, production of mucilages, glycoprotein, glomalin (the soil organic matter pool), polysaccharides, and other extracellular compounds [48]. The AMF extraradical hyphal length significantly decreased total soil loss by increasing soil cohesion [66]. This improves plant absorption efficacy for water and nutrients of the low mobility in soil [10]. They also alter the rates of above- and below-ground litter decomposition due to chemical changes in the roots and interactions with the decomposer fungi [15]. Furthermore, increased soil stability can reduce soil erosion, loss of nutrients and organic matter leading to increased aeration and water-holding capacity. This improved soil structure is not only influencing the behavior of organic contaminants in soil, but also helps to maintain high microbial activity, accelerating the biodegradation process, and ultimately enhancing crop safety.

5.2 AMF symbiosis for uptake and translocation of phosphate

Even though phosphorus is a major essential nutrient for plants growth and development, its excess application causes eutrophication of water [67]. It is also bound to Ca and Mg in alkaline soils and readily complexed to charged Al and Fe oxides in acidic soils [68], resulting in a decline of directly absorbed Pi (inorganic phosphate) by the plant root. The arbuscular mycorrhizal fungi (AMF) increase its exploitation of organic P or orthophosphate ions through the organic ions and siderophore production [32] and the changes in sorption balance of soil solution caused by microbial biomass turnover in the rhizosphere [69] besides increasing the absorbing surface area. AMF-induced enhancement in phosphatase activity could possibly mediate the release of organically bound phosphorous, hence increasing transport and uptake of phosphorous in AMF inoculated plants [59]. The mycorrhizal plants displayed a larger phosphorus inflow than the non-mycorrhizal controls [51]. Likewise, AMF fungi possess high-affinity inorganic phosphate (Pi) transporters, which are localized to the extraradical hyphae of G. *versiforme*, the site of phosphate uptake from the soil (**Figure 3**) [33]. Similarly, Smith & Smith [19] summarize the possible mechanisms for increased uptake of P by arbuscular mycorrhizal plants: (1) extensive physical exploration of soil by fungal hyphae, thus decreasing the distance that P ions must diffuse before uptake; (2) faster P uptake from solution; (3) production of exudates by AM plants and change of rhizosphere pH. Variation may also occur among AMF fungi in their ability to utilize various inorganic or organic P sources, or in their ability to proliferate in nutrient-rich patches of soil [21]. In AMF-colonized rice, about 70% of the overall Pi acquired is delivered by the symbiotic fungi [60]. In some cases, AMF may be responsible for 100% of host nutrients and have a significant impact on plant growth, health, and subsequently on plant biodiversity and ecosystem productivity [34]. It is evident that such reductions in phosphate application have important economic and environmental benefit [22].

5.3 Transport and assimilation of nitrogen in AMF symbiosis

Plants take up more NH_4^+ via the AM fungal from soils in very low concentrations than NO_3^- in agricultural soils due to its more efficient energetic and relatively immobile in the soil solution [19, 38, 39, 50], which is highly influenced by agricultural management decision [70]. This may be due to A high-affinity ammonium (NH_4^+) transporter (AMT2;2) is also localized in the AMF [17]. There are two possible reasons why N delivery by the fungus was negligible for nitrate-N; (1) the hyphal ability to acquire nitrate was very low due to poor development of the extraradical hyphae in comparison with the ammonium supply related to high mobility of nitrate; and (2) the assimilation and deliver rate of nitrate was low [38]. After uptake of N from the soil it is assimilated into Arginine (Arg) in ERM and translocated to IRM, then Arg is broken down to NH_4^+ via the catabolic arm of the urea cycle and reassimilated by plant through mycorrhiza-inducible transporters (**Figure 4**) [50, 71, 72]. The experiments of [71] show that fungi directly take up inorganic nitrogen and incorporate into amino acids and then transfer N from extraradical mycelium to the host plant without carbon.

Even though contribution of the AMF symbiosis to crop N nutrition is often thought to be small, as inorganic N(NO_3^+ or NH_4^+) in soil is much more mobile, and unlike ecto- and ericoid mycorrhizas organic N is not available for AMF due to lack of saprotrophic ability [18, 60]; several studies confirmed that AMF are able to take up organic N sources from the soil mainly in the form of amino acids using *RiPTR2* transporter [50]. AMF also play vital role in accelerating the decomposition of organic material and nutrient cycling in soils through providing decomposers with plant-derived carbon inputs and transfer decomposition products to the plant [50]. Thus, AMF can play an indirect stimulatory role in the uptake of N from organic sources [36] by influencing organic matter decomposition through bacterial activity [49].

In sustainable system, mineral N levels should be kept low as NO_3^+ can be reduced to N_2O, a potent greenhouse gas, or lost from the soil-plant system through leaching [46] and negatively impacting surface and ground water quality. Therefore, presence of AMF fungi in the plant-soil system can enhance mineralization of N from organic residues in soil, and the N released can be better used by plants tapping AMF networks located in the vicinity of mineralizing residues [46]. In AM root organ cultures, more than 21% of the total N in the roots was taken up by the ERM [73]. Similarly, Tanaka and Yano [38] demonstrated that maize leaves obtain up to 75% of the N taken up by AMF due to high expression levels of *GintAMT1* in the ERM, which primarily

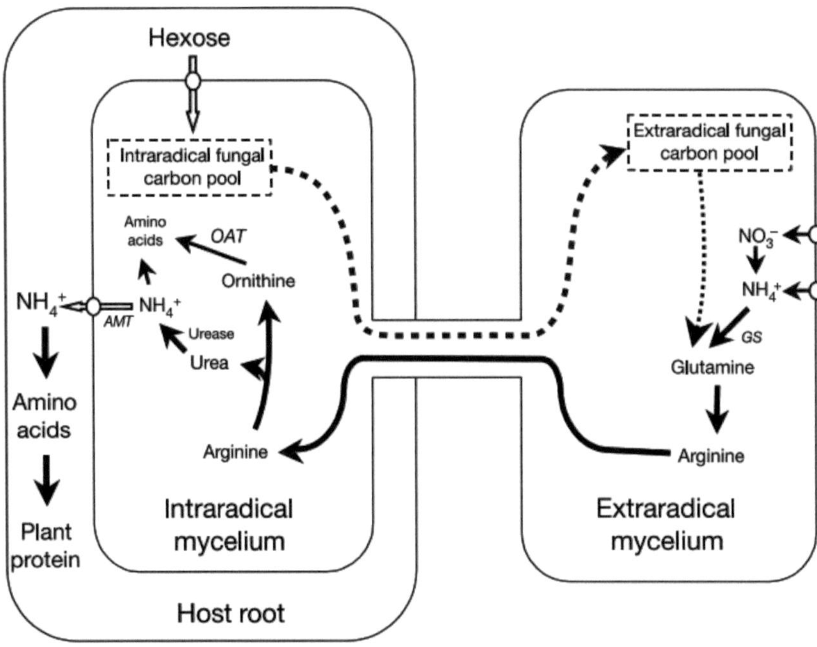

Figure 4.
Model of nitrogen movement in the arbuscular mycorrhizal symbiosis (source: [71]).

involved in the NH_4^+ acquisition of fungal hyphae from the soil. They conclude that the mycorrhizal fungus can rapidly deliver ammonium-N to the plants but lacks the capacity to transfer nitrate to the plant.

5.4 Metal ions transfer in AMF symbiosis

Besides providing resistance to disease and drought, AMF supply a range of limiting nutrients including copper, iron, and zinc to the plant in exchange for carbon [43]. Even though the roles of arbuscular mycorrhizal in the uptake of K, Ca, Mg, and S are poorly understood, numerous studies also reported that AMF increase uptake of K, Ca, and Mg [74]. AMF fungi can increase host plant uptake of K and modulate plant responses to long-term K limitation, by preventing ROS production and upregulating specific genes, including an ortholog of the plasma membrane K^+/H^+ exchanger *AtCHX20* (cation/H^+ exchanger) in *Medicago truncatula* [51].

Arbuscular mycorrhizal fungi (AMF) enhance the nutritional state of their host plant through acquiring and delivering a proportion of resource to their hosts and play potential role in sustainable crop production [75]. The utilization of microbial products has several advantages over conventional chemicals for agricultural purposes [2]: (i) microbial products are considered safer than many of the chemicals now in use; (ii) neither toxic substances nor microbes themselves will be accumulated in the food chain; (iii) self-replication of microbes circumvents the need for repeated application; (iv) target organisms seldom develop resistance as is the case when chemical agents are used to eliminate the pests harmful to plant growth; and (v) properly developed biocontrol agents are not considered harmful to ecological processes or the environment. Rini et al. [76] concluded that AMF application reduced 50% of compound fertilizer needed for oil palm seedlings. Hence, AMF is considered to be a good replacement for inorganic or chemical fertilizer in the future.

5.5 How AMF symbiosis reduces use of pesticides and other chemicals?

5.5.1 Bioremediation

A wide range of toxic pollutants including heavy metals are disposed of daily to the soil and water, which are the most are the most serious disaster affecting humans, plants, and the environment negatively [77, 78]. The conventional remediation methods cause high cost, intensive labor, irreversible changes in soil properties, and disturbance of native soil microflora. Thus, researchers obtain a novel approach called phytoremediation, which is an economically feasible and sustainable option to clean up heavy-metal-contaminated sites. It refers to a green technology that uses plants and associated soil microbes to reclaim HM and radionuclides from the environment by various detoxification mechanisms including phytoextraction, rhizofltration, phytostabilization, phytodesalination, photodegradation, and phytovolatilization; [77–79], which is one of the low-cost remediation techniques used by microorganisms to remove heavy metals [42].

AMF alleviate heavy metal toxicity due to the ability of fungal hyphae to bind heavy metals outside and inside the roots and restrict their uptake to upper parts [51, 59]. This binding ability is due to AMF hyphae being capable of secreting a glycoprotein called glomalin, which can bind heavy metals and subsequently remove heavy metals absorbed by the plants from contaminated soil. Glomalin can develop the properties and structure of the soil, which helps to enhance soil fertility by linked with soil carbon [42]. This binding of heavy metal to the cell walls of the fungal hyphae in roots and not release to the shoot as the presence of free amino acids, hydroxyl, carboxyl, and other groups are containing the cell wall of fungi representing as act as a filtration barrier against the transfer of heavy metals to plant shoots. Furthermore, metal dissolution by fungi may take place through proton-promoted or ligand-promoted mechanisms and organic acids provide both a source of protons for solubilization and metal-chelating anions to complex the metal cations [32]. The work of [40] suggested that mycorrhizal symbiosis becomes a promising and suitable as phytostabilizers of Cd stressed soil. Similarly, three arbuscular mycorrhizal fungi (AMF) from *Glomus*, *Acaulospora*, and *Scutellospora*, and four plant-growth-promoting rhizobacteria isolates related to genera *Streptomyces*, *Azotobacter*, *Pseudomonas*, and *Paenibacillus* were found to be effective in phytoremediation of Fe_3C contaminated soil [80]. In general, metal-tolerant fungi grew and solubilized toxic metal minerals better than non-tolerant isolates. Mycorrhizal colonization protects roots against metal such as Cd, Zn, Cu, Pb, Mn, and Ni, toxicity [40, 59, 81]. AMF that promote biodegradation will decrease contaminant residues in both crops and the environment.

5.5.2 As biocontrol agents (biopesticides, bioherbicides)

To prevent environmental pollution and occupational diseases, proactive preventative actions are needed [5]. AMF are particularly important in organic and/or sustainable farming systems that rely on biological processes rather than agrochemicals to control plant diseases. AMF can act as bio-controllers that allow host plants to grow healthier and decrease the application amount and frequency of pesticides and other environmentally harmful agrochemical products [1]. These processes protect plants against soil-borne pathogens and assist in contaminant removal [7] and reduce pesticide application via enhanced crop resistance to pathogens and improved competition with weeds [1]. The use of AMF as safe and sustainable biostimulators and bio-protectants is very promising alternative, as it does not harm the environment, resulting in sustainable food production [82].

The complex interactions between plant, pathogen, and AMF symbiosis can enhance host plant resistance or tolerance to root pathogenesis [83]. The major mechanisms related to bio-protection role of AMF are indirect mechanisms such as enhanced mineral nutrition status of plants (enhance resistance and tolerance), changes in root architecture (e.g., increase in lateral branches and cell wall lignification), competition for infection sites (the pathogen never penetrated arbuscule-containing cells), change in rhizodeposition (alter the chemotaxis to the roots by the pathogens), and activation of plant defense responses [83]. The mechanisms of biocontrol exercised by most microbial inoculants could be attributed to release of extracellular hydrolytic enzymes, competition for nutrients and secondary metabolites toxic to plant pathogens at very low concentrations [84].

AMF plants may be more appealing to herbivores, reduce disease symptoms for some disease, and more attractive to pollinators due to their altered architecture and improved nutrient status [12]. An interesting issue in mycorrhizal plant-parasite interaction is a recently reported potential of mycorrhiza to provide partly protection and control of Striga [85, 86]. They observed as AMF negatively impacted on Striga seed germination, reduced the number of Striga seedlings attaching and emerging, and delayed the emergence time of Striga both in pot and field experiments. Exudates from mycorrhizal sorghum plants resulted in much lower germination of Striga seeds than exudates of non-mycorrhizal sorghum plants. AMF also have an indirect beneficial effect by enhancing the fitness of the cereal host and thereby allowed the host to withstand Striga damage better. Therefore, AMF might be able to substitute for reduced fertilizer and biocide inputs in organic systems, which is not permitted in organic systems [74], while increasing productivity in the range of 16–78% by gaining more N, P, and other less mobile nutrients increased [24].

5.6 Factors affecting AMF-soil-environmental symbiosis

Mycorrhizal symbioses in agro-ecosystems are affected by various factors including species compatibility with the target environment, the degree of spatial competition with other soil organisms in the target niche, and the timing of inoculation [26, 87]. AMF symbiosis and growth are also influenced by the host plant genotype (for instance, legumes more mycotrophic than grasses [39]), AMF species, soil fertility, and percentage of root infection [13, 37, 56]. A very challenging for large-scale production of AMF is its obligate symbionts behaviors [26, 12] and having high degree of genetic and lack of a uninucleate cell stage in the life cycle of AMF [33]. Similarly, equally or even better performance of indigenous AMF [10] than commercial or culture collection isolates is another headache for producers [26], which may cause potential environmental risks without providing higher plant benefits [27]. This promotes encouraging farmers to autonomously produce their AMF to make technology affordable and sustainable.

Application of agrochemicals also reduces network complexity and the abundance of AMF particularly, the nutritional status of the plant and surrounding environment such as P [28, 51]. Because high levels of P repressed the expression of genes encoding carotenoid and strigolactone biosynthesis enzymes in plants, suggesting that high levels of P directly inhibit spore germination by reducing strigolactone biosynthesis then reducing the Pi flux from AM fungi, which in turn reduces the amount of carbohydrates transferred to AM fungi [51]. This reduces abundance of AMF in soil (e.g., by fertilization) could reduce the importance of AM fungi for ecosystem functioning, including their ability to reduce nutrient leaching losses [46].

Furthermore, land uses, crop rotations, and soil features affect the AMF diversity and their community functioning due to changes in soil physical and chemical characteristics [88]. Less-intensive tillage is a viable strategy for enhancing root colonization by indigenous AMF across soil types and crop species [27]. Reduced tillage, manures, and cover crops are recognized as a practical way to maintain high population of functional effective AMF eventually to support sustainable crop production [89]. This is true if you did not use herbicide such as glyphosate that hinders and detrimental for subsequent AMF recovery. This specificity suggests that crop species in rotations may influence the quality of the AMF population in the soil of a following crop [10]. Once AMF biodiversity is restored and well-established with AMF-friendly management before and after cultivation mycorrhizal hyphal network will remain unaltered and infective in the future [26].

Generally, for increasing effectiveness and success AMF inoculation reducing tillage rate, bare fallows, agrochemicals usage, and use of non-mycorrhizal crops in rotation vital for enhancing abundance and contribution of AMF fungi as ecological services in agriculture [22]. The review of [60] summarize the management of the AMF symbiosis is achievable through a variety of agronomic practices, in particular: (1) tillage, (2) crop nutrition, (3) grazing, and (4) integrated pest management, as well as by (5) the selection of crop genotypes and crop rotation sequences, (6) the use of AMF inoculants, and (7) the use of biotechnologies that enhance the AMF symbiosis of crop plants. Thus, understanding best management practice of AMF in crop production would insure good extraradical mycelium development leading to soil quality improvement and reduced activity of soil-borne pathogens [10].

6. Concluding remarks

Harnessing natural resources including beneficiary microorganisms is one of the most effective approaches to improving farm productivity and food quality in a sustainable way. Microbial inoculant technology will ensure healthy food security for the future population. AMF efficiently use for agricultural productivity in sustainable manner in which the diversity and function of soil microorganism is the decisive issue in the agrarian activities and ecosystem service [23]. AMF might be able to substitute chemical fertilizer and biocide inputs in organic systems [74], while increasing productivity in the range of 16–78% by gaining more nitrogen (N), phosphorus (P), and other less mobile nutrients increased [24]. This role of AMF can reduce the catastrophic effect indiscriminate use of agrochemicals in agricultural sectors and promotes. Hence, reintroducing AMF into agro-systems may also improve nutrient use efficiency, water-use efficiency, and tolerance to pathogens and herbivores [12]. Finlay [32] suggests more research on AMF as its relevant species have not yet been investigated since there has been a general concentration on agricultural systems in which additions of inorganic fertilizers, pesticides, and plant breeding may have selected against arbuscular mycorrhizal fungi with the capacity to mobilize organic substrates.

Mycorrhizal symbioses in agro-ecosystems are affected by various factors including all biotic, abiotic, and agronomic management factors related to complex interaction of soil-AMF and host plant. Thus, selection of new crop varieties giving yields on poor soils and in low fertilization conditions and understanding compatible management practice in crop production that would ensure good extraradical mycelium development leading to improvement of soil quality and crop productivity in holistic manner is paramount importance for successful exploitation of arbuscular mycorrhizal fungi.

Acknowledgements

The authors are highly thankful to researchers whose findings are included directly or indirectly in preparing this manuscript.

Funding

The authors received no direct funding for this research.

Conflicts of interest

The authors declare no conflict of interest.

Data availability

All data generated are included in this article reference's part.

Abbreviations

AMT	ammonium transporters
AMF	arbuscular mycorrhizal fungi
SYM	common symbiosis
ECM	ectomycorrhizal
ERM	extraradical mycelium
GHG	greenhouse gas
IRM	intraradical mycelium
PT4	Pi transporters
ROS	reactive oxygen species

Author details

Addisu Ebbisa
Faculty of Agriculture, Department of Plant Science (Agronomy), Haramaya University, Dire Dewa, Ethiopia

*Address all correspondence to: saebbisa@gmial.com; aebbisa@gmail.com

References

[1] Wang F, Adams CA, Yang W, Sun Y, Wang F. Technology benefits of arbuscular mycorrhizal fungi in reducing organic contaminant residues in crops: Implications for cleaner agricultural production cleaner agricultural production. Critical Reviews in Environmental Science and Technology. 2019;**0**:1-33. DOI: 10.1080/10643389.2019.1665945

[2] Alori ET, Babalola OO. Microbial inoculants for improving crop quality and human health in Africa. Frontiers in Microbiology. 2018;**9**:1-12. DOI: 10.3389/fmicb.2018.02213

[3] Bhardwaj D, Ansari MW, Sahoo RK, Tuteja N. Biofertilizers function as key player in sustainable agriculture by improving soil fertility, plant tolerance and crop productivity. Microbial Cell Factories. 2014;**13**(66):1-10

[4] Teklewold H, Kassie M, Shiferaw B, Köhlin G. Cropping system diversi fi cation, conservation tillage and modern seed adoption in Ethiopia: Impacts on household income, agrochemical use and demand for labor. Ecological Economics. 2013;**93**:85-93. DOI: 10.1016/j.ecolecon.2013.05.002

[5] Wimalawansa SA, Wimalawansa SJ. Agrochemical-related environmental pollution: Effects on human health. Global Jourmal of Biology, Agriculture & Health Science. 2014;**3**(3):72-83

[6] Rajbhandari BP, Bassi A, Ramyil C, Regmi R, Rijal J. Related papers an overview of agrochemicals and their effects on environment in Nepal. Applied Ecology and Environmental Sciences. 2014;**2**(2):66-73. DOI: 10.12691/aees-2-2-5

[7] Rathore KS, Singh S, Sharma AK. Arbuscular mycorrhizal fungi: Next generation bioagents for sustainable agriculture. In: et al, Saxena AK, editors. Everyman's Science. 6th ed. Vol. LIII(2). Kolkata: INDIAN; 2018. pp. 85-89

[8] Thirkell TJ, Charters MD, Elliott AJ, Sait SM, Katie J. Are mycorrhizal fungi our sustainable saviours? Considerations for achieving food security. Journal of Ecology. 2017;**105**:921-929. DOI: 10.1111/1365-2745.12788

[9] Javaid A. Role of arbuscular mycorrhizal fungi in nitrogen fixation in legumes. In: Khan MS, Musarrat J, Zaidi A, editors. Germay: Springer Wien New York; 2010. pp. 409-426

[10] Hamel C, Strullu D. Arbuscular mycorrhizal fungi in field crop production: Potential and new direction. Canadian Journal of Plant Science. 2006;**86**:941-950

[11] Dobo B. Effect of arbuscular mycorrhizal fungi (AMF) and rhizobium inoculation on growth and yield of *Glycine max* L . varieties. International Journal of Agronomy. 2022;**2022**:1-10

[12] Hart MM, Trevors JT. Microbe management: Application of mycorrhyzal fungi in sustainable agriculture. Frontiers in Ecology and the Environment. 2005;**3**(10):533-539

[13] Piotrowski JS, Rillig MC. Succession of arbuscular mycorrhizal fungi: Patterns, causes, and considerations for organic agriculture. Advances in Agronomy. 2008;**97**(07):111-130. DOI: 10.1016/S0065-2113(07)00003-X

[14] Qiu B, Wang Y. Phylogenetic distribution and evolution of mycorrhizas in land plants. Mycorrhiza. 2006;**16**:299-363. DOI: 10.1007/s00572-005-0033-6

[15] Soka G, Ritchie M. Arbuscular mycorrhizal symbiosis and ecosystem processes: Prospects for future research in tropical soils. Open Journal of Ecology. 2014;**4**(1):11-22

[16] Smith SE, Read D. Mycorrhizal Symbiosis. 3rd ed. Elsevier; 2008

[17] Bücking H, Liepold E, Ambilwade P. "The role of the mycorrhizal symbiosis in nutrient uptake of plants and the regulatory mechanisms underlying these transport processes," in Plant Science, Dhal NK Sahu.S.C., London: Intechopen; 2012. pp. 107-138

[18] Smith FA, Smith SE. Roles of arbuscular mycorrhizas in plant nutrition and growth: New paradigms from cellular to ecosystem scales. Annual Review of Plant Biology. 2011;**62**:227-250. DOI: 10.1146/annurev-arplant-042110-103846

[19] Smith FA, Smith SE. What is the significance of the arbuscular mycorrhizal colonisation of many economically important crop plants? Plant and Soil. 2011;**348**:63-79. DOI: 10.1007/s11104-011-0865-0

[20] Lurthy T, Pivato B, Lemanceau P, Mazurier S. Importance of the rhizosphere microbiota in iron biofortification of plants. Frontiers in Plant Science. 2021;**12**(744445):1-24. DOI: 10.3389/fpls.2021.744445

[21] Koide RT. Commentary functional complementarity in the arbuscular mycorrhizal symbiosis. The New Phytologist. 2000;**147**:233-235

[22] Gianinazzi S, Gollotte A, Binet M-N, van Tuinen D, Redecker D, Wipf D. Agroecology: The key role of arbuscular mycorrhizas in ecosystem services. Mycorrhiza. 2010;**20**:519-530. DOI: 10.1007/s00572-010-0333-3

[23] Mulugeta TE. The role of arbuscular mycorrhizal fungi on agricultural crop productivity and ecosystem service: A review. International Journal of Agroforestry and Silviculture. 2021;**9**(2):1-9

[24] Rivera R, Fernández F, Fernández K, Ruiz L, Sánchez C, Riera M. Advances in the management of effective arbuscular mycorrhizal symbiosis in tropical ecosystems. In: Chantal H, Christian P, Editors. Mycorrhizae in Crop Production. Madison: Taylor & Francis; 2007. pp. 151-195. DOI: 10.1300/5425

[25] Igiehon NO, Babalola OO. Biofertilizers and sustainable agriculture: Exploring arbuscular mycorrhizal fungi. Applied Microbiology and Biotechnology. 2017;**101**:4871-4881. DOI: 10.1007/s00253-017-8344-z

[26] Berruti A, Lumini E, Balestrini R, Bianciotto V. Arbuscular mycorrhizal fungi as natural biofertilizers: Let's benefit from past successes. Frontiers in Microbiology. 2016;**6**:1-13. DOI: 10.3389/fmicb.2015.01559

[27] Schaefer DA, Gui H, Mortimer PE, Xu J. Arbuscular mycorrhiza and sustainable agriculture. Circular Agricultural Systems. 2021;**1**(6):1-7

[28] Franz S, Der Heijden V, Marcel GA, Franz S, Der Heijden V, Marcel GA. Establishment success and crop growth effects of an arbuscular mycorrhizal fungus inoculated into Swiss corn fields. Agriculture, Ecosystems and Environment. 2019;**273**:13-24

[29] Hinsinger P, Betencourt E, Bernard L, Brauman A, Plassard C, Shen J. P for two, sharing a scarce resource: Soil phosphorus acquisition in the rhizosphere of intercropped species. Plant Physiology. 2011;**156**:1078-1086. DOI: 10.1104/pp.111.175331

[30] Fianko JR, Donkor A, Lowor ST, Yeboah PO. Agrochemicals and the Ghanaian Environment, a Review. Journal of Environmental Protection. 2011;**2**:221-230. DOI: 10.4236/jep.2011.23026

[31] Bender SF, Wagg C, Van Der Heijden MGA. An underground revolution: Biodiversity and soil ecological engineering for agricultural sustainability. Trends in Ecology & Evolution. 2016;**xx**(yy):1-13. DOI: 10.1016/j.tree.2016.02.016

[32] Finlay RD. Ecological aspects of mycorrhizal symbiosis: With special emphasis on the functional diversity of interactions involving the extraradical mycelium. Journal of Experimental Botany. 2008;**59**(5):1115-1126. DOI: 10.1093/jxb/ern059

[33] Bonfante P, Genre A. Mechanisms underlying beneficial plant—Fungus interactions in mycorrhizal symbiosis. Nature Communications. 2010;**1**(48):1-11. DOI: 10.1038/ncomms1046

[34] Harrison MJ. Signaling in the arbuscular mycorrhizal symbiosis. Annual Review of Microbiology. 2005;**59**:19-42. DOI: 10.1146/annurev.micro.58.030603.123749

[35] Pepe A, Giovannetti M, Sbrana C. Different levels of hyphal self-incompatibility modulate interconnectedness of mycorrhizal networks in three arbuscular mycorrhizal fungi within the Glomeraceae. Mycorrhiza. 2016;**26**:325-332. DOI: 10.1007/s00572-015-0671-2

[36] Hodge A, Helgason T, Fitter AH. Nutritional ecology of arbuscular mycorrhizal fungi. Fungal Ecology. 2010;**3**(4):267-273. DOI: 10.1016/j.funeco.2010.02.002

[37] Giovannini L et al. Arbuscular mycorrhizal fungi and associated microbiota as plant biostimulants: Research strategies for the selection of the best performing Inocula. Agronomy. 2020;**10**(106):1-14

[38] Tanaka Y, Yano K. Nitrogen delivery to maize via mycorrhizal hyphae depends on the form of N supplied. Plant, Cell and Environment. 2005;**28**:1247-1254

[39] Ingraffia R, Amato G, Frenda AS, Giambalvo D. Impacts of arbuscular mycorrhizal fungi on nutrient uptake, N2 fixation, N transfer, and growth in a wheat/faba bean intercropping system. PLoS One. 2019;**14**(3):1-16

[40] Abdelhameed RE, Metwally RA. Alleviation of cadmium stress by arbuscular mycorrhizal symbiosis. International Journal of Phytoremediation. 2019;**0**:1-9. DOI: 10.1080/15226514.2018.1556584

[41] Turrini A, Avio L, Giovannetti M, Agnolucci M. Functional complementarity of arbuscular mycorrhizal fungi and associated microbiota: The challenge of translational research. Frontiers in Plant Science. 2018;**9**(1407):10-13. DOI: 10.3389/fpls.2018.01407

[42] Herath B, Kwa M, Jpd L, Pn Y. Arbuscular mycorrhizal fungi as a potential tool for bioremediation of heavy metals in contaminated soil. WJARR. 2021;**10**(03):217-228

[43] van der Heijden MGA, Bardgett RD, van Straalen NM. The unseen majority: Soil microbes as drivers of plant diversity and productivity in terrestrial ecosystems. Ecology Letters. 2008;**11**:296-310. DOI: 10.1111/j.1461-0248.2007.01139.x

[44] Kuila D, Ghosh S. Current research in microbial sciences aspects, problems and utilization of arbuscular mycorrhizal (AM) application as bio-fertilizer in sustainable agriculture. Current Research in Microbial Sciences. 2022;**26**:325-332. DOI: 10.1016/j.crmicr.2022.100107

[45] Leifheit EF et al. Multiple factors influence the role of arbuscular mycorrhizal fungi in soil aggregation—A meta-analysis. Plant and Soil. 2014;**374**(1/2):523-537

[46] van der Heijden MGA. Mycorrhizal fungi reduce nutrient loss from model grassland ecosystems. Ecology. 2010;**91**(4):1163-1171. DOI: 10.1890/09-0336.1

[47] Bender SF et al. Symbiotic relationships between soil fungi and plants reduce N2O emissions from soil. The ISME Journal. 2014;**8**:1336-1345. DOI: 10.1038/ismej.2013.224

[48] Rillig MC, Mummey DL. Mycorrhizas and soil structure. The New Phytologist. 2006;**171**:41-53

[49] Saia S, Benítez E, Settanni L, Amato G. The effect of arbuscular mycorrhizal fungi on total plant nitrogen uptake and nitrogen recovery from soil organic material. The Journal of Agricultural Science. 2013;**1**:1-9. DOI: 10.1017/S002185961300004X

[50] Bücking H, Kafle A. Role of arbuscular mycorrhizal fungi in the nitrogen uptake of plants: Current knowledge and research gaps. Agronomy. 2015;**5**:587-612. DOI: 10.3390/agronomy5040587

[51] Wang W, Shi J, Xie Q, Jiang Y, Yu N, Wang E. Nutrient exchange and regulation in arbuscular mycorrhizal symbiosis. Molecular Plant. 2017;**492**:1-31. DOI: 10.1016/j.molp.2017.07.012

[52] Shao Y et al. Mycorrhiza-induced changes in root growth and nutrient absorption of tea plants. Plant, Soil and Environment. 2018;**64**(6):283-289

[53] Fan L, Dalpé Y, Fang C, Dubé C, Khanizadeh S. Influence of arbuscular mycorrhizae on biomass and root morphology of selected strawberry cultivars under salt stress. Botany. 2011;**403**:397-403. DOI: 10.1139/B11-028

[54] Allen MF. Linking water and nutrients through the vadose zone: A fungal interface between the soil and plant systems. Journal of Arid Land. 2011;**3**(3):155-163. DOI: 10.3724/SP.J.1227.2011.00155

[55] Nicolás E et al. Effectiveness and persistence of arbuscular mycorrhizal fungi on the physiology, nutrient uptake and yield of crimson seedless grapevine. The Journal of Agricultural Science. 2014;**2014**:1-13. DOI: 10.1017/S002185961400080X

[56] Marschner H, Dell B. Nutrient uptake in mycorrhizal symbiosis. Plant and Soil. 1994;**159**(1):89-102

[57] Khan MS, Zaidi A, Musarrat J. Microbial Strategies for Crop Improvement. Dordrecht Heidelberg London New York: Springer-Verlag Berlin Heidelberg; 2009

[58] Lenoir I, Fontaine J, Sahraoui AL. Phytochemistry arbuscular mycorrhizal fungal responses to abiotic stresses: A review. Phytochemistry. 2016;**xxx**:1-12. DOI: 10.1016/j.phytochem.2016.01.002

[59] Allah EFA, Abeer H, Alqarawi AA, Hend AA. Alleviation of adverse impact

of cadmium stress in sunflower (*Helianthus annuus* L.) by arbuscular mycorrhizal fungi. Pakistan Journal of Botany. 2015;**47**(2):785-795

[60] Yang C et al. Management of the arbuscular mycorrhizal symbiosis in sustainable crop production. In: Solaiman ZM et al., editors. Mycorrhizal Fungi: Use in Sustainable Agriculture and Land Restoration, No. 41. Berlin Heidelberg: Springer-Verlag Berlin Heidelberg; 2014. pp. 89-122

[61] Ahmad P, Rasool S. Emerging Technologies and Management of Crop Stress Tolerance. Vol. 2. UK: Elsevier Inc.; 2014

[62] Adesemoye AO, Kloepper JW. Plant-microbes interactions in enhanced fertilizer-use efficiency. Applied Microbiology and Biotechnology. 2009;**85**:1-12. DOI: 10.1007/s00253-009-2196-0

[63] Sadhana B. Arbuscular mycorrhizal fungi (AMF) as a biofertilizer—A review. International Journal of Current Microbiology and Applied Sciences. 2014;**3**(4):384-400

[64] Meng L, Zhang A, Wang F, Han X, Wang D, Li S. Arbuscular mycorrhizal fungi and rhizobium facilitate nitrogen uptake and transfer in soybean/maize intercropping system. Frontiers in Plant Science. 2015;**6**:1-10. DOI: 10.3389/fpls.2015.00339

[65] Ríos-ruiz WF, Torres-chávez EE, Torres-delgado J, Rojas-garcía JC, Bedmar EJ, Valdez-nuñez RA. Rhizosphere Inoculation of bacterial consortium increases rice yield (*Oryza sativa* L.) reducing applications of nitrogen fertilizer in San Martin region, Peru. Rhizosphere. 2020;**14**:100200. DOI: 10.1016/j.rhisph.2020.100200

[66] Mardhiah U, Caruso T, Gurnell A, Rillig MC. Arbuscular mycorrhizal fungal hyphae reduce soil erosion by surface water flow in a greenhouse experiment. Applied Soil Ecology. 2016;**99**:137-140. DOI: 10.1016/j.apsoil.2015.11.027

[67] Conley DJ et al. Controlling eutrophication: Nitrogen and phosphorus. Ecology. 2009;**323**: 1014-1015

[68] Hinsinger P. Bioavailability of soil inorganic P in the rhizosphere as affected by root-induced chemical changes: A review. Plant and Soil. 2001;**237**:173-195

[69] Raimi A, Adeleke R, Roopnarain A. Soil fertility challenges and Biofertiliser as a viable alternative for increasing smallholder farmer crop productivity in sub-Saharan Africa. Cogent Food & Agriculture. 2017;**9**(1):1-26. DOI: 10.1080/23311932.2017.1400933

[70] Thirkell T, Cameron D, Hodge A. Contrasting nitrogen fertilisation rates alter mycorrhizal contribution to barley nutrition in a field trial. Frontiers in Plant Science. 2019;**10**(1312):1-9. DOI: 10.3389/fpls.2019.01312

[71] Govindarajulu M et al. Nitrogen transfer in the arbuscular mycorrhizal symbiosis. Nature. 2005;**435**:819-823. DOI: 10.1038/nature03610

[72] Tian C, Kasiborski B, Koul R, Lammers PJ, Bucking H, Shachar-Hill Y. Regulation of the nitrogen transfer pathway in the arbuscular mycorrhizal symbiosis: Gene characterization and the coordination of expression. Plant Physiology. 2010;**153**:1175-1187. DOI: 10.1104/pp.110.156430

[73] Jin H, Pfeffer PE, Douds DD, Piotrowski E, Lammers PJ,

Shachar-Hill Y. The uptake, metabolism, transport and transfer of nitrogen in an arbuscular mycorrhizal symbiosis. The New Phytologist. 2005;**168**:687-696

[74] Gosling P, Hodge A, Goodlass G, Bending GD. Arbuscular mycorrhizal fungi and organic farming. Agriculture, Ecosystems and Environment. 2006;**113**:17-35. DOI: 10.1016/j.agee.2005.09.009

[75] Göhre V, Paszkowski U. Contribution of the arbuscular mycorrhizal symbiosis to heavy met phytoremediation. Planta. 2006;**223**(6):1115-1122. DOI: 10.1007/s00425-006-0225-0

[76] Rini MV, Yansyah MP, Arif MA. The application of arbuscular mycorrhizal fungi reduced the required dose of compound fertilizer for oil palm (*Elaeis Guineensis* Jacq.) in nursery. IOP Conference Series: Earth and Environmental Science. 2022;**1012**:1-7. DOI: 10.1088/1755-1315/1012/1/012011

[77] Mohebbi AH, Harutyunyan SS, Chorom M. Phytoremediation potential of three plant grown in monoculture and intercropping with date palm in contaminated soil. International Journal of Agriculture and Crop Sciences. 2012;**4**(20):1523-1530

[78] Ali H, Khan E, Sajad MA. Chemosphere phytoremediation of heavy metals—Concepts and applications. Chemosphere. 2013;**91**:869-881

[79] Liu X, Li X, Ong SMC, Chu Z. Progress of phytoremediation: Focus on new plant and molecular mechanism. Journal of Plant Biology & Soil Health. 2013;**1**(1):1-5

[80] Mishraa V et al. Synergistic effects of arbuscular mycorrhizal fungi and plant growth promoting rhizobacteria in bioremediation o ... Synergistic effects of arbuscular mycorrhizal. International Journal of Phytoremediation. 2016;**18**(7):697-703. DOI: 10.1080/15226514.2015.1131231

[81] Kumar A, Verma JP. Does plant-microbe interaction confer stress tolerance in plants: A review? Microbiological Research. 2018;**207**: 41-52. DOI: 10.1016/j.micres.2017.11.004

[82] Albuquerque M. Bioprotection by arbuscular mycorrhizal fungi in plants infected with Meloidogyne nematodes: A sustainable alternative. Crop Protection. 2020;**135**:105203. DOI: 10.1016/j.cropro.2020.105203

[83] Harrier LA, Watson CA. The potential role of arbuscular mycorrhizal (AM) fungi in the bioprotection of plants against soil-borne pathogens in organic and/or other sustainable farming systems †. Pest Management Science. 2004;**60**:149-157. DOI: 10.1002/ps.820

[84] Rani A, Singh R, Kumar P, Shukla G. Pros and cons of fungicides: An overview. IJESRT. 2017;**6**(1):112-117

[85] Lendzemo VW. The Tripartite Interaction between Sorghum, Striga Hermonthica, and Arbuscular Mycorrhizal Fungi. Wageningen; 2004

[86] Lendzemo VW, Van Ast A, Kuyper TW. Can arbuscular mycorrhizal fungi contribute to Striga management on cereals in Africa? Outlook on Agriculture. 2006;**35**(4):307-311

[87] Verbruggen E, Van Der Heijden MGA, Rillig MC, Kiers ET. Mycorrhizal fungal establishment in agricultural soils: Factors determining inoculation success. The New Phytologist. 2013;**197**:1104-1109

[88] Ontivero RE et al. Impact of land use history on the arbuscular

mycorrhizal fungal diversity in arid soils of Argentinean farming fields. FEMS Microbiology Letters. 2020;**367**:1-11

[89] Gupta MM, Chourasiya D, Sharma MP. Diversity of arbuscular mycorrhizal fungi in relation to sustainable plant production systems. In: Satyanarayana T, et al, editors. Microbial Diversity in Ecosystem Sustainability and Biotechnological Applications. Madhya Pradesh, India: Springer Nature Singapore Pte Ltd; 2019. pp. 167-187

Chapter 10

Ectomycorrhizal Fungi as Biofertilizers in Forestry

Muhammad Hanif, Zubaria Ashraf, Samar Bashir, Fatima Riaz, Rizwan Amanat, Nousheen Yousaf and Samina Sarwar

Abstract

Ectomycorrhiza (ECM) is an association of fungi with the roots of higher plants in which both the species are equally benefited and appears to be important for the survival of both parties, and no doubt this association offered extensive benefits in the restoration of forest and ecosystem soil aggregation and stabilization. The most important and vital role of ECM fungi, which is analyzed globally, is that they are best and environment friendly biofertilizers. ECM fungi considered as a biotechnological tool in forest management because their role in reforestation, bioremediation, control of soil pathogen, and restoration of ecosystem is reviewed extensively. ECM fungi not only increase the biomass of edible fruiting bodies but also improve soil structure, nutrient cycle, and also produce phytohormones, which increase the growth and survival of seedlings and enhance the photosynthetic rate of plants and also maintain their tolerance level against environmental stresses in forest nursery. Ectomycorrhizas also reduce fertilization expenses in an environment friendly manner. The importance of ECM fungi and mycorrhizal helper bacteria for the growth enhancement of the economically important trees and significantly important role in restoration of sites degraded by forestry operation.

Keywords: biofertilizer, phytopathogens, agriculture, biome, mycorrhization

1. Introduction

Fungi are the most valued and understudied group of microorganisms and are widely distributed in different ecological habitats that inhabit and the consequential requirement to contend against an assorted group of other microorganisms, that is, bacteria, fungi, and viruses also have a strong defense mechanism for survival as many scientists have studied their role industrially, ecologically, and biotechnologically; now, fungal association with trees, which termed as mycorrhiza, is becoming the center of focus for its role in forestry and ecosystem as a biofertilizer. Fungi as biofertilizers are considered the basic core of the ecosystem, and their impact will be felt more in the future; hence, the importance of fungi as biofertilizers cannot be denied. The term mycorrhiza originated from two Greek words "mykes and rhiza," which mean fungus and roots, respectively, was first coined in 1885 by Albert Bernhard

Frank who was a forest pathologist [1]. Frank also explained that mycorrhiza is an association of ectomycorrhiza (ECM) and roots of higher plant, and this relationship is symbiotic in nature; he further subdivided it into two subcategories, that is, ectomycorrhizal fungi (do not penetrate cortical cells) and endomycorrhizal fungi (penetrate cortical cells). A lot of research has been conducted on symbiotic relationships of fungi and plants by many scientists who showed that approximately 86% of territorial plants get their nutrients through mycorrhizal roots [2]. The number of fungal species that take part in ectomycorrhiza formation is about more than 7000, and, predominantly, species from basidiomycetes class contribute toward ectomycorrhiza formation. In ectomycorrhiza formation, a special network named as Hartig's is a bridge for the metabolic altercation of fungi and the roots through which it plays its role in the mobilization, translocation, and transportation of soil nutrients to the roots of plants through mycorrhizal mantle connection extended into the soil [3].

Many pieces of research also supported an assumption that ectomycorrhiza fungi developed polyphyletic assistance from multiple saprophytic species, as their diversity was initially based on the studies of the reproductive part of fungi, and now they are categorized based on their morph anatomical characteristics. With many significant and viable trees, such as poplar, pine, birch, and oak, ectomycorrhizal associations are being observed [4]. Mycorrhizal fungi (MF) can be helpful for the identification of the structure of the plant community. So, the determination of fungal companion, symbiotic relationship and to understand its structure, function and fundamental importance in ecology terms [5].

2. Role of microbes as biofertilizers

In the recent era, the agricultural sector has mostly relied on synthetic and chemical fertilizers for their betterment, but the excessive use of chemicals has caused various environmental issues, such as increase in temperature, destruction of habitat, unavailability of nutrients due to change in soil structure or profile, and environmental pollution causing health hazards. Hence, researchers are working on different agricultural tools that are considered to be effective or environment friendly with less consumption of energy. Thus, natural ecofriendly microbes (algae, fungi, and bacteria) have been recommended as practicable solutions for extensive agricultural applications economically and also support soil structure as well as various forms of agricultural land, support plant growth by enhancing its nutrient absorbing capacity, and reduce chances of soil-borne diseases [6]. Bacteria as biofertilizer ensure the fixation of atmospheric nitrogen and its availability to plants by synthesizing plant-growth-promoting substances and increasing the solubility of phosphorous. Ectomycorrhizal fungi are an important tool for the absorption of different mineral nutrients such as phosphorus. Therefore, all microbes play an important role in the agricultural sector by providing different services, such as disease resistance, drought tolerance, and increasing and maintaining the nutrient quality of soil [7].

The forest biomes offer vital ecosystem amenities, which include providing habitats for organisms and acting as a sink for different nutrients such as carbon, sodium, and potassium, controlling different harsh factors, such as erosion of soil, extenuating climate change, and manufacturing vital assets, for example, wood timber, fuel, and bioproducts. Owing to different human actions, the efficiency of forests has declined significantly over the passage of time. To overcome the shortage of nutrients and phytopathogens, different chemically synthesized products are being used in forest development or for diversity enhancement in different sectors. But, this frequently leads to nutrient losses through various factors such as leaching, gaseous losses, and many other harmful factors. The exploitation of biofertilizers instead of relying on different chemicals may increase the growth and development of plant species that enhance the productivity in a more maintainable way. Biofertilizers are mostly used in horticulture and agriculture sectors with less emphasis on forestry. It is mandatory to explore and exploit numerous mechanisms of action, viz., enabling nutrient uptake, phytoprotection, and modulation of phytohormone of biofertilizers that would have supportive role in promoting the ecosystem services of forest biomes. All these factors (i.e., the mechanisms of action of effective microorganisms in biofertilizers, applications of biofertilizers in the forestry sector, and factors influencing the effectiveness of biofertilizer application) will be considered in the given review.

S. No	Groups	Examples
List of nitrogen-fixing biofertilizers		
1	Free-living	*Azotobacter, Beijerinkia, Clostridium, Klebsiella, Anabaena*, and *Nostoc*
2	Symbiotic	*Frankia, Rhizobium, Anabaena*, and *azollae*
3	Associative symbiotic	*Azospirillum*
List of phosphorus-soluble biofertilizers		
1	Bacteria	Some phosphorous solubilizing bacteria are *Bacillus subtilis, Bacillus megaterium, Pseudomonas striata* and *Bacillus circulans*
2	Fungi	*Penicillium* sp. and *Aspergillus awamori*
List of phosphorus-mobilized biofertilizers		
1	Arbuscular mycorrhiza	*Glomus* sp., *Gigaspora* sp., *Acaulospora* sp., *Scutellospora* sp. and *Sclerocystis* sp.
2	Ectomycorrhiza	*Laccaria* sp., *Amanita, Pisolithus* sp. and *Boletus* sp.
3	Ericoid mycorrhiza	*Pezizella ericae*
4	Orchid mycorrhiza	*Rhizocotina solani*

S. No	Name	Crop suited	Benefits
List of commonly produced biofertilizers			
1	Rhizobium strains	Legumes (i.e., pulses, groundnut, soybean)	It can cause 10–35% increase in yield 50–200 kg N ha-
2	*Azotobacter*	Soil treatment for non-legume crops (including dry land crops)	Results in 10–15% increase in yield with addition of 20–25 kg N. It is also helpful in control of certain diseases.
3	*Azospirillum*	Nonlegumes (like maize, barley, oats, sorghum, millet, Sugarcane, rice, etc.)	It causes about 10–20% increase in yield. It can be applied to legumes as a co-inoculant.
4	Phosphate solubilizes two bacterial and two fungal species in this group	Soil application for all crops	5–30% increase in yield
5	Blue-green algae, Azolla	Rice/wetland	20–30 kg N ha^{-1}, Azolla can give biomass up to 41–52 tones and fix 32–100 kg N.
6	Mycorrhizas	Trees, crops, and few ornamental plants	It causes 30–50% increase in yield and promotes the availability of Zn, P, S, and water.

3. Fungal biofertilizer and its role in agriculture sustainability

Internationally, agricultural production and food manufacturing has to increase twofold in 2050 or the coming years so as to nourish the global growing population while reducing dependence on conventional chemical fertilizers and pesticides. In the past few decades, the increase in extreme chemical fertilizer and pesticide application for crop production has produced a difference in the soil ecosystem. In the present situation of global warming and a disturbed ecosystem, soil microflora regulates how healthy a generation of plants, animals, and humans would be. In light of this, the usage of fungal biofertilizers as a maintainable solution has gained importance over the years [8].

Pesticides, on the other hand, are a double-edged sword. Though targeted on exact targets, they constrain nontarget organisms including the soil mycoflora, which eases the growth of plants. Research shows how agricultural sustainability and economic stability depend on fungal biofertilizers [9].

4. Mycorrhizal fungi and other soil organisms

Mycorrhizal fungi interrelate with a vast variety of other soil microorganisms. This interaction may occur in the rhizosphere or in the loose soil. All these interactions may act as an inhibitor or may play a role as a stimulator; most of the time, they play a role of a symbiotic or a competitor. Effects can be observed evidently at all the phases of the life cycle of MF from the spore population dynamic role of mycorrhizal helper bacteria (MHB) in symbiosis interaction; it is very important for bacteria and grazing animals to maintain and sustain this interaction or symbiotic relationship. Mycorrhizal fungi are also capable to amend the associations of plants with other soil organisms—both pathogens, for example, root-inhabiting nematode, fungi, and particularly nitrogen-fixing bacteria. These connections may be significant in natural environment where microbes play regulatory roles, and in agricultural systems, mycorrhizas may be important in the formation of integrated systems of pest control and growth stimulation [10].

5. Mycorrhizas as nutritional mutualisms

Mycorrhiza is a symbiotic relationship between plants and fungi in which both get to benefit from each other; fungi provide all nutrient availability to plants and in return get food or photosynthetic material. The union of many dissimilar systems of mycorrhizas is a testimony to the shared benefits of these enterprises. Resource exchange in mycorrhizas can be observed, and for the inspection of exchanged resources among both partners, different tools are being used. MF is considered more efficient in exploring soil and uptake of nutrients for plants as compared to plant roots. The large size of soil is inhibited by MF that spreads far away from the nutrient reduction zone that develops around roots [11]. About external hyphae of ECM, it is assessed that it may increase the surface area up to 60-fold on average. Nutrients from soil pore spaces are extracted by the small diameter of fungal hyphae that is too small for plant roots to perform its activity [12].

6. Nitrogen fluxes in the plant and fungus symbiosis

Mycorrhizal fungi can use nitrogen in two forms, that is, organic or inorganic, from soil matrix and make their availability to plants. Different research studies carried out by researchers show that ammonium, nitrates, and urea marked with N^{15}. This form of nitrogen is translocated in the hyphae in the form of arginine, though it is quickly converted and then transformed into ammonium ion immediately, while the carbon skeletons created throughout the excruciating of the arginine are reincorporated again into the fungal groups. Now, it is easy to understand how the external hyphae of the ECM incarcerate inorganic nitrogen as nitrate ion, NH_4^+, and organic nitrogen as amino acids and distribute occasionally a huge fraction of them to the plants.

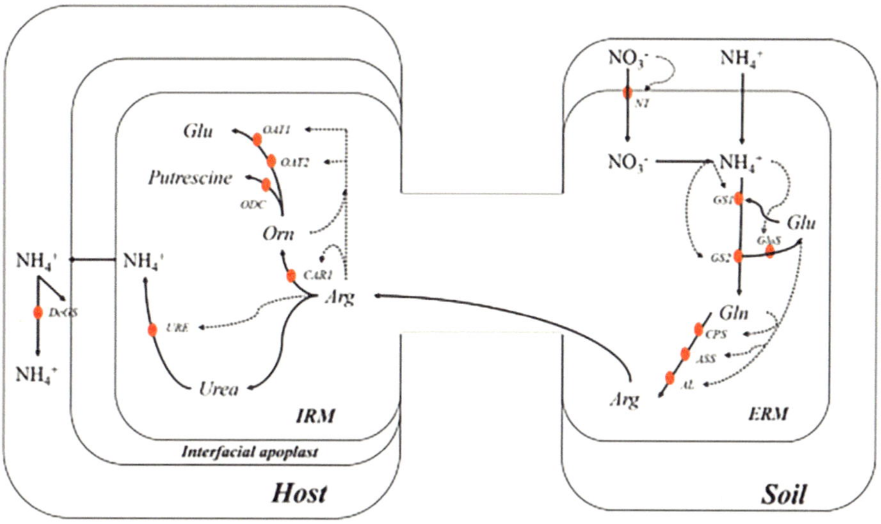

7. Forest biome and associated microorganisms

Microorganisms are almost ubiquitous in both the terrestrial and aquatic habitats, and regardless of their microscopic proportions, they play an important part in global nutrient cycling. All the multicellular organisms have a close relationship with different microbes, and those organisms that are living in natural environment cannot be considered free from germs. Mutually associated microorganisms use an extensive variety of effects on their hosts from helpful to harmful. Forest biomes are among the species-rich terrestrial systems, and this equally applies to microbes living in forests. Forest soil host not only has rich microbe groups, but also animals and plants that provide them a habitat. Recent developments significantly widened our consideration of the taxonomic and functional variety of microbes living in forest biomes, which can collectively be termed "forest micro biota." Still, many phases of the dynamics of microorganism values of their assemblage and their contacts with hosts remain unclear and require more research [13].

8. ECM fungi as a source of phosphorous mobilization in forestry

About 97% of phosphorous present in agricultural soil is not available to plants due to its organic or inorganic forms, so fungi are a major source of phosphorous; they provide phosphorous by producing a large amount of acid phosphate, alkaline, and other forms of acid. Different groups of fungi perform these functions and enhance the growth of plants such as *Gliocladium*, *Penicillium*, *Trichoderma*, and *Chaetomium*. Using 45–70 kg phosphorous enhances the yield of different crops by 15–30% with fungi as a mobilizer. The small concentration of organic acid mandatory to mix phosphorous ranges from 0.5 to 0.6%. The influence of microbes was greater in the fungal inoculated plants as compared to plant input. Fungal extracellular enzymes were more effective than their intracellular complement [14].

9. Role of ECM fungi in vegetative propagation in forest nurseries

Vegetative propagation is a vital tool for the enhancement of productivity of economically important horticultural and agricultural plants. Microorganism also plays a dynamic role in the horticultural sector apart from phytohormones, that is, bacteria, fungi, and most important arbuscular mycorrhizal (AM) fungi, because the symbiotic relationship between fungi and plant controls many factors, enhances the ability to endure harsh condition or harsh environmental factors, and triggers hormone production and adventitious root formation [15]. The initial inoculation of arbuscular mycorrhizal fungi onto the medium of roots improves the rate of vegetative-propagated plant species growth after establishing a mutual association with the plant. Furthermore, a series of successive signaling measures are known to happen between AM fungi and plant during the growth and development of roots [7].

10. Ectomycorrhizal fungi and forest restoration

Temperate and boreal forests are affected by anthropogenic activities and destruction. Ectomycorrhizal fungi play a significant role in the re-establishment of forest through building a symbiotic relationship with roots of higher plants of temperate and boreal forests, providing nutrients to their hosts and surface area for gaseous exchange to plant. ECM fungi are involved in woody plant existence and development and help them to tolerate severe environmental conditions [16]. Much research has been conducted on the restoration of forest ecosystems by using ECM fungi in the site where heavy metals and soil erosion destroy plants drastically. The result has proven that ECM fungi restore sites occupied by plant species that are not native. Moreover, boundaries, knowledge gaps, and possible unwanted results of the use of ectomycorrhizal fungi (EMF) in forest reestablishment and proposed for the further incorporation of this fungal group into forest management. Ectomycorrhizal fungi (ECMF) host connections could progress the chances of success of future reestablishment programs in different forests [10].

11. Drought tolerance and ECM fungi

ECMF increase the capacity of plants to bear damaging effects triggered by the shortage of water. This symbiotic relationship affects different metabolic mechanisms of the host plant and improves the gas exchange mechanism in leaves, photosynthetic rate, and direct assimilation of water from the soil, and allocation to the host plant increases the role of enzymes involved in antioxidant defense, uptake of nitrates, increase application of water through enhanced root hydraulic conductivity osmotic modification, and variations in the flexibility of cell wall. ECMF are useless in severe drought situations [17]. Drought is very common and natural disaster affecting both crop development and livelihood in many ways. It has been assessed that drought damages many plants and crops [18]. Since ECMF have the ability to make the plants which may tolerate water shortage and used to lessen such crop losses. Different research studies have shown that adding ECMF can enhance the growth and production of plants under water-stressed environments. Observations showed that the soil moisture content is higher in ECMF-protected treatments, which shows that ECMF increase soil moisture [19].

12. Disease resistance and ECM fungi

Much research has confirmed that ECMF increase disease resistance in plants against soil-borne plant pathogens. But the efficiency of biocontrol achieved by ECMF is reliant on the species complex, substrate, and host plants. Also, the defense given by ECMF is not effective against all the pathogens controlled by soil and other environmental conditions [20]. This symbiotic relationship protects plants from plant pathogens by using different mechanisms such as changing root development and morphological character creating variations in the tissues of host root and also manufacturing biochemical and physiological variations within the host, varying host nutrition, adapting mycorrhizal sphere that affects microbial populations, challenging for colonization spots and foods, and triggering defense mechanisms and parasitism on nematodes [17].

13. Nitrogen-fixating bacteria and ECM fungi

Recently, research has been conducted to assess the outcomes of biofertilizers, that is, the role of bacteria involved in nitrogen fixation and MF collectively on lenticels growth under rainfed conditions associated with plants with no application of *Azotobacter* bacteria and mycorrhizal fungi. Results showed highly effective improvement in leaf moisture, increased amount of chlorophyll, reduced protein amount, and increased biomass production and seed production of plants compared with the control plant [18]. The valuable consequence of mycorrhizal fungi and *Azotobacter* applications showed positive results on photosynthesis in which plant growth was enhanced due to better water uptake and nutrient supply mainly under the shortage of water. Results provide an innovative indication about the effects of biological fertilizers on the growth of lentils and biomass production under rainfed environments. The focus should be on increasing the production of lentil in arid and semiarid area [21].

14. The mycorrhiza helper bacteria

Mycorrhiza derived from word "*mycorrhization*" was first given by two scientist Duponnois and Garbaye in 1991. They described that these bacteria are helpful in the formation of the plant root and fungus symbiotic relationship. MHB helps to increase the efficiency of mycorrhizal fungi, that is, absorption of different nutrients from soil, and to protect plant roots from pathogens attack and enhance the capacity of plants to uptake a growth factor. This would certainly offer a new measurement of the ecology, evolutionary, and physiology biology of mycorrhiza association. MHB may increase the efficacy of fungal inoculum with a less budget because bacteria are easily grown in profitable numbers than most MF. This research shows that more mycorrhiza helper bacteria work should be devoted to mycorrhizal fungi for commercial concerns as well as used as a laboratory models in research. These fungi include *Pisolithus* spp., *Laccaria bicolor*, and *Glomus intraradices* genomes that are being sequenced as arbuscular fungi ectomycorrhizas. In addition, growing concern about the pollution of soil and the resulting tendency toward reducing the input of chemicals in plant production should substitute eco-friendly practices such as controlled microbial bioremediation, for example, by using ECM fungi as carriers of depolluting bacteria [22]. Different investigation and research supported by the genomic development may be a great opportunity to place MHB in the top prior list of future mycorrhiza research and to increase general field of fungal-bacterial interactions in ecosystems.

15. Role of ECM fungi in the agriculture sector

Uses of different chemicals for acute and chronic disease inhibition may cause environmental pollution and major human health problems [23]. For the sustainability of agriculture, different methods are established by fungal species [21]. Members in the genera *Alternaria, Aspergillus, Chaetomium, Fusarium, Penicillium, Serendipita, Phoma*, and *Trichoderma* are commonly identified as plant development stimulating fungi and also play a vital role as biofertilizer [24]. Different research studies have proven that fungi act as opponents and destroy soil-borne plant pathogens and help in the control of plant disease [25]. Entophytic fungi can be efficiently used as plant defenders, growth stimulators, and competitors of microorganisms, which have capacity for consumption in an extensive diversity of medical agricultural and industrial fields. This is mostly due to their abundant dispersion as symbionts with plants. ECMF recover soil structure, and nutrient source defends plants against root pathogens and also enhances plant growth by making different plants hormones, which may increase the growth rate and also enhance the photosynthesis rate of plants [26]. These fungi are important for the growth improvement of commercially important crops including different trees from the genera *Castanopsis, Dipterocarpus, Eucalyptus, Fagus, Picea, Pinus, Quercus*, and *Shorea*. *Scleroderma* and *Thelephora* are famous ectomycorrhiza genera that upsurge the rate of existence and growth of eucalyptus, oak, and pine seedlings in reestablishment and replantation programs [24].

16. Conclusion

The ectomycorrhiza group of fungi mainly belongs to Basidiomycota and Ascomycota members, and higher plants have mutual association. Different nutrients and water are translocated, absorbed, and utilized by roots of plants; ectomycorrhizal fungi are frequently involved in it. Most of the identified species of eatable fungal are saprophytes, and few are ectomycorrhizal fungi, and in the past few years, ectomycorrhizal fungi has been a topic for research because the role of ectomycorrhizal fungi in the ecosystem is unbeatable; ectomycorrhizal fungi play a vital role in different environments, for example, in terrestrial ecosystem and frequently involved in nutrient cycling, that is, universal carbon and nutrient cycles in the fungus plant interface the role of carbon and nutrient seems significant. Researchers should focus on molecular and functional tools involved in relationship between fungi, plant, and soil. For eras, scientists were considering that ectomycorrhizal fungi are only effective in forest ecosystem, but latest research has shown that ectomycorrhizal fungi play an important role in all the environments, but further investigation is required to specify its role in other environments; some examples justify that ectomycorrhizal fungi should be considered a basic tool in ecosystem restoration as in the nursery the application of ectomycorrhiza is very important, to yield ectomycorrhizal forest seedlings and also has vital impact in forest restoration. Now, scientists should focus on identifying suitable tools for the viable methods of development and increase in mycorrhizal fungi application on large scale under maintained environmental conditions. The part of ECM as biofertilizers in bioremediation and biocontrol in reestablishment, replantation, and environmental renovation has been important so far, and its significance role in the sustainability of the environment can be huge, enhancing the acceptance of plants against living and abiotic stress. The role of ECM fungi in present drawbacks of ecosystem, such as the vulnerable decline in plants such as oaks and pines and the process of plant remediation of polluted soils, seems favorable. Further research is going on to choose a new group mutual association for terrestrial reestablishment and replantation.

Author details

Muhammad Hanif[1*], Zubaria Ashraf[1], Samar Bashir[1], Fatima Riaz[1], Rizwan Amanat[1], Nousheen Yousaf[1] and Samina Sarwar[2]

1 Department of Botany, GC University, Lahore, Pakistan

2 Department of Botany, LCW University, Lahore, Pakistan

*Address all correspondence to: dr.mhanif@gcu.edu.pk

References

[1] Smith SE. Mycorrhizal Symbiosis. 3rd ed. London: Academic Press; 2008. p. 787

[2] Peterson RL, Massicotte HB, Melville LH. Mycorrhizas: Anatomy and Cell Biology. Wallingford, Oxon: CABI Publishing, CAB International; 2004

[3] Sharma R. Ectomycorrhizal mushrooms: Their diversity, ecology and practical applications. In: Varma A, Prasad R, Tuteja N, editors. Mycorrhiza—Function, Diversity, State of the Art. Switzerland: Springer International Publishing; 2017. pp. 99-131

[4] Gheda SF, Ahmed DA. Improved soil characteristics and wheat germination as influenced by inoculation of Nostoc kihlmani and Anabaena cylindrica. Rend Lincei. 2015;**26**:121-131

[5] Prieto I, Roldán A, Huygens D, Alguacil Navarro-Cano M. Species-specific roles of ectomycorrhizal fungi in facilitating interplant transfer of hydraulically redistributed water between Pinus halepensis saplings and seedlings. Plant and Soil. 2016;**2016**:15-27

[6] Miransari M. Soil microbes and plant fertilization. Applied Microbiology and Biotechnology. 2011;**2011**:875-885

[7] Lebin S, Singh R. Microbial bio fertilizer: Types and application. Biofertlizer for Sustainable Agriculture and Environment. 2019;**2019**:1-19

[8] Abdallah M, Ameye M, De Saeger S, Audenaert K, Haesaert G. Biological control of mycotoxigenic fungi and their toxins: An update for the pre-harvest approach. In: Mycotoxins - Impact and management Strategies. 2018. DOI: 10.5772/intechopen.72353

[9] Ali AF, Sohail FM, Salim HA, Abed AH. *Azotobacter chroococcum, Azospirillum brasilense: Trichoderma harzianum* and Tri-calcium phosphate on hydroponically grown barley grains. IOP Publishing; 2021;**735**:012056 DOI: 10.1088/1755-1315/735/1/012056

[10] Fitter AH, Gabraye J. Interaction between mycorrhizal fungi and other soil organisms. Plant and Soil. 1994;**1994**:123-132

[11] Simard SW, Durall D, Jones M. Carbon and nutrient fluxes within and between mycorr hizal plants. In: van der Heijden MGA, Sanders I. editors. Mycorrhizal Ecology. Berhn, Heidelberg, New York: Springer; 2002. pp. 33-74

[12] van Breemen N, Finlay R, Lundstrom U, Jongmans AG, Giesler R, Olsson M. Mycorrhizal weathering: A true case of mineral plant nutrition? Bio geochemistry. 2000;**49**:53-67

[13] Bhardwaj D, Ansari MW, Sahoo RK, Tuteja N. Biofertilizers function as key player in sustainable agriculture by improving soil fertility, plant tolerance and crop productivity. Microbial Cell Factories. 2014;**13**:66

[14] Tarafdar JC. Fungal inoculants for native phosphorus mobilization. Biofertlizer for Sustainable Agriculture and Environment. 2019;**2019**:21-40

[15] Azul AM, Nunes J, Ferreira I, Coelho AS, Verissimo P, Trovao J. Valuing native ectomycorrhizal fungi as a mediterranean forestry component for sustainable and innovative solutions. Botany-Botanique. 2014;**2014**:161-171

[16] Adeleke RA, Nunthkumar B, Roopnarain A, Obi L. Applications of plant–microbe interactions in agro-ecosystems. In: Microbiome in Plant

Health and Disease. Singapore: Springer; 2019. pp. 1-34

[17] Bonfante P. The future has roots in the past: The ideas and scientists that shaped mycorrhizal research. New Phytologist. 2018;**2018**:982-995

[18] Marcel G, Heijden V, Martin FM, Selosse MA, Sanders IR. Mycorrhizal ecology and evolution: The past, the present, and the future. New Phytologist. 2015;**2015**

[19] Chalote M, Plassard C. Ectomycorrhiza and nitrogen provision to the host tree. In: Polacco JC, Todd CD, editors. Ecological Aspects of Nitrogen Metabolism in Plants. John Wiley & Sons; 2011. pp. 69-94

[20] Bonfante P, Genre A. Mechanisms underlying beneficial plant-fungus interactions in mycorrhizal symbiosis. Nature Communications. 2010;**2010**:1-48

[21] Bamisile BS, Akutse KS, Siddiqui J, Xu Y. Model application of entomopathogenic fungi as alternatives to chemical pesticides: Prospects, challenges, and insights for next-generation sustainable agriculture. Frontiers in Plant Science. 2021;**2021**

[22] Sarand I, Timonen S, Nurmiaho-Lassila E-L, Koivula T, Haatela K, Romantschuk M, et al. Microbial biofilms and catabolic plasmid harbouring degradative fluorescent pseudomonads in Scots pine mycorrhizospheres developed on petroleum contaminated soil. FEMS Microbiology and Ecology. 1998;**27**:115-126

[23] Aktar MW, Sengupta D, Chowdhury A. Impact of pesticides use in agriculture: Their benefits and hazards. Interdisciplinary Toxicology. 2009;**2009**:1-12

[24] Hyde KD, Xu J, Rapior S, Jeewon R, Lumyong S, Niego AGT. The amazing potential of fungi: 50 ways we can exploit fungi industrially. Fungal Diversity. 2019;**2019**:1-136

[25] Pirttilä AM, Tabas MP, Baruah N, Koskimäki J. Biofertilizers and biocontrol agents for agriculture: How to identify and develop new potent microbial strains and traits. Microorganisms. 2021;**9**:817

[26] Santoyo G, Guzmán-Guzmán P, Parra-Cota FI, Santos-Villalobos SDL, Orozco-Mosqueda M, Glick BR. Plant growth stimulation bymicrobial consortia. Agronomy. 2021;**11**:219. DOI: 10.3390/agronomy11020219

Section 3

Mycorrhizae Biology and Development

Chapter 11

Plant Microbiome and Mycorrhizal Fungi

YingWu Shi, Hongmei Yang, Ming Chu, XinXiang Niu, Huifang Bao, Ning Wang, Faqiang Zhan, Xuanqi Long, Rong Yang, Qing Lin and Kai Lou

Abstract

In this paper, the research results on the synergy between mycorrhizal fungi and plant microorganisms in China and abroad were summarized. The purpose of this paper was to elaborate the effects of the synergy mechanism between mycorrhizal fungi and plant microorganisms on crop growth and stress resistance, soil physical and chemical properties, and soil microbial diversity and to analyze the contribution of the interaction between mycorrhizal fungi and plant microorganisms in agriculture and forestry, so as to provide theoretical basis for the further preparation of composite microbial agents, the healthy and green improvement of crop yield, and the ecological restoration of forestry stress resistance. The main directions of future research in this field were also analyzed.

Keywords: symbiosis, endophytic microbe, rhizosphere microorganism, interaction

1. Introduction

Mycorrhizal fungi can infect plant roots to form mycorrhiza [1, 2]. In recent years, there have been many studies on the symbiotic relationship between mycorrhizal fungi and plants [3, 4]. The dynamic interaction between plants and mycorrhizal fungi can promote the effective absorption of minerals by plants from soil and provide protection for various environmental stresses [5, 6]. However, some studies have found that this interaction is closely related to mycorrhizal helper bacteria (MHB). Mycorrhizal helper microorganism is a kind of special rhizosphere and endophyte that can specifically bind to mycorrhizal fungi, promote mycorrhizal fungi infection to the host, and promote plant growth and development [7, 8]. The interaction of mycorrhizal fungi and auxiliary microorganisms can promote plant growth, reduce the occurrence of soil-borne diseases, and improve the availability of nutrients in soil and plant absorption rate [9–11]. However, there are still insufficient studies on the synergistic effect of mycorrhizal fungi and auxiliary microorganisms on agricultural crops and forests.

The purpose was to elaborate the effects of the synergy mechanism between mycorrhizal fungi and mycorrhizal fungi, plant microflora, rhizosphere microorganisms and mycorrhizal relationships, and the relationship between endophytic bacteria and mycorrhizal fungi.

2. Mycorrhiza and mycorrhizal fungi

Mycorrhiza refers to the reciprocal symbiosis formed by fungi and plant roots, according to different morphological structures. Mycorrhiza is divided into seven types: arbuscular mycorrhiza, ecto mycorrhiza, endo mycorrhiza, orchid mycorrhiza, arbutoid mycorrhiza, monotropoid mycorrhiza, and ericoid mycorrhiza. Among them, arbuscular mycorrhiza, ecto mycorrhiza, orchid mycorrhiza, and ericoid mycorrhiza are four common types [12]. The characteristic is that fungi do not invade the inner cortex and can form a typical structure of hyphal ring in root cortex cells. Aging hyphal rings are eventually digested and absorbed by plant cells. Studies have shown that the biological binding force of orchid mycorrhiza (OM) symbionts to each other is not symmetrical, and plants are the beneficiaries of symbiotic relationships [13]; some OM fungi are not specifically associated with orchids [14]; fungal heterotrophic orchids have more diverse mycorrhizal fungal lineages than autotrophic orchids [15].

Mycorrhizal fungi are fungi that can infect plant roots to form mycorrhiza [16]. Mycorrhizal fungi and terrestrial plants have formed mutually beneficial, mutually conditional physiological wholes and symbionts with different morphological characteristics 500 million years ago. The core of symbiosis is the bidirectional exchange of plant carbon sequestration and nutrients obtained by fungi. Mycorrhizal fungi mycelium can interact directly or indirectly with other beneficial organisms in soil interior and surface and participate in nutrient transformation, absorption, and recycling processes in soil ecosystems [17]; it plays a role in maintaining the balance of atmospheric composition, regulating ecosystems, increasing biodiversity, and stabilizing and maintaining sustainable productivity of ecosystems [18]. It can also improve the absorption of nutrients and water and salt resistance and disease resistance of plants by combining with the roots of host plants, releasing plant hormones and enzymes to promote plant growth [19].

In recent years, there have been many studies on the symbiotic relationship between mycorrhizal fungi and plants. The dynamic interaction between plants and mycorrhizal fungi can promote the effective absorption of minerals by plants from soil and provide protection for various environmental stresses [20]. However, some studies have found that this phenomenon is also closely related to the mycorrhizal helper bacteria (MHB) [21], which is less studied. MHB is a kind of special rhizosphere bacteria that can specifically bind to mycorrhizal fungi, promote the infection of mycorrhizal fungi to the host, and promote the growth and development of plants [22]. Studies have shown that MHB not only promotes mycorrhizal symbiosis by triggering plant growth factors but also promotes spore germination, root colonization, metabolic diversity, and biological control of soil-borne diseases [23]. The interaction of mycorrhizal fungi and MHB can promote plant growth, reduce the occurrence of soil-borne diseases, and improve the availability of nutrients in soil and plant absorption rate [24, 25]. However, there are still insufficient studies on the synergistic effect of mycorrhizal fungi and MHB on crops in China and abroad.

3. Plant microbiome

Plant microbial groups include microbial communities that generally interact with plants. They can survive within or outside plant tissues and carry out various beneficial activities, including inhibiting potential plant pathogens and promoting plant growth. Plant microflora mainly includes rhizosphere microbial community, leaf microbial community, and endogenous microbial community. The microbial community composition of different flora is complex, and its diversity, preference, and abundance are affected by host plants and environment, which is the medium of plant-soil-atmosphere interaction. They communicate material and energy through plant body and have formed a highly close symbiotic relationship in the long-term co-evolution [26]. During plant growth, microorganisms are actively recruited from the surrounding microbial pool. In soil, plant roots provide a unique niche for soil microbial communities in addition to fixing plants and as organs for absorbing water and nutrients, attracting various microbial communities to distribute in rhizosphere, root, and to a certain extent above ground parts [27].

Rhizosphere microbial communities can promote plant phenotypic plasticity, such as the flowering time of plants may be affected by them. Studies have found a molecular interaction network linking nitrogen cycle, tryptophan (Trp) synthesis of plant hormone IAA and flowering time, which is of great significance to the study of plant phenotype in climate change and the improvement of crop yield [28].

Endophytes and above-ground microbial communities are known for their potential to promote plant growth, improve disease resistance, and alleviate stress tolerance. Endophytes include fungi and bacteria, archaea, and the rarely explored viral world, among which there are abundant studies on endophytic fungi and their metabolites.

4. Relationship between rhizosphere microorganism and mycorrhiza

4.1 Mycorrhiza and auxiliary bacteria

Garbaye first proposed the concept of MHB by analyzing the isolation, identification, and symbiosis of bacteria in mycorrhiza [24]. MHB acts on mycorrhiza through root sensitivity to fungi, rhizosphere soil improvement, and fungal reproductive germination. At present, in the study of exogenous mycorrhizal, MHB can promote spore germination and mycelial growth by producing growth factors, detoxification antagonistic substances, and inhibiting competitors. The change of mycelium growth represents the adaptability of MHB to fungi and the close relationship between fungi and host plants [29]. MHB can improve the infection rate of exogenous mycorrhizal fungi on plant seedlings, enhance their colonization, and then promote plant growth two categories [30, 31]: the first category is through the production of enzymes that break down the spore wall and other volatile substances such as terpenoids, thereby promoting spore germination and bacterial formation [32]. The second category is by changing the absorption of inorganic salt by mycorrhizal fungi to affect spore germination, the establishment of symbiotic system [33, 34]. The MHB isolated and identified mainly includes *Agrobacterium*, *Burkholderia*, *Pseudomonas*, *Bacillus*, *Paenibacilus*, and *Streptomyces* [35].

4.2 Mycorrhizal and rhizosphere microorganisms

Mycorrhiza fungi treatment can significantly increase the species of beneficial fungi and reduce the species of pathogenic bacteria. Mycorrhizal fungi and some rhizosphere beneficial microorganisms have synergistic promotion relationship [36], which can significantly improve plant resistance [37–39]. Arbuscular mycorrhizal (AM) fungi have effects on rhizosphere microbial community structure and activity under atrazine stress, but there are few reports on the composition and diversity of rhizosphere bacterial community. Arbuscular mycorrhizal fungi (AMF) increased the phospholipid fatty acid analysis (PLFA) biomass of AM fungi but decreased the phospholipid fatty acid analysis (PLFA) biomass of microbial bacteria and fungi in rhizosphere soil.

Under the same site conditions, the number of fungi in rhizosphere soil of larch mycorrhizal fungi was significantly different from that in non-rhizosphere soil, and the species composition and dominant population were also significantly different. Since the relationship between tree roots, mycorrhizal fungi, and soil is very complex, many studies have been reported. However, the research field of mycorrhizal rhizosphere fungi has just started, and there are still many contents to be further studied. At the same time, mycorrhizal fungi have direct or indirect effects on plant growth and development. Enzymes, hormones, and toxins produced by mycorrhizal fungi affect the mineral nutrition, water absorption, and root development of plant roots in soil environment.

5. Relationship between endophytic bacteria and mycorrhizal fungi in plants

Endophytic bacteria, as a unique form of arbuscular mycorrhizal fungi (AMF), exist in the host AMF. At present, studies have shown that endophytic bacteria can enhance the ecological adaptability of the host AMF and improve its environmental stress resistance [40]. At present, it can only be preliminarily proved that the growth promotion effect and defense ability are related to endophytic bacteria. However, AMF without endophytic bacteria symbiosis, such as *R. intraradices*, can still produce the effect of disease resistance and injury stress on host plants without endophytic bacteria symbiosis. The interaction mechanism and disease resistance of this kind of AMF with host plants are different from those of the above endophytic bacteria. There are few studies on the interaction mechanism between symbiotic endophytic bacteria and host AMF. Strengthening the research in this direction has important innovative value and scientific significance for further exploration and evaluation of AMF fungal resources.

The unique arbuscular mycorrhizal structure of AMF and its diverse ecological functions formed by its interaction with host plants play a positive role and influence in the restoration of ecological environment vegetation, the restoration of ecological areas with moderate to strong alkaline effects [41], the restoration of wasteland containing metal mines [42], the phytoremediation of heavy metals in sewage-contaminated soil [43], and the rapid restoration of soil vegetation after interference [40]. The interaction between improving crop growth and mineral nutrition [44], plant productivity, and nutrient absorption is still an important research direction at present and in the future [45].

Ectomycorrhizal fungi (ECMF) usually selects bacterial communities from the surrounding soil according to symbiotic function or habitat requirements, and the short-term contact between soil bacteria and their fruiting bodies can also directly or indirectly promote the endobacteria (EB) community composition of fruiting bodies [46, 47]. According to the ecological function or habitat requirements, ECMF selects specific bacteria to colonize its body, which lays the foundation for the establishment and maintenance of symbiotic relationship. However, at present, it is still necessary to further explore the extent to which these specific species affect ECMF microbial communities and how they interact. Bacteria colonize selectively in ECMFs suitable for their own growth and form a new symbiotic relationship under severe environmental stresses [48].

It has been proved that whether ECMF selects specific bacterial groups according to symbiotic function or habitat requirements, or bacteria colonize in ECMF due to their own nutrient needs or niche expansion, the symbiotic mode formed between the two is more excellent in adapting to environmental changes.

In the symbiotic interaction between bacteria and ECMF, bacteria can loosely interact with the mycelium surface; it can also colonize mycelia and fruiting bodies, or show some symbiotic species specificity, as well as some potential metabolic complementation and ecological functions, thereby promoting mycelial growth, biomass increase, fruiting body formation of host fungi, and providing nitrogen sources for host fungi through nitrogen fixation [40, 46, 49]. At the same time, the mycelia and fruiting bodies of fungi create a suitable habitat for EB by providing different carbon sources for EB and protecting it from environmental stresses [50].

6. Contribution of mycorrhizal fungi to agriculture

The effect of AM fungi inoculation combined with appropriate agronomic measures on improving soil fertility, promoting plant growth, and increasing plant yield is greater than any single effect. For example, rotation, intercropping, or grass orchard inoculation of AM fungi are more obvious. At present some countries in the world are not rich in land resources, continuous cropping cultivation, especially the protection land contact for many years, resulting in soil fertility, soil quality, and ecological comprehensive service function decline, it is difficult to meet the needs of production. Therefore, it is worthy of systematic research and development to establish a new system of soil and seedling management dominated by mycorrhizal fungi inoculation to overcome continuous cropping obstacles under the combined cropping pattern of multiple crops mixed planting structure and rotation and intercropping. It is foreseeable that the implementation of AM fungi inoculation combined with appropriate agricultural measures will continue to develop and become an important emerging agricultural technology.

6.1 Application of AMF technology in agriculture

Most crops show some mycorrhizal dependence on AMF. AMF can promote crop uptake of soil nutrients, promote crop growth, improve crop yield, improve fertilizer use efficiency, and reduce fertilizer consumption, which is conducive to maintaining sustainable agricultural development. At present, there are research institutions and companies in the world that can produce AMF fungicides on a large scale [51].

In practical application, selecting crops with strong mycorrhizal dependence and high economic value to promote the application of mycorrhizal biotechnology has great ecological and economic benefits.

Many studies have shown that AMF inoculation can increase the yield of maize, wheat, soybean, and other major grain crops, improve crop nutritional status, and enhance crop resistance to drought, salinity, and other stresses [52, 53]. The commercial fungicide MYKEPROSG2 (Canada) was used to inoculated maize, and it was found that the P content of host plants could still be increased in the presence of indigenous AMF [54]. AMF combined with chemical fertilizer can improve the quality of rice and increase the contents of Fe and Zn in rice [55]. Ceballos inoculated cassava (Manihot esculenta), an important food crop in tropical regions, with *Rhizophagus irregularis*, which greatly increased its yield [56]. The effectiveness and economy of AMF inoculation could be taken into account when 2500 spores were inoculated per cassava.

6.2 Application of AMF technology in horticulture

Most horticultural crops can form symbiotic systems with AMF, such as Solanaceae, Onionaceae, most fruit trees, herbs, herbs, and ornamental flowers. Horticultural crops have high economic value and are used to nursery and container seedling, which provides convenient conditions for mycorrhizal plants at seedling stage. After mycorrhizal inoculation, it can shorten the nursery period of seedlings, improve the survival rate of transplanting, promote seedling growth, enhance disease resistance, ultimately improve product yield and quality, increase the economic added value of horticultural crops, and obtain greater benefits with less investment.

Under the premise of halving fertilizer application rate, Ziane (2020) inoculated tomatoes with commercial fungicide Symbivit, indigenous fungicide *Aspergillus moxicanus*, and *Septoglomus constrictum*, which could improve plant biomass and achieve the effect of 100% fertilizer application rate without inoculation [57]. The inoculation of horticultural crops such as fruit trees and vegetables can improve the yield and quality of fruits and vegetables. For example, the inoculation of *Glomus trufemii* on cucumber improved the quality of single fruit and the total yield per plant, while the contents of soluble protein and Vc in fruit were significantly increased [58]. AMF can also promote the formation of secondary metabolites and active substances in herb spices and medicinal plants, such as *Salvia japonica* inoculated with commercial bacteria and *Septoglomus viscosum*, the content and quality of essential oil in plants were significantly improved, and the proportion of taxol increased [59]. Xie (2018) found that the content of glycyrrhizic acid and liquiritin in roots of *Glycyrrhiza uralensis* increased significantly after inoculation with AMF [60]. Mycorrhizal plants have stronger adaptability under adverse conditions such as drought and salinity. The activities of antioxidant enzymes such as superoxide dismutase (SOD), catalase (CAT), peroxidase (POD), and guaiacol peroxidase (GPOX) in vivo are significantly increased, which can slow down the oxidative damage of plants and make plants have stronger resistance to water deficit [61, 62]. Selecting appropriate AMF for inoculation of plants can not only promote the absorption of phosphorus (P) by plants but also increase the contents of Zn, Cu, Mn, and other trace elements in plants [63, 64]. Therefore, mycorrhizal biotechnology can substantially reduce the use of fertilizers and pesticides, avoid pesticide residues in fruit and vegetable products, and improve the quality of fruits and vegetables.

In the international production of fruit trees and flowers, AMF has been widely used as biological protective agent, biological accelerator, and biological fertilizer.

7. Application of mycorrhizal fungi in forestry

Once a plant enters a new growth environment, the introduction fails due to lack of symbiotic mycorrhizal fungi or suitable mycorrhizal fungi. This phenomenon is common in all continents. For example, Puerto Rico introduced foreign pine, and annual introduction failed repeatedly until the introduction of mycorrhiza fungi from the origin of the successful introduction. The survival rate of imported pine reached above by mycorrhizal fungi inoculation and afforestation with mycorrhizal seedlings in China. Mycorrhizal seedling afforestation can not only improve the ability of trees to absorb and utilize nutrients but also improve the survival rate of afforestation and timber yield. Moreover, inoculation at seedling stage can also greatly reduce the amount of fungicides. To this end, the United States also established mycorrhizal technology companies. Mycorrhizal biological agents were specially provided for mycorrhizal forest trees. The mycorrhizal research and development center of Chinese Academy of Forestry has also carried out research on the production and application technology of mycorrhizal biological agents, has been popularized in various provinces and regions throughout the country, and achieved remarkable results. A large number of studies have shown that mycorrhizal fungi not only prevent the invasion of pathogenic bacteria in structure but also secrete antibiotics to inhibit the growth and development of pathogenic bacteria, which can improve the disease resistance of forest trees and play an important role in the biological control of forest root diseases. At present, the inoculation of *Larix gmelinii* with Boletus has been successfully used to prevent and control root rot caused by *Pythium fulvus*.

7.1 Application of mycorrhizal fungi in tree breeding

When introducing new tree species, corresponding mycorrhizal species should be introduced, especially pine and other specialized mycorrhizal species. As early as in the early 1950s, after the failure of introducing eucalyptus, Iraq introduced undisinfected soil and finally introduced it successfully and significantly promoted seedling growth [65]. In 2001, Hua et al. put forward the concept of forest mycorrhizal bioengineering or forest mycorrhizal biotechnology [66], that mycorrhizal trees and the whole forest ecosystem is of great significance, that not only can improve forestry productivity but also provide a sustainable development path for forestry development.

In the current study on the effects of mycorrhizal fungi on forest growth, most of the experiments inoculated mycorrhizal fungi on forest seedlings to observe the changes in growth indicators, namely mycorrhizal forest. Zhang et al. test showed that the survival rate of mycorrhizal fungi in *Pinus elliottii* was above 90% [67]. The ground diameter and height of mycorrhizal *P. sylvestris* seedlings were 37% and 39% higher, respectively, and the survival rate of seedlings under drought conditions was improved [68]. Zhang found that inoculation with mycorrhizal fungi could promote seedling height growth, ground diameter growth, and biomass growth of poplar through experiments [69]. Some studies also inoculated seven mycorrhizal fungi of Russula on *Pinus massoniana* seedlings [70], which could not only form mycorrhizal fungi with *P. massoniana* seedlings but also promote the growth of seedlings to varying degrees. He et al. treated the rhizosphere of *P. massoniana* seedlings with mycorrhizal fungi by non-sterile inoculation method [71]. The height, ground diameter, and crown width of mycorrhizal *P. massoniana* seedlings were 29%, 51.7%, and 46.6% higher than those of the control group, respectively. Wu et al. conducted pot experiments after soil sterilization, and the results showed that exogenous mycorrhizal fungi could promote the growth and development of *P. tabulaeformis* seedlings and the accumulation of phosphorus *in vivo* and could significantly increase the basal diameter and leaf length of plants [72]. At the same time, it was concluded that the difference between natural forest and artificial forest in the natural regeneration of seedlings was due to the composition and diversity of exogenous mycorrhizal communities. Therefore, mycorrhizal fungi played an important role in the natural regeneration and growth of *P. tabulaeformis*.

Changes in plant water use efficiency, stomatal conductance, and CO_2 absorption efficiency can also affect the distribution of nitrogen content. Mycorrhiza can change the photosynthetic efficiency and thus affect the growth of trees through the above methods [73]. Experiments show that mycorrhizal *P. elliottii* can increase chlorophyll content, thereby increasing photosynthesis [67]. Zhu et al. found that the net photosynthetic rate, intercellular CO_2 concentration, stomatal conductance, seedling height, and ground diameter of *Zelkova schneideriana* were significantly improved after inoculation with mycorrhizal fungi [74].

Mycorrhiza promoted root growth of seedlings. Wang (2013) inoculated exogenous mycorrhiza on *P. massoniana* seedlings [75]. The results showed that the main root length, lateral root length, and lateral root number of mycorrhizal seedlings were higher than those of control group. Song et al. found that the establishment of symbiotic relationship between *P. cathayana* seedlings and VA mycorrhizal fungi could promote the development of seedling roots, increase the root surface absorption area, increase the ratio of active root absorption area, and enhance the activity of polyphenol oxidase in roots [76]. Mycorrhiza can indirectly promote forest growth by promoting root absorption of nutrients in soil. Lu found that after AM fungi infected roots, the growth of host plants changed significantly, and seedling height and root length changed significantly in the early stage of seedlings [77]. Jin (2019) inoculated different strains of *Quercus* spp., and the inoculated mycorrhizal fungi showed different degrees of growth-promoting effects on the host and the ability to significantly regulate root architecture [78]. Ditengou et al. found that *C. bicolor* could secrete sesquiterpenoid small-molecule substances, which had a significant regulatory effect on poplar roots, resulting in an increase in the number of secondary roots, expanding the root area of poplar, thus making poplar absorb nutrients [79]. Wu inoculated *R. irregularis* in arbuscular mycorrhizal fungi on *Populus deltoides* and found that inoculation with mycorrhizal fungi could increase the contents of iron, copper, manganese, and phosphorus in roots at high nitrogen levels [80]. It was speculated that *R. irregularis* could alleviate the nutritional imbalance caused by high nitrogen stress by promoting the absorption of nutrients and the content of elements in roots of *P. deltoides*. At the same time, this experiment found that mycorrhizal poplars might absorb nitrogen through mycorrhizal pathways and downregulate the expression of nitrogen transport-related genes in roots.

7.2 Application of mycorrhiza in forestry nursery

With the continuous understanding and in-depth study of mycorrhiza, mycorrhiza seedling technology is more and more widely used in forestry. From introduction, mycorrhizal seedling to afforestation in ecologically fragile areas, the application of mycorrhizal has achieved initial results. For mycorrhizal plants, countries or regions should introduce corresponding mycorrhizal fungi while introducing a new plant. There are many examples of failed introduction due to the lack of mycorrhizal fungi. For example, Puerto Rico in South America introduced pine tree species from all countries in the world in the 1930s, which failed. It was not until the introduction of mycorrhizal fungi from the origin that the afforestation was successful in 1955. Iraq's migration to different tree species was also unsuccessful. In the early 20th century, Iraq 's move to different eucalyptus species was also unsuccessful. Later, the unsterilized soil in the introduced eucalyptus forest was introduced and applied to the roots of the seedlings. Finally, it was successful and significantly promoted seedling growth. Finally, it succeeded and significantly promoted the growth of seedlings. Three kinds of pine trees, *P. elliottii*, *P. caribbeanensis*, and *Pinus taeda*, were introduced by Guangdong Forestry Research Institute in China. However, due to the lack of sufficient mycorrhizal fungi, afforestation was declared to be unsuccessful. Subsequently, the seedlings inoculated with mycorrhizal fungi were used for afforestation so that the survival rates of these three pines reached, and a large number of *P. yunnanensis* trees were introduced into Hainan Island in China. Due to the adoption of mycorrhizal fungi inoculation measures, the seedlings grew vigorously, and the introduction and afforestation were also successful. Chen (2013) reported the role of Australian and domestic exogenous mycorrhizal fungi in the introduction of Australian tail leaves, and the results showed that the inoculation of exogenous mycorrhizal fungi could improve the plant growth [81].

There are many advantages of mycorrhizal seedlings compared with ordinary afforestation technology, which can greatly improve the survival rate of tree afforestation, promote the growth of seedlings, increase the absorption capacity of plants to equal greenhouse gases, improve the carbon fixation ability of plants, improve the ability of seedlings to absorb nutrients such as nitrogen, phosphorus, and potassium, and secrete a variety of enzymes to activate soil components to form soil aggregates and improve soil rhizosphere microenvironment. At present, many countries and regions in the world stipulate that afforestation in some ecologically fragile areas must adopt mycorrhizal seedling technology.

American mycorrhizal development has been in the forefront of the world and set up a special mycorrhizal technology company to provide special mycorrhizal agents for forest mycorrhizal. The relevant departments of the United States have stipulated that seedling cultivation and afforestation in wet grassland areas must be inoculated before the Soviet Union stipulated in the forest steppe zone, and the mycorrhizal inoculation technology must be used to establish nursery for the seedlings. Mycorrhizal inoculation of roots at the early stage of seedlings was not only beneficial to mycorrhizal formation but also had less usage and obvious effect.

Forestry scientists in China have been working on mycorrhizal seedling technology in recent years. Mycorrhizal agent developed by China Academy of Forestry is mainly used for seedling cultivation of *P. massoniana*, which can increase seedling height, ground diameter, biomass, lateral root number, yield of qualified seedlings, and survival growth rate of afforestation. Therefore, it can solve the problems of low survival rate and slow growth in the early stage of *P. massoniana* and has been popularized in various provinces, cities, and regions nationwide, and remarkable results have been achieved. Using mycorrhizal technology to develop *P. elliottii*, studies on seedling cultivation and seedling ectomycorrhizal rate up to the above indicated that mycorrhizal treatment of seedlings of *P. massoniana* and *P. elliottii* could promote root growth of seedlings, increase seedling growth, increase ground diameter by one, and grow vigorously.

In areas where conventional afforestation is difficult, selecting suitable mycorrhizal fungi and adopting mycorrhizal technology according to the principle of suitable land and tree can greatly improve the survival rate of seedlings and afforestation effect, accelerate the recovery ability of vegetation, prevent further deterioration of the environment, and promote ecological balance. Mycorrhizal technology will play an important role in the restoration and reconstruction of ecological environment in China and other countries.

In summary, mycorrhizal technology has been applied in a variety of tree species, and pine mycorrhizal research has also been widely concerned. Many scientists have done some statistical analysis on the mycorrhizal formation of a tree species or a mycorrhizal fungi, described the mycorrhizal fungi of American pine seedlings, and described the necessary species of ectomycorrhizal roots for radiation pine seedlings. The exophytic mycorrhizas of P. radiata were divided into subclasses and subsubclasses. A series of morphological, histological, and cytological studies on the mycorrhiza were carried out. A key to the mycorrhiza of *Platycladus erinaceus*, *Picea norwayensis*, and Pine New Year's Eve was developed, from which the 10 genera of Boletus, such as Russula, Lactarius, Bacteroides, Boletus, Cephalophyllum, Chlamydomonas, and so on, which formed mycorrhiza, can be retrieved. It is a comprehensive reference for the identification of mycorrhiza. *P. tabulaeformis* is the second largest species of pine in planting area and quantity. The application value and prospect of fine mycorrhizal fungi breeding and propagation technology for *P. tabulaeformis* are quite extensive.

The inoculation of *P. tabulaeformis* with mycorrhizal technology can not only provide excellent nursery stocks but also improve the ecological environment in ecologically fragile areas, control soil erosion, and improve soil fertility. In landscaping, it can increase the absorption capacity of equivalent greenhouse gases, that is carbon sequestration capacity is very strong, and can slow down the heat island effect in the city. In today's global warming, without sacrificing the premise of economic development, this technology makes a certain contribution to the social energy conservation and emission reduction. Using mycorrhizal inoculation technology to cultivate fine *P. tabulaeformis* seedlings, carrying out industrial production, establishing nursery, and applying it to afforestation under difficult site conditions will greatly improve the survival rate of seedlings and forest growth, accelerate the pace of afforestation in obstacle areas, and promote a new step in afforestation in China. It plays an important role in improving the ecological environment, improving the productivity of forest land, controlling desertification land, and preventing soil erosion.

8. Future perspectives

With the continuous strengthening of people's awareness of forest resource protection, the ecological value of forestry has attracted more and more attention. Mycorrhizal fungi, as an important role in forestry, play a positive role in the growth and development of trees and the whole forestry ecological environment. It is mainly manifested in the introduction of new species of trees, breeding and afforestation, promoting the growth and development of trees, affecting the photosynthesis of trees, strengthening the absorption of soil nutrients by tree roots and the resistance to adverse environments, and improving the soil and the whole forestry ecology.

Although the research on mycorrhizal science is deepening, there are still shortcomings. In order to better apply mycorrhizal science to forestry development and provide a theoretical basis for the sustainable development of forestry economy, it is suggested to carry out research from the following aspects. At present, there are few statistical studies on the influence of geographical environment on mycorrhizal fungi. By analyzing the distribution characteristics of mycorrhizal fungi in various mountainous areas and forest areas under the influence of environmental factors, the dominant mycorrhizal fungi in various forest areas can be selected for investigation, analysis and in-depth study. The classification of mycorrhizal fungi needs to be further refined, which can be associated with peripheral factors such as symbiotic trees, surrounding undergrowth vegetation, and other microorganisms in soil and soil element content, to more comprehensively analyze the role of mycorrhizal fungi in forest land ecology. Molecular biology and omics are rarely used in mycorrhizal fungi, and the study on the dynamic changes of forest molecular level under the action of mycorrhizal fungi is rare, including omics such as molecules, proteins, and metabolism, which can strengthen the study on molecular level and further reveal the mechanism of symbiosis between mycorrhizal fungi and forest.

9. Conclusion

The interaction of mycorrhizal fungi and MHB can promote plant growth, reduce the occurrence of soil-borne diseases, and improve the availability of nutrients in soil and plant absorption rate. Plant microorganisms encode more genes than the host plant itself and affect the growth, development, and health of plants through the cooperation and competition between microorganisms and plants. The number of fungi in mycorrhizal rhizosphere is large, indicating that the rhizosphere microenvironment is conducive to the growth and development of fungi. The differences in species composition reflected that mycorrhiza and its metabolites changed the growth environment of soil fungi. The differences in dominant species indicated that some fungi in mycorrhizal rhizosphere adapted to the environment of mycorrhizal rhizosphere and survived. Endophytic bacteria symbiotic AMF significantly promoted the increase of chlorophyll content and promoted the utilization of photosynthesis in host plants. Endophytic bacteria symbiotic AMF significantly increased chitinase activity and activity of phenylalanine ammonia-lyase (PAL) and promoted host plant resistance to disease stress. These endophytic fungi are host plant mycorrhizal fungi. The effect of AM fungi inoculation combined with appropriate agronomic measures on improving soil fertility, promoting plant growth, and increasing plant yield is greater than any single effect.

Author details

YingWu Shi[1,2,3]*, Hongmei Yang[1,2,3], Ming Chu[1,2,3], XinXiang Niu[3,4], Huifang Bao[1,2], Ning Wang[1,2,3], Faqiang Zhan[1,2], Xuanqi Long[1,2], Rong Yang[1,2], Qing Lin[1,2] and Kai Lou[1,2]

1 Institute of Microbiology, Xinjiang Academy of Agricultural Sciences, Urumqi, Xinjiang, China

2 Xinjiang Laboratory of Special Enviromental Microbiology, Urumqi, Xinjiang, China

3 Key Laboratory of Agriculture Environment in Northwest Oasis of Agriculture, Urumqi, Xinjiang, China

4 Institute of Soil and Fertilizer and Agricultural Water Conservation, Xinjiang Academy of Agricultural Sciences, Urumqi, Xinjiang, China

*Address all correspondence to: syw1973@126.com

References

[1] Wu C, Qu J, Liu L, Kang H, Sun H, Zhang Y, et al. Quo vadis: Signaling molecules and small secreted proteins from mycorrhizal fungi at the early stage of mycorrhiza formation. Symbiosis. 2021;**85**(2):123-143

[2] Zhou Y, Li X, Gao Y, Ren A. Plant endophytes and arbuscular mycorrhizal fungi alter the decomposition of Achnatherum sibiricum litter. Applied Soil Ecology. 2022;**180**:104616

[3] Liang SM, Zheng FL, Wu QS. Elucidating the dialogue between arbuscular mycorrhizal fungi and polyamines in plants. World Journal of Microbiology and Biotechnology. 2022;**38**(9):1-12

[4] Helber N, Wippel K, Sauer N, Schaarschmidt S, Hause B, Requena N. A versatile monosaccharide transporter that operates in the arbuscular mycorrhizal fungus Glomus sp is crucial for the symbiotic relationship with plants. The Plant Cell. 2011;**23**(10):3812-3823

[5] Oliveira TC, Cabral JSR, Santana LR, Tavares GG, Santos LDS, Paim TP, et al. The arbuscular mycorrhizal fungus Rhizophagus clarus improves physiological tolerance to drought stress in soybean plants. Scientific Reports. 2022;**12**(1):1-15

[6] Liang Y, Pan F, Jiang Z, Li Q, Pu J, Liu K. Accumulation in nutrient acquisition strategies of arbuscular mycorrhizal fungi and plant roots in poor and heterogeneous soils of karst shrub ecosystems. BMC Plant Biology. 2022;**22**(1):1-12

[7] Lies A, Delteil A, Prin Y, Duponnois R. Using mycorrhiza helper microorganisms (MHM) to improve the mycorrhizal efficiency on plant growth. In: Role of Rhizospheric Microbes in Soil. Singapore: Springer; 2018. pp. 277-298

[8] Kulczyk-Skrzeszewska M, Kieliszewska-Rokicka B. Influence of drought and salt stress on the growth of young Populus nigra 'Italica' plants and associated mycorrhizal fungi and non-mycorrhizal fungal endophytes. New Forests. 2022;**53**(4):679-694

[9] Weng W, Yan J, Zhou M, Yao X, Gao A, Ma C, et al. Roles of arbuscular mycorrhizal fungi as a biocontrol agent in the control of plant diseases. Microorganisms. 2022;**10**(7):1266

[10] Gomes SIF, Fortuna MA, Bascompte J, Merckx VSFT. Mycoheterotrophic plants preferentially target arbuscular mycorrhizal fungi that are highly connected to autotrophic plants. The New Phytologist. 2022;**235**(5):2034-2045

[11] Lu Y, Liu X, Zhou S. Nitrogen addition altered the plant-arbuscular mycorrhizal fungi network through reducing redundant interactions in an alpine meadow. Soil Biology and Biochemistry. 2022;**171**:108727

[12] Sun G, Jie Wg HW, Zhang YZ, Qiao W, Wei L, Jiang YT, et al. Effects of mycorrhizal fungi and mycorrhizal helper bacteria on crop development: A review. Chinese Agricultural Science Bulletin. 2022;**38**(09):88-92

[13] Martos F, Munoz F, Pailler T, Kottke I, Gonneau C, Selosse MA. The role of epiphytism in architecture and evolutionary constraint within mycorrhizal networks of tropical orchids. Molecular Ecology. 2012;**21**:5098-5109

[14] Veldre V, Abarenkov K, Bahram M, Martos F, Selosse MA, Tamm H, et al. Evolution of nu-tritional modes of Ceratobasidiaceae (Cantharellales, Ba-sidiomycota) as revealed from publicly available ITS sequences. Fungal Ecology. 2013;**6**:256-268

[15] Dearnaley JDW, Martos F, Selosse M-A. Orchid mycorrhizas: molecular ecology, physiology, evolution and conservation aspects. In: Hock B, editor. Fungal Associations. 2nd ed. Berlin: Springer-Verlag; 2012. pp. 207-230

[16] Dos Santos LL, Neto AR, de Moura JB, de Souza RF, Santos MEF, de Moura LF, et al. Mycorrhizal fungi arbuscular in forage grasses cultivated in Cerrado soil. Scientific Reports. 2022;**12**(1):1-6

[17] Hodge A, Fitter AH, Diaz SM. Substantial nitrogen acquisition by arbuscular mycorrhizal fungi from organic material has implications for N cycling. Proceedings of the National Academy of Sciences of the United States of America. 2010;**107**(31):13754-13759

[18] Liang QQ, Li M, Liu RJ, Guo SX. Function and functioning mechanisms of mycorrhizal fungi under global changes. Acta Ecologica Sinica. 2014;**34**(21):6039-6048

[19] Cao BF, Jiang HX, Lu YG, Liu L, Wang MS. Research advances in symbiotic relationship between tobacco (Nicotiana tabacum L.) and Arbuscular mycorrhizal fungi. Soil and Fertilizer Sciences in China. 2021;**1**:327-338

[20] Guo YQ, Chen XT, Wu YY, Zhang L, Cheng JM, Wei GH, et al. Natural revegetation of a semiarid habitat alters taxonomic and functional diversity of soil microbial communities. Science of the Total Environment. 2018;**635**(1):598-606

[21] Sangwan S, Prasanna R. Mycorrhizae helper bacteria: Unlocking their potential as bioenhancers of plant–arbuscular mycorrhizal fungal associations. Microbial Ecology. 2022;**84**(1):1-10

[22] Garbaye J. Tansley review No. 76 helper bacteria: A new dimension to the mycorrhizal symbiosis. New Phytologist. 1994;**128**(2):197-210

[23] Seemakram W, Suebrasri T, Khaekhum S, Ekprasert J, Aimi T, Boonlue S. Growth enhancement of the highly prized tropical trees siamese rosewood and burma padauk. Rhizosphere. 2021;**19**:100363

[24] Rocha I, Ma Y, Souza-Alonso P, Vosatka M, Freitas H, Oliveira RS. Seed coating: A tool for delivering beneficial microbes to agricultural crops. Frontiers in Plant Science. 2019;**10**(6):1357-1373

[25] Yao YX, Jie WG, Hu W, Zhao DM, Jiang YT. Research progress on the effects of microbial fertilizer on the growth and development of soybean and the properties of rhizosphere soil. Hubei Agricultural Sciences. 2019;**58**(20):21-38

[26] Trivedi P, Leach JE, Tringe SG, Sa T, Singh BK. Plant–microbiome interactions: From community assembly to plant health. Nature Reviews Microbiology. 2020;**18**(11):607-621

[27] Compant S, Samad A, Faist H, Sessitsch A. A review on the plant microbiome: Ecology, functions, and emerging trends in microbial application. Journal of Advanced Research. 2019;**19**:29-37

[28] Lu T, Ke M, Lavoie M, Jin Y, Fan X, Zhang Z, et al. Rhizosphere microorganisms can influence the timing of plant flowering. Microbiome. 2018;**6**(1):1-12

[29] Adeleke R, Dames JF. Kalaharituber pfeilii and associated bacterial interactions. South African Journal of Botany. 2014;**90**(90):68-73

[30] Rigamonte TA, Pylro VS, Duarte GF. The role of mycorrhization helper bacteria in the establishment and action of ectomycorrhizae associations. Brazilian Journal of Microbiology. 2010;**41**(4):832-840

[31] Miransari M. Interactions between arbuscular mycorrhizal fungi and soil bacteria. Applied Microbiology Biotechnology. 2011;**89**(4):917-930

[32] Ratti N, Kumar S, Verma HN, Gautam SP. Improvement in bioavailability of tricalcium phosphate to Cymbopogon martinii var. motia by rhizobacteria, AMF and *Azospirillum* inoculation. Microbiological Research. 2001;**156**(2):145-149

[33] Tan SP, Tang C, Guo SX, Wang WH, Liu RJ. Advances in study of interactions between mycorrhizal fungi and bacteria. Journal of Qingdao Agricultural University (Natural Science). 2013;**30**(4):240-246

[34] Hestrin R, Hammer EC, Mueller CW, Lehmann J. Synergies between mycorrhizal fungi and soil microbial communities increase plant nitrogen acquisition. Communications Biology. 2019;**2**(1):1-9

[35] Li LL, Yang HY. Research progress on interaction between mycorrhizal fungi and mycorrhizal helper bacteria. Heilongjiang Agricultural Sciences. 2020;**9**:121-124

[36] Cao QQ, Feng YQ, Liu YF, Guo XP, Zhang GQ, Qin L. Advance of plant phosphorus uptake improved by mycorrhiza fungi. Chinese Bulletin of Life Sciences. 2011;**4**:407-413

[37] Zhang YW, Chai LW, Wang DW, Wang J, Huang Y. Effect of ectomycorrhizae on heavy metals sequestration by thermostable protein in rhizosphere of *Pinus tabulaeformis* under Cu and Cd stress. Environmental Science. 2014;**35**(3):169-175

[38] Sheng JM, Wu XQ. Interaction between mycorrhizal fungi and rhizosphere microorganisms. Journal of Northwest Forestry University. 2007;**22**(5):104-108

[39] Garbaye J. Mycorrhization helper bacteria: A new dimension to the mycorrhizal symbiosis. Acta Botanica Gallica. 1994;**141**(4):517-521

[40] Salvioli A, Ghignone S, Novero M, Navazio L, Venice F, Bagnaresi P, et al. Symbiosis with an endobacterium increases the fitness of a mycorrhizal fungus, raising its bioenergetic potential. The ISME Journal. 2016;**10**(1):130-144

[41] Parihar M, Rakshit A, Bahadur HS, Rana K. Diversity of arbuscular mycorrhizal fungi in alkaline soils of hot sub humid eco-region of middle gangetic plains of India. Acta Agriculturae Scandinavica, Section B-Soil Plant Science. 2019;**69**(5):386-397

[42] Zhan FD, Li B, Jiang M, Li T, He Y, Li Y, et al. Effects of arbuscular mycorrhizal fungi on the growth and heavy metal accumulation of bermudagrass [*Cynodon dactylon* (L.) pers.] grown in a lead zinc mine wasteland. International Journal of Phytoremediation. 2019;**21**(9):849-856

[43] Khalid ME, Fahad A, Al-Mana SE-H, Selwey AW, Elgorban A. Arbuscular mycorrhizal fungi mitigates heavy metal toxicity adverse effects in sewage water contaminated soil on Tagetes erectal. Soil Science and Plant Nutrition. 2018;**64**(5):662-668

[44] Rifa A, Tatsuhiro E. Nestedness in arbuscular mycorrhizal fungal communities in a volcanic ecosystem: selection of disturbance-tolerant fungi along an elevation gradient. Microbes and Environments. 2019;**34**(3):327-333

[45] Muchane MN, Pulleman MM, Vanlauwe B, Jefwa J, Kuyper TW. Impact of arbuscular mycorrhizal fungi and earthworms on soil aggregate stability, glomalin, and performance of pigeon pea, Cajanus Cajan. Soil Research. 2018;**57**(1):53-65

[46] Pent M, Põldmaa K, Bahram M. Bacterial communities in boreal forest mushrooms are shaped both by soil parameters and host identity. Frontiers in Microbiology. 2017;**8**:836

[47] Antony-Babu S, Deveau A, Van Nostrand JD, Zhou JZ, Le Tacon F, Robin C, et al. Black truffle-associated bacterial communities during the development and maturation of *Tuber melanosporum* ascocarps and putative functional roles. Environmental Microbiology. 2014;**16**(9):2831-2847

[48] Nazir R, Zhang MZ, De Boer W, Van Elsas JD. The capacity to comigrate with *Lyophyllum* sp. strain Karsten through different soils is spread among several phylogenetic groups within the genus *Burkholderia*. Soil Biology and Biochemistry. 2012;**50**:221-233

[49] Rangel-Castro IJ, Danell E, Taylor AF. Use of different nitrogen sources by the edible ectomycorrhizal mushroom *Cantharellus cibarius*. Mycorrhiza. 2002;**12**(3):131-137

[50] Schulz-Bohm K, Tyc O, De Boer W, Peereboom N, Debets F, Zaagman N, et al. Fungus-associated bacteriome in charge of their host behavior. Fungal Genetics and Biology. 2017;**102**:38-48

[51] Vosatka M, Latr A, Gianinazzi S, Albrechtova J. Development of arbuscular mycorrhizal biotechnology and industry: Current achievements and bottlenecks. Symbiosis. 2012;**58**(1):29-37

[52] Selvakumar G, Shagol CC, Kim K, Han S, Sa T. Spore associated bacteria regulates maize root K+/Na+ion homeostasis to promote salinity tolerance during arbuscular mycorrhizal symbiosis. BMC Plant Biology. 2018;**18**(1):1-13

[53] Tian H, Drijber RA, Zhang JL, Li XL. Impact of long-term nitrogen fertilization and rotation with soybean on the diversity and phosphorus metabolism of indigenous arbuscular mycorrhizal fungi within the roots of maize (*Zea mays* L.). Agriculture, Ecosystems Environment. 2013;**164**:53-61

[54] Antunes PM, Koch AM, Dunfield KE, Hart MM, Downing A, Rillig MC, et al. Influence of commercial inoculation with *Glomus intraradices* on the structure and functioning of an AM fungal community from an agricultural site. Plant and Soil. 2009;**317**(1):257-266

[55] Zhang X, Chen B, Ohtomo R. Mycorrhizal effects on growth, P uptake and Cd tolerance of the host plant vary among different AM fungal species. Soil Science and Plant Nutrition. 2015;**61**:359-368

[56] Ceballos I, Ruiz M, Fernández C, Peña R, Rodríguez A, Sanders IR. The in vitro mass-produced model mycorrhizal fungus, *Rhizophagus irregularis*, significantly increases yields of the globally important food security crop cassava. PLoS One. 2013;**8**(8):e70633

[57] Schubert R, Werner S, Cirka H, Rödel P, Tandron Moya Y, Mock HP, et al. Effects of arbuscular mycorrhization

on fruit quality in industrialized tomato production. International Journal of Molecular Sciences. 2020;**21**(19):7029

[58] Wu YS, Guo SR, Zhang J, Du NS, Sun J. Effects of exogenous spermidine and arbuscular mycorrhizal fungi on plant growth of cucumber. Chinese Journal of Applied Ecology. 2018;**29**(3):891-898

[59] Tarraf W, Ruta C, Tagarelli A, De Cillis F, De Mastro G. Influence of arbuscular mycorrhizae on plant growth, essential oil production and phosphorus uptake of Salvia officinalis L. Industrial Crops and Products. 2017;**102**:144-153

[60] Xie W, Hao Z, Zhou X, Jiang X, Xu L, Wu S, et al. Arbuscular mycorrhiza facilitates the accumulation of glycyrrhizin and liquiritin in *Glycyrrhiza uralensis* under drought stress. Mycorrhiza. 2018;**28**(3):285-300

[61] Ahmad H, Hayat S, Ali M, Liu T, Cheng Z. The combination of arbuscular mycorrhizal fungi inoculation (*Glomus versiforme*) and 28-homobrassinolide spraying intervals improves growth by enhancing photosynthesis, nutrient absorption, and antioxidant system in cucumber (*Cucumis sativus* L.) under salinity. Ecology and Evolution. 2018;**8**(11):5724-5740

[62] Domokos E, Jakab-Farkas L, Darko B, BIrO-Janka B, Mara G, Albert C, et al. Increase in Artemisia annua plant biomass artemisinin content and guaiacol peroxidase activity using the arbuscular mycorrhizal fungus *Rhizophagus irregularis*. Frontiers. Plant Science. 2018;**9**:478

[63] Lehmann A, Veresoglou SD, Leifheit EF, Rillig MC. Arbuscular mycorrhizal influence on zinc nutrition in crop plants – A meta-analysis. Soil Biology and Biochemistry. 2014;**69**:123-131

[64] Lehmann A, Rillig MC. Arbuscular mycorrhizal contribution to copper, manganese and iron nutrient concentrations in crops: A meta-analysis. Soil Biology and Bio-chemistry. 2015;**81**:147-158

[65] Zhong K, Liu HX. The new characters and application of mycorrhizal studies. Ecological Science. 2008;**3**:169-178

[66] Hua XM. Mycorrhizal biotechnology of forest trees. World Forestry Research. 2001;**1**:22-29

[67] Zhang Y, Zeng LQ, Ye JR, Wu XQ, Zhang LP. Effects of ectomycorrhizal fungi inoculation on growth and rhizosphere microorganisms of *Pinus elliottii* seedlings. Journal of Nanjing Forestry University (Natural Sciences Edition). 2014;**38**(5):57-61

[68] Lei ZY. Effect of inoculating mycorrhizal fungi on seedling of *Pinus sylvestris*. Journal of Liaoning Forestry Science Technology. 1996;**4**:8-9

[69] Zhang Y. Inoculation Effect of Four Ectomycorrhizal fungi on Populus×Canadensis cv. 'Zhongliao 1'. Shenyang: Shenyang Agricultural University; 2018

[70] Jiang QB, Zhong CL, Chen Y, Zhang Y, Chen Z. Study on inoculating Russula fungi with *Pinus massoniana* seedling. Journal of Central South University of Forestry Technology. 2016;**36**(8):6-38

[71] He HZ, Xv CL, Luo LX, Feng SB, Li LX, Zhang YF, et al. Effect of *Cantharellus* mycorrhizal fungi on the masson growth. Journal of Anhui Agricultural Sciences. 2010;**38**(25):13709-13710

[72] Wu YH. Studies on Ectomycorrhizal Fungi Biodiversity and Its Impacts on

Seedling Growth Of *Pinus tabulaeformis* at the Huanglong forest region. Yangling: Northwest A F University; 2019

[73] Heinonsalo J, Juurola E, Linden A, Pumpanen J. Ectomycorrhizal fungi affect Scots pine photosynthesis through nitrogen and water economy, not only through increased carbon demand. Environmental and Experimental Botany. 2015;**109**:103-112

[74] Zhu LJ, Fu ZY, Zhang JC, Wang JP, Lin J, Yuan ZM, et al. Effects of mycorrhizal fungi on photosynthetic characteristics of *Zelkova serrata* (Thunb.). Journal of Nanjing Forestry University (Natural Sciences Edition). 2018;**42**(6):121-127

[75] Wang Y, Ding GJ. Effects of ectomycorrhizal on growth, physiological characteristics and nutrition in *Pinus massoniana* seedlings. Journal of Nanjing Forestry University (Natural Sciences Edition). 2013;**37**(2):97-102

[76] Song FQ. Ecological and Physiological Research on VA Mycorrhizas of Populus Ussuriensis. Harbin: Northeast Forestry University; 2002

[77] Lu LL. Community characteristics of arbuscular mycorrhizal (AM) fungi in the rhizosphere of *Cunninghamia lanceolata* and Its Effect on Seedling Growth. Beijing: Beijing Forestry University; 2020

[78] Jin W. Diversity of ectomycorrhizal fungi associated with several introduced and native *Quercus* Species and their effects on seedling growth and nutrient uptake. Beijing: China Academy of Forestry Sciences; 2019

[79] Ditengou FA, Müller A, Rosenkranz M, Felten J, Lasok H, Van Doorn MM, et al. Volatile signalling by sesquiterpenes from ectomycorrhizal fungi reprogrammes root architecture. Nature Communications. 2015;**6**(1):1-9

[80] Wu F. Effects of nitrogen and arbuscular mycorrhizal fungi on the growth of *Populus×canadensis* 'Neva'. Yangling: Northwest A F University; 2018

[81] Chen YL, Gong MQ, Xu DP, Zhong CL, Wang FZ, Chen Y. Screening and inoculant efficacy of Australian ectomycorrhizal fungi on *Eucalyptus urophylla* in field. Forest Research. 2000;**13**(6):569-576

Chapter 12

Role of Non-Coding RNAs in Plant Nutrition through Mycorrhizal Interactions

Nidhi Verma, Yeshveer Singh, Anupam Patra and Tanvi Singh

Abstract

In nature, many plants rely on symbiotic interaction with mycorrhizae for their nutrition and survival. For instance, nitrogen-fixing nodules and mycorrhizae are well established mutualistic biotic interactions between plants and bacterial/fungal partners under nitrogen limiting environment. Many small regulatory components of RNA like micro-RNAs play a critical role in establishment of these symbioses. These regulatory components are also crucial for balancing hormone levels, and synchronization of plant defenses and development pathways. However, functions of various sRNAs are still need to be addressed. This chapter will detailed out various important parts these regulatory components (sRNA, miRNA and siRNA) are playing during mycorrhizal interactions for plant growth, development and nutrition.

Keywords: miRNA, siRNA, mycorrhiza, nodulation, symbiosis, nutrient uptake

1. Introduction

During course of co-evolution since millions of years, plants have established symbiotic associations with the fungi and bacteria. Established mycorrhizal and rhizobia symbiosis with the plants are the best illustrated examples of such interactions. These symbiotic associations are entrenched by the molecular cross-talk including correct recognition and specific activation/repression of signaling pathways. Legume-rhizobia interactions are specific in terms of molecular cross talk, as the host plant secretes flavonoids which are perceived by compatible rhizobia for the induction, expression and activation of *Nod* genes in the bacteria, necessary for the nodule formation in the host plant. The secreted Nod factors once recognized by host specific intra-cellular kinase and extra-cellular LysM domain containing receptors, a cascade of cytoplasmic events starts within root epidermal cells. Depolarization of the membrane, alteration in calcium levels and induction of calmodulin based kinase signaling makes favorable environment for rhizobia infection thread formation and successive penetration of plant host cell through branching. Subsequently, 'Bacteroids' formation and nitrogen fixation initiates in host cell cytoplasm. In contrast, mycorrhizal interactions are not specific in terms of host range as they can colonize almost all terrestrial plants [1]. Although the signaling pathway for

IntechOpen

mycorrhizal symbiosis activation shares some attributes of rhizobial symbiosis events, induction and reprogramming of the host cells starts after the recognition of myc-LCO (mycorrhizal lipo-chito-oligosaccharide), which leads to altered metabolic cascade in host and hyphae as well [2]. This molecular cross-talk establishes nutrients and mineral transport through specialized and branched structures called 'Arbuscules' from Arbuscular mycorrhizal (AM) partner and photo-synthetically fixed carbon sources mobilization from host plant in exchange. To bear an invasion of microorganisms, plants must have some specialized mechanisms to distinguish beneficial microbes from harmful ones. Since last few decades, we are learning about regulators of fine tuning among symbiotic associations and plant immunity [3, 4]. Contributions of non-coding RNAs (ncRNAs) in this regulation of host defenses to establish symbiosis are indispensible according to recent studies [5].

In this chapter, we have summarized the genesis of various important classes of non-coding RNAs and their role in nutrient uptake, transport, assimilation and homeostasis in plants via mycorrhizal symbiosis, and discuss the recent discoveries of cross-kingdom RNA interference (RNAi) during plant-fungus interactions. We also provide the insights and future perspectives for improved understanding of mycorrhizal associations, which would aid in the development of innovative strategies for enhancing the crop yield.

2. Genesis of non-coding RNAs and classification

Currently, a large number of endogenously formed ncRNAs involved in different regulatory functions have been discovered and functionally characterized in various plant species [6, 7]. On account of their average size, the regulatory ncRNAs can be classified into sRNAs (small RNAs of typically 18–30 nt in size), medium-sized ncRNAs (broad range of 31–200 nt), and more than 200 nt sized Long-non-coding RNAs (lncRNAs). Furthermore, depending on their morphology, theycan be classified as linear or circular (circRNA). Recently, small regulatory RNAs (sRNAs), miRNAs and small interfering (si)RNAs, have been well characterized with respect to plant immunity and symbiosis. Although thought to be small, they play vital functions in response to the biotic, abiotic stress and environmental fluctuations by regulation/modulation of target genes expression [8–11]. Similarly, lncRNAs were considered transcriptional noises, but later attracted attention for the heterogeneous groups of ncRNAs and long range [12]. Remarkably, unlike other linearly regulated ncRNAs, the newly discovered circRNAs belong to a novel class, which lacks free 5′ and 3′ end [13]. In addition, many small ncRNAs, which are derivatives of tRNAs, which are identified and characterized in plants typically comprised of 15–42 nt, termed as tsRNAs [14–16]. The tsRNAs are also classified as regulatory ncRNAs for multiple functions. Generally, the functions of certain ncRNA are similar, but some differ and overlap in silencing signaling pathways [17].

As reviewed by Chao et al. [18], the biogenesis of miRNAs is a multi-step procedure which involves transcription, processing, alterations, and then RNA-induced silencing (RISC) complex assembly. First, a pri-miRNA (primary miRNA) is transcribed from RNA Polymerase II containing a hairpin RNA secondary structure. Next, the base pri-miRNA hairpin is then cleaved by a DICER Like RNase-III family enzyme (usually DCL1). To release miRNA-miRNA* duplex, these hairpins are cleaved again and subsequently methylated (at 2′O- position) by HUA Enhancer 1 (HEN1) nuclear protein for the stability. Finally, in nucleus the mature miRNA strand enters into AGO1 to form miRNA-AGO1 complex, which are then transported to cytoplasm leaving behind cleaved miRNA* fragment for the induction of post-transcriptional gene silencing.

Depending on their mechanism of action, siRNAs can further be classified into three major sub-categories: (1) native antisense siRNAs (nat-siRNAs), (2) heterochromatin siRNAs (hc-siRNAs), and (3) trans-acting siRNAs (ta-siRNAs). ta-siRNA is generated from the TAS gene which is transcribed from RNA Pol II into single-stranded RNA and loses its cap and poly-A tail during miRNA-AGO1 complex-controlled cleavage [8, 19]. Later, the 5′ or 3′ cleaved fragments are end protected by the suppressor of gene silencing 3 (SGS3) protein and transformed into double-stranded RNA (dsRNA) via RDRP-VI [20]. Finally, by HEN1 and DCL activities they are methylated and processed to form ta-siRNAs (21–24 nt). To participate in post-transcriptional modulation/silencing of target genes by pairing with its complementary mRNAs, these 21–24 nt sized strands are integrated with AGO1/AGO7 present in the cytoplasm, whereas a few ta-siRNAs are loaded onto AGO4/6 for assisting methylation of TAS genes via RNA Pol V.

tsRNAs, with a wide size range (15–42 nt), represent a unique ncRNA class that can be sub-categorized based on their cleavage sites: (1) tRF-1 s, (2) tRF-2 s, (3) tRF-3 s, (4) tRF-5 s, and (5) tiRNAs. However, plant research is still in its infancy and many questions remain unanswered in reference to its existence. For instance, the biosynthetic pathway for tsRNAs and their regulatory or physiological roles in plants are still very limited [21].

CircRNAs are known as RNA biomolecules which are circular, covalently closed and single-stranded [22]. They were first identified and characterized from plant viruses by Sanger and colleagues in 1976. The organization of circRNAs can be divided into three groups [23]. (1) The Exon circRNAs are generated by the circularization of lariat-derived and intron pairing, (2) the intronic circRNAs are formed by the partial intron degradation after lasso structure formation; and (3) exo-syntronic circRNAs are composed of exons as well as introns, and circularized during the splicing process.

lncRNAs biogenesis can be categorized into five major types in accordance to the sites being transcribed via RNA-Pol II. (1) The antisense lncRNA is transcribed over the complementary strand of the exon; whereas (2) sense lncRNA is transcribed on the same strand as the exon. As name indicates, (3) Intron lncRNA is transcribed into an intron. (4) The Inter-genic lncRNAs are situated between two different genes and (5) the enhancer lncRNA mostly arises from the enhancer region of the protein-encoding gene [24]. They can control target regulation in a variety of ways, including chromatin re-modeling, transcriptional repression, splicing of RNA and its transcriptional enhancers. Additionally, lncRNAs can code for certain small peptides required for various cellular processes [25]. Notably, several lncRNAs are regulated under abiotic/biotic stresses in the plants.

3. Role of sRNAs in plant nutrition

Induction of miRNAs regulates the expression of an array of genes and promotes plant nutritional homeostasis. Owing high-throughput RNASeq techniques and target prediction tools the role of ncRNAs in nutrition and stress signaling has been investigated in recent past. Majorly the role of ncRNAs in nitrogen (N), phosphate (Pi) and sulfur (S) homeostasis has been discussed below:

3.1 Nitrogen

Evidence for miRNAs controlling nitrogen responses in plants has been illustrated [26, 27]. Up-regulation of pri-miRNA156, pri-miRNA447c and down-regulation of pri-miRNA169 and pri-miR398a has been characterized in Arabidopsis under nitrogen-deficient conditions [28]. Expression of nitrogen responsive miRNA, like miRNA160, miRNA167, miRNA168 in the maize roots and miRNA164, miRNA171 in shoots whereas, miRNA169 in both are reported under nitrogen-limiting conditions [27]. Similarly, several nitrogen-responsive miRNAs have been investigated in legumes, for instance, a total of altered expressions of 168 miRNAs are reported in limiting-nitrogen-tolerant and limiting-nitrogen-sensitive genotype of soybean using RNASeq [26]. A down-regulation of miRNA2606a/b-3p and up-regulation of miRNA1512a-5p was found in limiting-nitrogen-tolerant and limiting-nitrogen-sensitive genotype respectively. Moreover, mRNA transcripts encoding Cathepsin and E3-Ubiquitin ligase protein were found to be targeted and degraded by miRNA396b/g-5p and miRNA156b/6f-5p respectively under nitrogen stress.

3.2 Phosphate

Phosphate-responsive sRNA involved in Pi-uptake, transport, assimilation and homeostasis through targeting mRNA transcripts are extensively studied and identified in plant including rice, maize, tomato, soybean and Arabidopsis [29–33]. Among these plant species, common set of plant miRNA families are characterized modulating signaling networks, including miRNA156, miRNA159, miRNA166, miRNA319, miRNA395, miRNA398, miRNA399, miRNA447, and miRNA827 are demonstrated in response to Pi-limiting environment [34, 35]. An elevated level of miRNA156, miRNA399, miRNA778, miRNA827, miRNA2111 and suppressed miRNA169, miRNA395 and miRNA398 levels are observed under Pi stress [28]. Role of miRNA2111 has been illustrated under N and Pi limiting conditions [36]. The expression of phosphate-responsive PHO2 transporter was altered by miRNA827 and miRNA399 [37]. Moreover, miRNA827 targets the Major Facility Superfamily (MFS)-XPS proteins which are involved in Pi sensing and transport [38]. Common response against nutrient starvation includes anthocyanin accumulation in plants. MYB TF regulated anthocyanin biosynthesis pathway genes are targeted by siRNAs produced by ta-siRNA4 under the regulation of Pi-responsive miRNA828, post-transcriptionally [39]. The major regulatory role in maintenance of mineral homeostasis in host plant under N, Pi and C limiting environments is performed by miRNA398a [40]. Among all the characterized Pi-responsive miRNAs, altered levels of different alleles of miRNA399 were found conserved and pre-dominant under Pi-limiting conditions [41]. miRNAs and siRNAs induction was reported upon *Candidatus liberibacter* infection in citrus plants and interestingly, miRNA399 level was found elevated in infected plants than healthy host under Pi-limiting conditions [33]. These facts demonstrate the critical role of sRNAs in post-transcription regulation of Pi-responsive transcripts enabling host adaption under nutrition stress.

3.3 Sulfate

Sulfate transporters located on root epidermal and cortical cell membrane are the key components in sulfur uptake and transport to the plants in $SO4^{2-}$ form. Based on their substrate affinity, sequence similarity and their location of expression, they are categorized into five major groups. Group 1, group 2 and group 3 are characterized as high affinity, low affinity, and moderate affinity transporters for sulfur substrate, respectively [42–44]. Group 4 and 5 transporters are characterized as efflux transporter on tonoplast and molybdenum transporters (for being actively involved in molybdenum transport across the plant) respectively [45, 46]. S uptake from soil to the root is carried out by group1 and 2 transporters, while root to shoot transport of S is done by group 4 transporters. Under S-limiting conditions plethora of miRNAs induced including miRNA66, miRNA67 and miRNA395 while suppression of miRNA14, miRNA20 and miRNA43 is associated with regulation of post-transcription modification of S signaling. miRNA395 is a S-specific ncRNA signal and has been characterized to function as a key regulator of the sulfate depletion pathway. Under S-deprivation, miRNA395 positively regulates the expression of the low-affinity transporter AtSul2;1 [47], supporting sulfate uptake and transport of cells to shoots and leaves in *Arabidopsis thaliana*. The initial step of S assimilation into cysteine is catalyzed by ATP dependent sulphurylases (APSs), which are the target for miRNA395 in plastids [35].

4. Nutrient uptake/exchange during mutualistic plant-fungus interactions

One of the characteristics of the beneficial mycorrhizal interactions is the bidirectional nutrient exchange between both the partners and to support the growth of plant host [48, 49]. In these relationships, the fungus provides nitrogen, phosphate and sulfate nutrients to the host, whereas, in return, the host plant transfers photosynthetically fixed carbon (4–20%) to the mycorrhizal fungus [50]. In AM roots, the fungus proliferates into the root cortex intercellularly as well as intracellularly, whereas in case of ECM roots, it only covers intercellular regions, indicating the differences in the mechanism of colonization and nutrient uptake/exchange. The uptake of nutrients by plants from soil is limited by the repressed mobility of nutrients. Importantly, during AM symbiosis, the plant phosphate (P) transporters are down-regulated [51, 52]. Under these conditions, the nutrient such as P uptake is predominantly achieved through mycorrhizal pathway [1, 53, 54]. It has been observed that contribution of the mycorrhizal pathway in nutrient uptake varies with the plant and fungal partners that are involved in the interaction and also on the nutrient type [48, 52, 55–57]. To facilitate the nutrient uptake via mycorrhizal interface, the peri-arbuscular membrane (PAM) harbors high affinity transporters that are specifically induced in mycorrhizal roots. For instance, Pt4 and AMT2 are the high affinity transporters for P and ammonium ($NH4^+$) that mediate transfer of respective nutrients from fungus to plant host [58–60]. Moreover, a few mycorrhiza-inducible sulfate transporters have also been reported in AM roots [61, 62]. Recently, a sulfate transporter (SiSulT) and iron transporter (PiFtr) from *Serendipita indica* (previously known as *Piriformospora indica*) has been characterized, which transfers sulfur to the maize and iron to the rice plants, respectively and improves its growth [63, 64]. These studies highlight the importance of sulfur transport via mycorrhizal associations.

On the other hand, the plant transfers photosynthates as sucrose from source to ECM roots that serve as a carbon sink, which is then converted to simpler sugars such as glucose or fructose by invertase enzyme of host. The glucose and fructose are taken up by fungal counterparts through mycorrhizal interface. For instance, an arbuscular membrane localized monosaccharide transporter (MST2) is involved in the uptake of glucose and xylose molecules by AM fungi [65]. Intriguingly, host carbon supply has been found to trigger the fungal gene expression, and P and N uptake during AM and ECM symbiosis [66–68]. This also leads to increased hydrolysis of polyP (an important source of P and N) and release of Pi and Arg in the fungal cytoplasm. The Arg is further broken down to $NH4^+$ and is transferred to host plant via mycorrhizal interface. Importantly, these transport processes, from host to fungus and from fungus to the host, involve diverse molecular players that mediate membrane transport for the nutrition exchange between mycorrhizal plants and fungus at the interaction interface. The membrane transporters and channels that mediate the transport of molecules such as P, N, S, K, sugar and water are collectively referred as the 'transportome' [69].

Moreover, several robust and tightly regulated signaling processes are involved in establishing the successful mycorrhizal colonization for the efficient exchange of nutrients between plant and fungus. Although, the regulatory processes of plant and fungus both are important during symbiosis, the major proportion of studies has focused on the regulation from plant's perspective. These regulators include a variety of non-coding RNAs, phytohormones, peptide signals, transcription factors such as *CYCLOPS* and *NODULATION SIGNALING PATHWAY* (NSP1 and NSP2) [70–72]. The *CYCLOPS*, NSP1 and NSP2 are conserved members of rhizobial and mycorrhizal symbiosis phenomenon. The detailed overview of transcriptional regulation of AM development has been provided by Pimprikar and Gutjahr [73]. The non-coding RNA mediated-regulation of mycorrhizal symbiosis is now gaining the scientist's attention and emerging as new area of research.

5. Plant's ncRNAs modulating interconnection networks with fungi

5.1 miRNA

In recent past decades, the focus has pushed beyond the traditional defense pathway's transcriptional control in establishing pathogenic or beneficial plant-microbe interactions to attempts to understand novel transcriptional regulators. In particular, many groups have started investigating the function of microRNAs (miRNAs) in regulating signaling processes accompanying the symbiotic interactions. Many plant miRNAs have been reported to be involved in modulating the plant-pathogenic microbe interactions. The majority of miRNAs investigated and characterized to date complements the alterations described well in the transcriptome. For example, one of the defenses against necrotrophic pathogens involves improving the physical tolerance of plant cells. *Arabidopsis* miRNA408 and miRNA160a induce physical cell reinforcement by positive regulation of lignification and callose deposition [74, 75]. In *Medicago* and *Oriza*, various miRNA targets the ET (Ethylene), JA (Jasmonic Acid), and SA (Salicylic Acid) biogenesis during pathogenesis at very early stages [76, 77]. During attack by biotrophic pathogens, many miRNAs also suppress routine cellular detoxification. One example is miRNA398, which enhances ROS generation in *Magnaporthe oryzae*-infected tissues [78]. On the other hand, in case of tomato, the necrotrophic association between plant and *Alternaria solani*, miRNAs target gene transcripts are reported to be actively involved in toxin detoxification [79]. This suggests that miRNAs have potential to modify plant cells into the toxic ecosystems, poisoning invading microbes before they spread further to strengthen host immunity.

In mutual bio-trophic interactions, a few recent investigations have shown that the a major proportion of miRNAs synthesized all through interaction establishment modulate hormone response pathways, protein methylation, and functions of innate immunity components [78, 80]. A well-studied example includes the miRNA (annotated as E4D3Z3Y01BW0TQ) which is reported to be induced during AM symbiosis progression and interferes with GA signaling. GA signaling pathway is known to inhibit symbiotic-association [81–83]. On the other hand, miRNA172c promotes nodule formation in many plants by repressing the translation of APETALA2 TF [84, 85]. In AM-colonized roots, miRNA171b hampers with GRAS TF-responsive transcripts targeting through miRNA171, which are necessary for both nodulation and symbiosis.

5.2 siRNA

Microorganisms can have a significant impact on how plants react to colonization; they are not just background actors in the process. Since past few decades, effector proteins and siRNAs have been the two main areas of research. A variety of plant signaling pathways are altered by microbial effector proteins, which are generally tiny secreted proteins that are substantially stimulated during the colonization process. Similar to effector proteins, siRNAs target essential plant transcripts, disrupt transcripts via the ARGONAUTE (AGO) pathway, or act in a manner resembling that of miRNAs by inhibiting tRNA binding and localization. The microorganism-secreted siRNAs are taken up by the host plant, and disrupt the key transcripts of host.

Botrytis cinerea, a fungal pathogen, has been demonstrated to alter plant physiology during colonization by secreting siRNAs [86]. *B. cinerea* siRNAs initially penetrate host plant cells during the pathogenesis of tomato and *Arabidopsis*, where they diminish the host's RNAi apparatus. These relatively tiny molecules can therefore be thought of as variants of conventional effector proteins. In case of *A. thaliana*, *Bc*-siRNA3.1, *Bc*-siRNA3.2, and *Bc*-siRNA5 collectively silenced the stress-related genes, such as PRXIIF, WAK, MPK1 and MPK2 to eliminate the plant defense [87]. However, *Bc*-siRNA37 AtPMR6, AtFEI2 and AtWRKY7 are selectively silenced by *Bc*-siRNA37 [88]. While this is the only example so far in which siRNAs have been formed and released by microbes and reached to the host plant cells, there can be another common means based on genomic analysis by which microbes can modulate host responses during colonization.

5.3 lncRNA

The role of lncRNAs has been well-investigated in the context of plant defense against fungal, bacterial and viral pathogens [89–94]. Furthermore, the functions of a number of plant lncRNAs intricated in plant defenses are experimentally validated. For instance, it has been observed that the lncRNA-ACOD1 is dispensable for viral entrance but not for viral replication in the host [95]. Moreover, when turnip crinkle virus infection occurs in *Arabidopsis*, the expression of the long intergenic ncRNA LINC-AP2 gene is negatively regulated [89]. LncRNA33732 has been characterized to function as a positive regulator in tomatoes, enhancing the expression of the respiratory burst oxidase gene and raising H_2O_2 build-up, thereby increasing tomato resistance to *Phytophthora infestans* [96]. Also, lncRNA23468 in tomato can compete with endogenous RNA to regulate the NBS-LRR gene by feeding on miRNA482b, thereby controling tomato resistance to *P. infestans* pathogenesis [90]. Numerous lncRNAs have been identified as being modulated in drought, nitrogen-stress and phosphate depletion in maize [97–99]. The maize inbred line B73 tissues were subjected more than 700 high-fidelity RNA-Seq studies, which identified nearly 18,165 maize lncRNAs [100]. Although lncRNAs are involved in the regulation of plant-microbe interactions, there are no available reports characterizing lncRNA responses to AM fungi, so far as the experimental evidence is considered.

6. miRNA in symbiosis: regulation through repression

Emphasis on existence and importance of small ncRNA, especially miRNA started with the discovery of its association with regulation of gene expression in Petunia [101, 102]. Owing genome wide studies and high-throughput sequencing efforts, till date thousands of miRNAs have been characterized throughout the kingdom of life. Most of the characterized miRNA associated with symbiosis are either involved in nutrient signaling, exchange and homeostasis or development of nodule/arbuscules or both [103, 104] are illustrated in **Figure 1**. Based on morphological analysis of host, AM Symbiosis (AMS) can be sub-categorized into four major stages: (1) pre-contact signaling, (2) contact establishment between plant root and fungal hyphae, 3) intra-radical proliferation, and 4) arbuscule formation [105]. A non-canonical form of miRNA171, which is found repressed under phosphate starvation, regulates an important transcription factor (TF) involved in common symbiotic signaling pathway, NSP1/2 (NODULE SIGNALING PATHWAY 1/2) during mycorrhizal symbiosis.

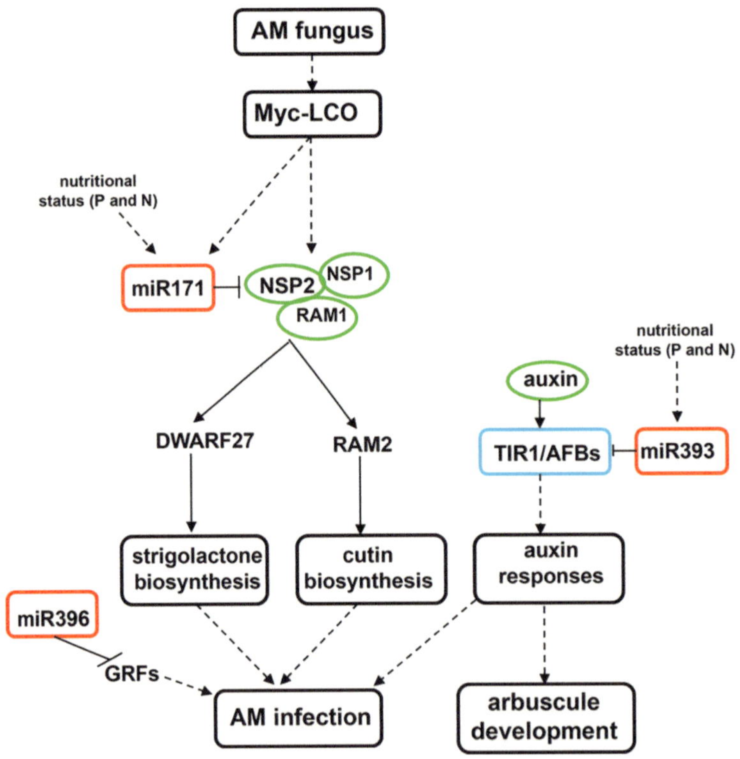

Figure 1.
Illustration of cross-talk between myc-LCO (mycorrhizal lipo-chito-oligosaccharide) factor induced common pathway signaling, miRNAs and regulatory hormones during AM symbiosis. Solid and dotted lines represents direct and indirect interconnections respectively.

The interaction of NSP2 with the mycorrhizal-specific GRAS TF, RAM1 (Required for Arbuscular Mycorrhization 1) regulates RAM2, and DWARF27 that are part of cutin and strigo-lactone biosynthetic pathway, respectively, as these two pathways facilitate AM inoculation. miRNA393, also identified as a N- and Pi-responsive miRNA, is involved in the homeostasis of auxin signaling and thus inhibits the auxin receptor TRANSPORT INHIBITOR RESPONSE1/AUXIN SIGNALING F-BOX PROTEIN (TIR1/ABF), also mediates the repression of root growth regulatory factors (GRFs) to affect fungal colonization and arbuscule development. miRNA396b also investigated to perform a significant role in root colonization and development during mycorrhizal symbiosis by targeting six GRFs and a TF in *Medicago truncaluta* [106].

For initiating rhizobia symbiosis, interactions between miRNAs, Nod factor signaling, and hormone regulation through NSP2 is controlled by a nodule-specific miRNA171. The NUCLEAR FACTOR (NF)-YA gene, a TF necessary for the nodule initiation and maintenance of meristem, is negatively regulated by miRNA169 [107]. The combination of cytokinin-responsive miRNA172 genes, Nod factor, and various

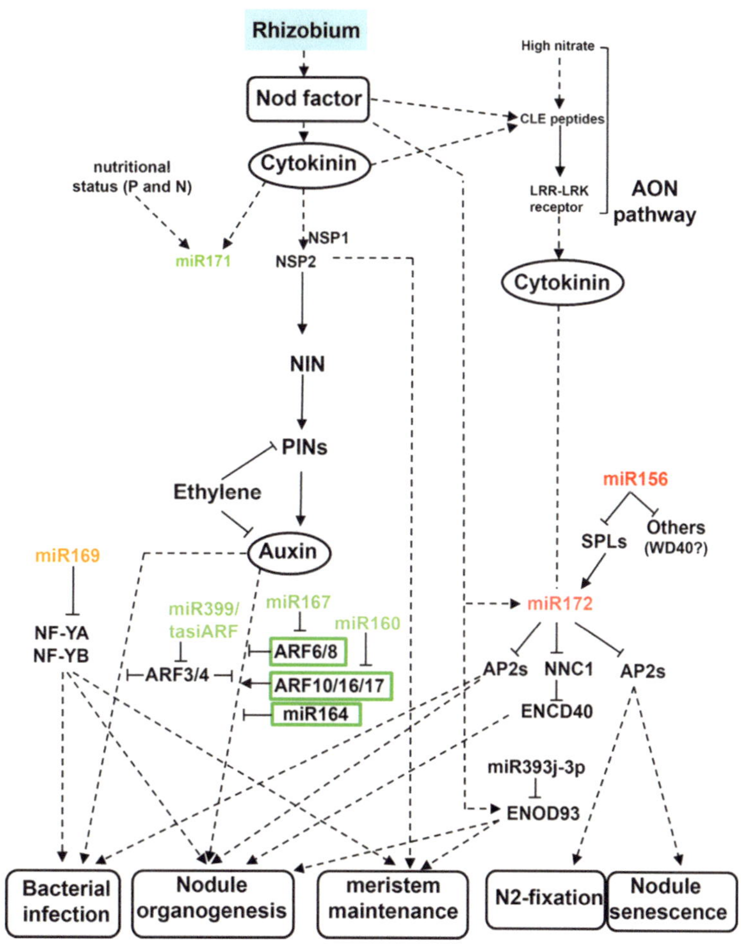

Figure 2.
Illustration of cross-talk between Nod factor induced common pathway signaling, miRNAs and regulatory hormones during rhizobial symbiosis. Solid and dotted lines represents direct and indirect interconnections respectively.

AP2 (APETALA2) targets has been implicated in rhizobia infection, stimulation of nodule organogenesis, N2 fixation, and delaying senescence in nodule cells.

The auto-regulation of the nodulation (AON) pathway, which governs the number of nodules that are formed in host plants, is one mechanism by which miRNA172 can work. In this pathway, leucine-rich repeat receptor-like kinase (LRR-LRK) receptors recognize NF- or nitrate-induced Clavata3/Embryo Surrounding Region-Related-peptides, resulting in an inhibitory signal (including CK) to cells, for establishing new nodules. By suppressing certain squamosa promoter-binding protein-like (SPL) TFs that stimulate miRNA172 production, miRNA156 antagonizes the action of miRNA172 [108].

Through the repression of nodulin gene (specifically ENOD93), miRNA393j-3p restricts nodules [109] whereas, miRNA1512 and miRNA1515 over-expression was discovered to be linked to increased nodule formation [75]. Finally, miRNA160 and miRNA167 cleave the transcripts of multiple auxin response factors (ARFs) that play key roles in the auxin response and pre-requisite for nodule initiation [110]. The detailed regulation mechanism is shown in **Figure 2**. miRNA390encourages the formation of a transacting small interfering (tasi)RNA, that represses ARF3 and ARF4during rhizobia colonization and nodule growth [78], it is known to combine auxin and ethylene signals. While, ARF10, ARF16, ARF17 and ARF6, ARF8 are directly regulated/targeted by miRNA160 and miRNA167, respectively [111] for auxin-responsive root development in both cases.

7. Mycorrhiza-derived non-coding RNAs and cross-kingdom signaling during symbiosis

Recent investigations have established non-coding RNAs as one of the central mediators of cross-kingdom communication between the plant and microbes [86, 88, 112–119]. These non-coding RNAs can move from donor organism to recipient organism, and target the specific host mRNAs for degradation. Sometimes sRNAs also trigger the production of secondary sRNA and thereby modulate the host defense and metabolic pathways [87, 120, 121]. Most of the studies focus on the plant-parasite or plant-pathogen (fungi and oomycetes) interactions [86, 88, 112, 120], however, such processes have been rarely explored in case of plant-mycorrhiza associations. Emerging body of evidences suggest that many plant miRNAs show differential expression patterns during AM symbiosis, nevertheless, their functions and cross-kingdom mobility remains unclear [80, 83, 111, 122, 123]. Mewalal et al. [124] identified several sRNAs from *Polpulus* spp. which were responsive to mutualistic/symbiotic interaction with mycorrhizal fungi like *Laccaria bicolor* and *Rhyzophagus irregularis*. Interestingly, they did not find any Populus RNAs interacting with *R. irregularis*, however, some of the miRNAs could interact with *L. bicolor*. Further the study revealed that these miRNAs can potentially target multiple host mRNAs encoding for vesicular transport and transcription regulatory proteins along with several uncharacterized proteins.

On the other hand, at present, very little information is available regarding the non-coding RNA biogenesis machinery and their functions in the development of arbuscular mycorrhizal (AM) and ectomycorrhizal (ECM) fungi or while interacting with the host plants. However, the successful application of host-induced gene silencing (HIGS) and virus-induced gene silencing (VIGS) approaches [65, 125–128] indicates that AMF, like pathogenic fungi, also possess functional RNAi machinery.

An *in silico* study identified putative RNAi machinery including a Dicer-like (DCL) gene, Argonaute-like (AGO-like) and RNA-dependent RNA polymerase (RdRp) gene families in *R. irregularis*, and validated their transcript-level expression [129]. An unusual expansion of AGO-like (5 members) and RdRp (21 members) gene families was observed in *R. irregularis.* Authors postulated that 15 out of 21RdRp genes, could be the product of a recent gene expansion event. The study also characterized the fungal sRNA and microRNA-like sequences, and predicted 237 transcripts of *Medicago truncatula* as their potential targets including a few known mRNAs that are modulated during AMF colonization. For instance, some of the *M. truncatula* mRNAs that are potentially targeted by *Rir*-sRNAs encode for the nuclear-binding leucine-rich repeat (*NBS-LRR*) type disease resistance gene, Non-specific phospholipase C4 (*NPC4*), *MtVapyrin* (Ankyrin repeat RF-like protein) and DREPP plasma membrane protein (*MtDREPP*) [129]. The homologs of NBS-LRR and NCP4 proteins from rice and arabidopsis, respectively, are involved in the plant immunity [130, 131], thus repression of these genes may allow AM colonization without triggering the robust host defense responses. MtVapyrin plays crucial role in arbuscule formation [132–134]. The down-regulation of *MtDREPP* has been reported in mycorrizal roots [135]. Though, further experimental validation is required, these findings indicate the possible existence of non-coding RNA-mediated post-transcriptional regulation and cross-kingdom gene silencing by AMF.

Another study by Silvestri et al. [136] identified the small RNA population from AMF *Gigaspora margarita* and showed their origin from different genetic sources such as endobacteria, RNA viruses and non-integrated DNA sequences from mitoviruses. Intriguingly, the extracellular vesicles (EVs), that are deployed in delivering the sRNA molecules to the other interacting partner [112, 120, 137], have also been observed in the peri-arbuscular interface of *R. irregularis* during the whole lifespan of arbuscules. This indicates the crucial role of EVs in cross-kingdom communication and nutrient exchange during AMF symbiosis [138]. More recently, a breakthrough discovery demonstrated that an ECM fungus *Pisolithus microcarpus* encodes 11 miRNAs, six of them were found induced during host colonization process. Notably, the miRNA (*Pmic_miR-8*) enters the plant cell and partakes in cross-kingdom gene silencing at some stage in symbiotic interaction with host plant *Eucalyptus grandis* [139]. The inhibition of *Pmic_miR-8* resulted in less developed Hartig nets, whereas, supplementation showed increased Hartig net depth in host tissue. Further the study showed that *Pmic_miR-8* may target the host NB-ARC (nucleotide-binding adaptor shared by APAF-1, R proteins, and CED-4) domain containing transcripts, indicating its potential role in modulating host signaling to stabilize the mutualistic association. As the CC (coiled-coil) nucleotide binding and leucine-rich repeat domain immune receptors (CC-NLR) are the largest category of NLRs, thus *Pmic_miR-8* may target several plant genes belonging to this class. Importantly, this is the first study which established the cross-kingdom gene silencing by mycorrhizal fungi and its role in beneficial interactions with host.

8. Conclusion and future prospects

In the last decade, the non-coding RNAs have emerged as one of the key regulators of diverse plant process including their development, response to abiotic/biotic stress, and nutrient uptake. A significant advancement has been made to understand the crucial roles of non-coding RNAs in plant-microbe interactions, particularly pathogenic interactions. Nutrient uptake via mycorrizal association is an important aspect of plants lifestyle and the studies suggest the extensive involvement of non-coding RNAs in regulating the plant nutrient status via affecting symbiosis. Notably, the sRNA-mediated regulatory mechanisms during mycorrhizal symbiosis have mainly focused on plant's perspective. These studies have provided crucial insights on understanding how the mycorrizal colonization proceeds and how the host plants fine-tune the extent of fungal colonization so that it does not turns pathogenic. On the contrary, the sRNA-mediated regulation of symbiosis from the fungal perspective remains infancy. Moreover, an integrated view of both the organisms (plant and fungus) will be required to appropriately comprehend the beneficial relationships. To understand ncRNA-molecular interaction networks occurring at plant host-AM symbiosis interface, experimental evidences and rewriting of dynamics of interaction, sensing, uptake, transport, assimilation and homeostasis of nutrients regulation are required. Extensive investigations for ncRNAs mediated regulation of cross-talk between AM and host plant are also needed for teeming knowledge voids. We anticipate that the recent discovery of cross-kingdom gene silencing by ECM fungus *Pisolithus microcarpus* and AM-like model EM fungus *Serendipita indica* would pave the way for future investigations of non-coding RNA-mediated regulatory networks in mycorriza growth and development as well as during host interactions, and would trigger novel research ideas among plant scientists. The better understanding of these regulatory circuits would aid in improving the nutritional status of plants in order to combat the elevating global quality-food demand.

Author details

Nidhi Verma*, Yeshveer Singh, Anupam Patra and Tanvi Singh
International Centre for Genetic Engineering and Biotechnology, New Delhi, India

*Address all correspondence to: vermanidhi5@gmail.com

References

[1] Smith SE, Smith FA. Roles of Arbuscular Mycorrhizas in plant nutrition and growth: New paradigms from cellular to ecosystem scales. Annual Review Plant Biology. 2011;**62**:227-250

[2] Zamioudis C, Pieterse CM. Modulation of host immunity by beneficial microbes. Molecular Plant-Microbe Interactions: MPMI. 2012;**25**(2):139-150

[3] López C, Szurek B, Perez-Quintero Á. Small non-coding RNAs in plant immunity. In: Dhal N, Sahu S, editors. Plant Science. London, UK: IntechOpen; 2012

[4] Plett JM, Martin FM. Know your enemy, embrace your friend: using omics to understand how plants respond differently to pathogenic and mutualistic microorganisms. The Plant Journal. 2018;**93**(4):729-746

[5] Xu L, Hu Y, Cao Y, Li J, Ma L, Li Y, et al. An expression atlas of miRNAs in Arabidopsis thaliana. Science China Life Sciences. 2018;**61**(2):178-189

[6] Bonnet E, Van de Peer Y, Rouze P. The small RNA world of plants. The New Phytologist. 2006;**171**:451-468

[7] Liu X, Hao L, Li D, Zhu L, Hu S. Long non-coding RNAs and their biological roles in plants. Genomics Proteomics Bioinformatics. 2015;**13**:137-147

[8] Chen X. Small RNAs and their roles in plant development. Annual Review of Cell and Developmental Biology. 2009;**25**:21-44

[9] D'Ario M, Griffiths-Jones S, Kim M. Small RNAs: Big impact on plant development. Trends in Plant Science. 2017;**22**:1056-1068

[10] Martinez G, Kohler C. Role of small RNAs in epigenetic reprogramming during plant sexual reproduction. Current Opinion in Plant Biology. 2017;**36**:22-28

[11] Tang J, Chu C. MicroRNAs in crop improvement: Fine-tuners for complex traits. Nat. Plants. 2017;**3**:17077

[12] Chekanova JA. Long non-coding RNAs and their functions in plants. Current Opinion in Plant Biology. 2015;**27**:207-216

[13] Sablok G, Zhao H, Sun X. Plant circular RNAs (circRNAs): Transcriptional regulation beyond miRNAs in plants. Molecular Plant. 2016;**9**:192-194

[14] Alves CS, Vicentini R, Duarte GT, Pinoti VF, Vincentz M, Nogueira FT. Genome-wide identification and characterization of tRNA-derived RNA fragments in land plants. Plant Molecular Biology. 2017;**93**:35-48

[15] Cognat V, Morelle G, Megel C, Lalande S, Molinier J, Vincent T, et al. The nuclear and organellar tRNA-derived RNA fragment population in Arabidopsis thaliana is highly dynamic. Nucleic Acids Research. 2017;**45**:3460-3472

[16] Megel C, Morelle G, Lalande S, Duchene AM, Small I, Marechal-Drouard L. Surveillance and cleavage of eukaryotic tRNAs. International Journal of Molecular Sciences. 2015;**16**:1873-1893

[17] Axtell MJ, Meyers BC. Revisiting criteria for plant microrna annotation in the era of big data. Plant Cell. 2018;**30**:272-284

[18] Chao H, Hu Y, Zhao L, Xin S, Ni Q, Zhang P, et al. Biogenesis, functions, interactions, and resources of non-coding RNAs in plants. International Journal of Molecular Sciences. 2022;**23**:3695

[19] Fei Q, Xia R, Meyers BC. Phased, secondary, small interfering RNAs in posttranscriptional regulatory networks. Plant Cell. 2013;**25**:2400-2415

[20] Peragine A, Yoshikawa M, Wu G, Albrecht HL, Poethig RS. SGS3 and SGS2/SDE1/RDR6 are required for juvenile development and the production of trans-acting siRNAs in Arabidopsis. Genes & Development. 2004;**18**:2368-2379

[21] Zhu L, Ow DW, Dong Z. Transfer RNA-derived small RNAs in plants. Science China. Life Sciences. 2018;**61**:155-161

[22] Kolakofsky D. Isolation and characterization of Sendai virus DI-RNAs. Cell. 1976;**8**:547-555

[23] Liu CX, Li X, Nan F, Jiang S, Gao X, Guo SK, et al. Structure and degradation of circular RNAs regulate PKR activation in innate immunity. Cell. 2019;**177**:865-880.e21

[24] Waititu JK, Zhang C, Liu J, Wang H. Plant non-coding RNAs: Origin, biogenesis mode of action and their roles in abiotic stress. International Journal Molecular Science. 2020;**21**:8401

[25] Matsumoto A, Nakayama KI. Hidden peptides encoded by putative noncoding RNAs. Cell Structure and Function. 2018;**43**:75-83

[26] Wang Y, Zhang C, Hao Q, Sha A, Zhou R, Zhou X, et al. Elucidation of miRNAs-mediated responses to low nitrogen stress by deep sequencing of two soybean genotypes. PLoS One. 2013;**8**:e67423

[27] Xu Z, Zhong S, Li X, Li W, Rothstein SJ, Zhang S, et al. Genome-wide identification of microRNAs in response to low nitrate availability in maize leaves and roots. PLoS One. 2011;**6**:e28009

[28] Hsieh LC, Lin SI, Shih ACC, Chen JW, Lin WY, Tseng CY, et al. Uncovering small RNA-mediated responses to phosphate deficiency in Arabidopsis by deep sequencing. Plant Physiology. 2009;**151**:2120-2132

[29] Gu M, Liu W, Meng Q, Zhang WQ, Chen AQ, Sun SB, et al. Identification of microRNAs in six solanaceous plants and their potential link with phosphate and mycorrhizal signalings. Journal of Integrative Plant Biology. 2014;**56**:1164-1178

[30] Lundmark M, Korner CJ, Nielsen TH. Global analysis of micro RNA in Arabidopsis in response to phosphate starvation as studied by locked nucleic acid-based microarrays. Physiologia Plantarum. 2010;**140**:57-68

[31] Pei L, Jin Z, Li K, Yin H, Wang J, Yang A. Identification andcomparative analysis of low phosphate tolerance associated microRNAs in two maize genotypes. Plant Physiology and Biochemistry. 2013;**70**:221-234

[32] Xu F, Liu Q, Chen L, Kuang J, Walk T, Wang J, et al. Genomewide identification of soybean microRNAs and their targets reveals their organ specificity and responses to phosphate starvation. BMC Genomics. 2013;**14**:e66

[33] Zhao X, Liu X, Guo C, Gu J, Xiao K. Identification and characterization of microRNAs from wheat (*Triticum aestivum* L.) under phosphorus

deprivation. Journal of Plant Biochemistry and Biotechnology. 2013;**22**:113-123

[34] Sun SB, Gu M, Cao Y, Huang XP, Zhang X, Ai PH, et al. A constitutive expressed phosphate transporter, OsPht1; 1, modulates phosphate uptake and translocation in phosphate-replete rice. Plant Physiology. 2012;**159**:1571-1581

[35] Sunkar R, Li YF, Jagadeeswaran G. Functions of microRNAs in plant stress responses. Trends in Plant Science. 2012;**17**:196-203

[36] Liang G, He H, Yu D. Identification of nitrogen starvation responsive miRNAs in *Arabidopsis thaliana*. PLoS One. 2012;7:e48951

[37] Zeng HQ, Zhu YY, Huang SQ, Yang ZM. Analysis of phosphorus-deficient responsive miRNAs and cis elements from soybean (*Glycine max* L.). Journal Plant Physiology. 2010;**167**:1289-1297

[38] Pacak A, Barciszewska-Pacak M, Swida-Barteczka A, Kruszka K, Sega P, Milanowska K, et al. Heat stress affects Pi-related genes expression and inorganic phosphate deposition/accumulation in Barley. Frontiers in Plant Science. 2016;7:e926

[39] Rajagopalan R, Vaucheret H, Trejo J, Bartel DP. A diverse and evolutionarily fluid set of microRNAs in *Arabidopsis thaliana*. Genes & Development. 2006;**20**:3407-3425

[40] Dugas DV, Bartel B. Sucrose induction of Arabidopsis miR398 represses two Cu/Zn superoxide dismutases. Plant Molecular Biology. 2008;**67**:403-417

[41] Kuo HF, Chiou TJ. The role of microRNAs in phosphorus deficiency signaling. Plant Physiology. 2011;**156**:1016-1024

[42] Howarth JR, Fourcroy P, Davidian JC, Smith FW, Hawkesford MJ. Cloning of two contrasting high-affinity sulfate transporters from tomato induced by low sulfate and infection by the vascular pathogen *Verticillium dahliae*. Planta. 2003;**218**:58-64

[43] Takahashi H. Regulation of sulfate transport and assimilation in plants. International Review of Cell and Molecular Biology. 2010;**281**:129-159

[44] Yoshimoto N, Inoue E, Watanabe-Takahashi A, Saito K, Takahashi H. Posttranscriptional regulation of high-affinity sulfate transporters in Arabidopsis by sulfur nutrition. Plant Physiology. 2007;**145**:378-388

[45] Shinmachi F, Buchner P, Stroud JL, Parmar S, Zhao FJ, McGrath SP, et al. Influence of sulfur deficiency on the expression of specific sulphate transporters and the distribution of sulfur, selenium and molybdenum in wheat. Plant Physiology. 2010;**153**:327-336

[46] Zuber H, Davidian JC, Wirtz M, Hell R, Belghazi M, Thompson R, et al. Sultr4;1 mutant seeds of Arabidopsis have an enhanced sulphate content and modified proteome suggesting metabolic adaptations to altered sulphate compartmentalization. BMC Plant Biology. 2010;**10**:e78

[47] Zeng H, Wang G, Hu X, Wang H, Du L, Zhu Y. Role of microRNAs in plant responses to nutrient stress. Plant and Soil. 2014;**374**:1005-1021

[48] Bucking H, Liepold E, Ambilwade P. The role of the mycorrhizal symbiosis in nutrient uptake of plants and the

regulatory mechanisms underlying these transport processes. In: Dhal NK, Sahu SC, editors. Plant Science. London, UK: IntechOpen; 2012

[49] Huey CJ, Gopinath SCB, Uda MNA, Zulhaimi HI, Jaafar MN, Kasim FH, et al. Mycorrhiza: A natural resource assists plant growth under varied soil conditions. 3 Biotech. 2020 May;**10**(5):204

[50] Wright DP, Read DJ, Scholes JD. Mycorrhizal sink strength influences whole plant carbon balance of trifolium repens L. Plant, Cell and Environment. 1998;**21**:881-891

[51] Chiou TJ, Liu H, Harrison MJ. The spatial expression patterns of a phosphate transporter (MtPT1) from Medicago truncatula indicate a role in phosphate transport at the root/soil interface. The Plant Journal. 2001;**25**(3):281-293

[52] Grunwald U, Guo WB, Fischer K, Isayenkov S, Ludwig-Müller J, Hause B, et al. Overlapping expression patterns and differential transcript levels of phosphate transporter genes in Arbuscular Mycorrhizal, Pi-Fertilised and phytohormone-treated medicago truncatula roots. Planta. 2009;**229**(5):1023-1034

[53] Li HY, Smith SE, Holloway RE, ZhuYG, SmithFA. Arbuscularmycorrhizal fungi contribute to phosphorus uptake by wheat grown in a phosphorus-fixing soil even in the absence of positive growth responses. New Phytologist. 2006;**172**:536-543

[54] Smith SE, Jakobsen I, Grønlund M, Smith FA. Roles of Arbuscular Mycorrhizas in plant phosphorus nutrition: interactions between pathways of phosphorus uptake in Arbuscular Mycorrhizal roots have important implications for understanding and manipulating plant phosphorus acquisition. Plant Physiology. 2011;**156**:1050-1057

[55] Smith SE, Smith FA, Jakobsen I. Mycorrhizal Fungi can dominate phosphate supply to plants irrespective of growth responses. Plant Physiology. 2003;**133**(1):16-20

[56] Tanaka Y, Yano K. Nitrogen delivery to maize via Mycorrhizal hyphae depends on the form of N supplied. Plant, Cell and Environment. 2005;**28**:1247-1254

[57] Toussaint JP, St-Arnaud M, Charest C. Nitrogen transfer and assimilation between the Arbuscular Mycorrhizal Fungus Glomus intraradices Schenck and Smith and Ri T-DNA roots of *Daucus carota* L. in an vitro compartmented system. Canadian Journal of Microbiology. 2004;**50**:251-260

[58] Guether M, Neuhauser B, Balestrini R, Dynowski M, Ludewig U, Bonfante P. A Mycorrhizal-specific ammonium transporter from lotus japonicus acquires nitrogen released by Arbuscular Mycorrhizal fungi. Plant Physiology. 2009;**150**(1):73-83

[59] Harrison MJ, Dewbre GR, Liu J. A phosphate transporter from Medicago Truncatula involved in the acquisition of phosphate released by Arbuscular Mycorrhizal Fungi. The Plant Cell. 2002;**14**:2413-2429

[60] Xu GH, Chague V, Melamed-Bessudo C, Kapulnik Y, Jain A, Raghothama KG, et al. Functional characterization of LePT4: A phosphate transporter in tomato with Mycorrhiza-enhanced expression. Journal of Experimental Botany. 2007;**58**(10):2491-2501

[61] Allen JW, Shachar-Hill Y. Sulfur transfer through an Arbuscular Mycorrhiza. Plant Physiology. 2009;**149**(1):549-560

[62] Casieri L, Gallardo K, Wipf D. Transcriptional response of Medicago truncatula Sulphate Transporters to Arbuscular Mycorrhizal symbiosis with and without sulphur stress. Planta. 2012. DOI: 10.1007/s00425-012-1645-7

[63] Narayan OP, Verma N, Jogawat A, Dua M, Johri AK. Sulfur transfer from the endophytic fungus Serendipita indica improves maize growth and requires the sulfate transporter SiSulT. The Plant Cell. 2021;**33**(4):1268-1285

[64] Verma N, Narayan OP, Prasad D, Jogawat A, Panwar SL, Dua M, et al. Functional characterization of a high-affinity iron transporter (PiFTR) from the endophytic fungus Piriformospora indica and its role in plant growth and development. Environmental Microbiology. 2022;**24**(2):689-706

[65] Helber N, Wippel K, Sauer N, Schaarschmidt S, Hause B, Requena N. A versatile monosaccharide transporter that operates in the Arbuscular Mycorrhizal fungus Glomus sp is crucial for the symbiotic relationship with plants. The Plant Cell. 2011;**23**:3812-3823

[66] Bücking H, Heyser W. Uptake and transfer of nutrients in ectomycorrhizal associations: interactions between photosynthesis and phosphate nutrition. Mycorrhiza. 2003;**13**:59-68

[67] Fellbaum CR, Gachomo EW, Beesetty Y, Choudhari S, Strahan GD, Pfeffer PE, et al. Carbon availability triggers fungal nitrogen uptake and transport in Arbuscular Mycorrhizal symbiosis. Proceedings of the National Academy of Sciences of the United States of America. 2012;**109**(7):2666-2671

[68] Hammer EC, Pallon J, Wallander H, Olsson PA. Tit for tat? a mycorrhizal fungus accumulates phosphorus under low plant carbon availability. FEMS Microbiology Ecology. 2011;**76**:236-244

[69] Garcia K, Doidy J, Zimmermann SD, Wipf D, Courty PE. Take a trip through the plant and fungal transportome of Mycorrhiza. Trends in Plant Science. 2016;**21**(11):937-950

[70] Diédhiou I, Diouf D. Transcription factors network in root endosymbiosis establishment and development. World Journal of Microbiology and Biotechnology. 2018;**34**(3):37

[71] Middleton H, Yergeau É, Monard C, Combier JP, El Amrani A. Rhizospheric plant-microbe interactions: miRNAs as a key mediator. Trends in Plant Science. 2021;**26**(2):132-141

[72] Müller LM, Harrison MJ. Phytohormones, miRNAs, and peptide signals integrate plant phosphorus status with arbuscular mycorrhizal symbiosis. Current Opinion in Plant Biology. 2019;**50**:132-139

[73] Pimprikar P, Gutjahr C. Transcriptional regulation of arbuscular mycorrhiza development. Plant & Cell Physiology. 2018;**59**(4):673-690

[74] Feng H, Zhang Q, Wang Q, Wang X, Liu J, Li M, et al. Target of tae-miR408, a chemocyanin-like protein gene (TaCLP1), plays positive roles in wheat response to high-salinity, heavy cupric stress and stripe rust. Plant Molecular Biology. 2013;**83**:433-443

[75] Li Y, Zhang Q, Zhang J, Wu L, Qi Y, Zhou JM. Identification of microRNAs involved in pathogen-associated molecular pattern-triggered plant innate immunity. Plant Physiology. 2010;**152**:2222-2231

[76] Baldrich P, Campo S, Wu MT, Liu TT, Hsing YI, San Segundo B. MicroRNA-mediated regulation of gene expression in the response of riceplants to fungal elicitors. RNA Biology. 2015;**12**:847-863

[77] Wong J, Gao L, Yang Y, Zhai J, Arikit S, Yu Y, et al. Roles of small RNAs in soybean defense against Phytophthora sojae infection. The Plant Journal. 2014;**79**:928-940

[78] Li Y, Lu YG, Shi Y, Wu L, Xu YJ, Huang F, et al. Multiple rice microRNAs are involved in immunity against the blast fungus Magnaporthe oryzae. Plant Physiology. 2014;**164**:1077-1092

[79] Sarkar D, Maji RK, Dey S, Sarkar A, Ghosh Z, Kundu P. Integrated miRNA and mRNA expression profiling reveals the response regulators of a susceptible tomato cultivar to early blight disease. DNA Research. 2017;**24**:235-250

[80] Formey D, Sallet E, Lelandais-Briere C, Ben C, Bustos-Sanmamed P, Niebel A, et al. The small RNA diversity from *Medicago truncatula* roots under biotic interactions evidences the environmental plasticity of the miRNAome. Genome Biology. 2014;**15**:457

[81] Floss D, Levy J, Lévesque-Tremblay V, Pumplin N, Harrison M. DELLA proteins regulate arbuscule formation in arbuscular mycorrhizal symbiosis. Proceedings of the National Academy of Sciences of the United States of America. 2013;**110**:E5025-E5034

[82] Martín-Rodríguez JA, Ocampo JA, Molinero-Rosales N, Tarkowská D, Ruíz-Rivero O, García-Garrido JM. Role of gibberellins during arbuscular mycorrhizal formation in tomato: New insights revealed by endogenous quantification and genetic analysis of their metabolism in mycorrhizal roots. Physiologia Plantarum. 2015;**154**:66-81

[83] Wu P, Wu Y, Liu CC, Liu LW, Ma FF, Wu XY, et al. Identification of arbuscular mycorrhiza (AM)-responsive microRNAs in tomato. Frontiers in Plant Science. 2016;7:429

[84] Holt DB, Gupta V, Meyer D, Abel NB, Andersen SU, Stougaard J, et al. microRNA 172 (miR172) signals epidermal infection and is expressed in cells primed for bacterial invasion in Lotus japonicus roots and nodules. The New Phytologist. 2015;**208**:241-256

[85] Wang L, Qin L, Liu W, Zhang D, Wang Y. A novel ethylene-responsive factor from Tamarix hispida, ThERF1, is a GCC-box- and DRE-motif binding protein that negatively modulates abiotic stress tolerance in Arabidopsis. Physiologia Plantarum. 2014;**152**:84-97

[86] Wang M, Thomas N, Jin H. Cross-kingdom RNA trafficking and environmental RNAi for powerful innovative pre- and post-harvest plant protection. Current Opinion in Plant Biology. 2017a;**38**:133-141

[87] Weiberg A, Wang M, Lin FM, Zhao H, Zhang Z, Kaloshian I, et al. Fungal small RNAs suppress plant immunity by hijacking host RNA interference pathways. Science. 2013;**342**:118-123

[88] Wang M, Weiberg A, Dellota E Jr, Yamane D, Jin H. Botrytis small RNA Bc-si R37 suppresses plant defense genes by cross-kingdom RNAi. RNA Biology. 2017b;**14**:421-428

[89] Gao R, Liu P, Irwanto N, Loh R, Wong SM. Upregulation of LINC-AP2 is negatively correlated with AP2 gene expression with Turnip crinkle virus

infection in Arabidopsis thaliana. Plant Cell Reports. 2016;**35**:2257-2267

[90] Jiang N, Cui J, Shi Y, Yang G, Zhou X, Hou X, et al. Tomato lnc RNA23468 functions as a competing endogenous RNA to modulate NBS-LRR genes by decoying miR482b in the tomato-Phytophthora infestans interaction. Horticulture Research. 2019;**6**:28

[91] Li WQ, Jia YL, Liu FQ, Wang FQ, Fan FJ, Wang J, et al. Genome-wide identification and characterization of long non-coding RNAs responsive to Dickeya zeae in rice. RSC Advances. 2018;**8**:34408-34417

[92] MuthusamyM,UmaS,SuthanthiramB, Saraswathi MS, Chandrasekar A. Genome-wide identification of novel, long non-coding RNAs responsive to Mycosphaerella eumusae and Pratylenchus co_eae infections and their differential expression patterns in disease-resistant and sensitive banana cultivars. Plant Biotechnology Report. 2019;**13**:73-83

[93] Yang Y, Liu T, Shen D, Wang J, Ling X, Hu Z, et al. Tomato yellow leaf curl virus intergenic siRNAs target a host long noncoding RNA to modulate disease symptoms. PLoS Pathogens. 2019;**15**:e1007534

[94] Zhang H, Hu W, Hao J, Lv S, Wang C, Tong W, et al. Genome-wide identification and functional prediction of novel and fungi-responsive lincRNAs in Triticum aestivum. BMC Genomics. 2016;**17**:238

[95] Runtsch MC, O'Neill LA. GOTcha: lncRNA-ACOD1 targets metabolism during viral infection. Cell Research. 2018;**28**:137-138

[96] Cui J, Jiang N, Meng J, Yang G, Liu W, Zhou X, et al. LncRNA33732-respiratory burst oxidase module associated with WRKY1 in tomato-Phytophthora infestans interactions. Plant Journal Cell Molecular Biology. 2019;**97**:933-946

[97] Du Q, Wang K, Zou C, Xu C, Li WX. The PILNCR1-miR399 regulatory module is important for low phosphate tolerance in Maize. Plant Physiology. 2018;**177**:1743-1753

[98] Lv Y, Liang Z, Ge M, Qi W, Zhang T, Lin F, et al. Genome-wide identification and functional prediction of nitrogen-responsive intergenic and intronic long non-coding RNAs in maize (*Zea mays* L.). BMC Genomics. 2016;**17**:350

[99] Pang J, Zhang X, Ma X, Zhao J. Spatio-temporal transcriptional dynamics of maize long non-coding rnas responsive to drought stress. Genes (Basel). 2019;**10**:138

[100] Han L, Mu Z, Luo Z, Pan Q, Li L. New lncRNA annotation reveals extensive functional divergence of the transcriptome in maize. Journal of Integrative Plant Biology. 2019;**61**:394-405

[101] Napoli C, Lemieux C, Jorgensen R. Introduction of a chimeric chalcone synthase gene into petunia results in reversible co-suppression of homologous genes in trans. The Plant Cell. 1990;**2**(4):279-289

[102] van der Krol AR, Mur LA, Beld M, Mol JN, Stuitje AR. Flavonoid genes in petunia: Addition of a limited number of gene copies may lead to a suppression of gene expression. The Plant Cell. 1990;**2**(4):291-299

[103] Bazin J, Bustos-Sanmamed P, Hartmann C, Lelandais-Briere C, Crespi M. Complexity of miRNA-dependent regulation in rootsymbiosis.

Philosophical Transactions of the Royal Society of London. Series B. Biological Sciences. 2012;**367**:1570-1579

[104] Simon SA, Meyers BC, Sherrier DJ. MicroRNAs in therhizobia legume symbiosis. Plant Physiology. 2009;**151**:1002-1008

[105] Gutjahr C, Parniske M. Cell and developmental biology of arbuscular mycorrhiza symbiosis. Annual Review of Cell and Developmental Biology. 2013;**29**:593-617

[106] Bazin J, Khan GA, Combier JP, Bustos-Sanmamed P, Debernardi JM, Rodriguez R, et al. miR396 affects mycorrhization and rootmeristem activity in the legume Medicago truncatula. The Plant Journal. 2013;**74**:920-934

[107] Laporte P, Lepage A, Fournier J, Catrice O, Moreau S, Jardinaud MF, et al. The CCAAT box-binding transcription factor NF-YA1 controls rhizobial infection. Journal of Experimental Botany. 2014;**65**:481-494

[108] Yan Z, Hossain MS, Wang J, Vald ´es-L´opez O, Liang Y, Libault M, et al. miR172 regulates soybean nodulation. Molecular Plant-Microbe Interactions: MPMI. 2013;**26**:1371-1377

[109] Etemadi M, Gutjahr C, Couzigou JM, Zouine M, Lauressergues D, Timmers A, et al. Auxin perception is required for arbuscule development in arbuscularmycorrhizal symbiosis. Plant Physiology. 2014;**166**:281-292

[110] Khan GA, Declerck M, Sorin C, Hartmann C, Crespi M, Lelandais-Bri'ere, C. MicroRNAs as regulators of root develop-ment and architecture. Plant Molecular Biology. 2011;77:47-58

[111] Lelandais-Brière C, Moreau J, Hartmann C, Crespi M. Noncoding RNAs, emerging regulators in root endosymbioses. Molecular Plant-Microbe Interactions. 2016;**29**(3):170-180

[112] Cai Q, He B, Kogel K-H, Jin H. Cross-kingdom RNA trafficking and environmental RNAi — nature's blueprint for modern crop protection strategies. Current Opinion in Microbiology. 2018a;**46**:58-64

[113] Chaloner T, van Kan JAL, Grant-Downton RT. RNA "information warfare" in pathogenic and mutualistic interactions. Trends in Plant Science. 2016;**21**:738-748

[114] Knip M, Constantin ME, Thordal-Christensen H. Trans-kingdom cross-talk: small RNAs on the move. PLoS Genetics. 2014;**10**(9):e1004602

[115] Wang M, Weiberg A, Lin F-M, Thomma BPHJ, Huang H-D, Jin H. Bidirectional cross-kingdom RNAi and fungal uptake of external RNAs confer plant protection. Nature Plants. 2016;2:16151

[116] Weiberg A, Bellinger M, Jin H. Conversations between kingdoms: Small RNAs. Current Opinion in Biotechnology. 2015;**32**:207-215

[117] Weiberg A, Jin H. Small RNAs-the secret agents in the plant-pathogen interactions. Current Opinion in Plant Biology. 2015;**26**:87-94

[118] Weiberg A, Wang M, Bellinger M, Jin H. Small RNAs: A new paradigm in plant-microbe interactions. Annual Review of Phytopathology. 2014;**52**:495-516

[119] Zhou G, Zhou Y, Chen X. New insight into inter-kingdom communication: Horizontal transfer

of mobile small RNAs. Frontiers in Microbiology. 2017;**8**:1-9

[120] Cai Q, Qiao L, Wang M, He B, Lin F-M, Palmquist J. Plants send small RNAs in extracellular vesicles to fungal pathogen to silence virulence genes. Science. 2018b;**360**(6393):1126-1129

[121] Shahid S, Kim G, Johnson NR, Wafula E, Wang F, Coruh C, et al. MicroRNAs from the parasitic plant Cuscuta campestris target host messenger RNAs. Nature. 2018;**553**:82-85

[122] Couzigou JM, Lauressergues D, André O, Gutjahr C, Guillotin B, Bécard G, et al. Positive gene regulation by a natural protective miRNA enables arbuscular mycorrhizal symbiosis. Cell Host & Microbe. 2017;**21**:106-112

[123] Devers EA, Branscheid A, May P, Krajinski F. Stars and Symbiosis: MicroRNA and microRNA*-mediated transcript cleavage involved in arbuscular mycorrhizal symbiosis. Plant Physiology. 2011;**156**:1990-2010

[124] Mewalal R, Yin H, Hu R, Jawdy S, Vion P, Tuskan GA, et al. Identification of populus small RNAs responsive to mutualistic interactions with mycorrhizal fungi laccaria bicolor and rhizophagus irregularis. Frontiers in Microbiology. 2019;**10**:515

[125] Kemppainen M, Duplessis S, Martin F, Pardo AG. RNA silencing in the model mycorrhizal fungus Laccaria bicolor: Gene knock-down of nitrate reductase results in inhibition of symbiosis with populus. Environmental Microbiology. 2009;**11**(7):1878-1896

[126] Kikuchi Y, Hijikata N, Ohtomo R, Handa Y, Kawaguchi M, Saito K, et al. Aquaporin-mediated long-distance polyphosphate translocation directed towards the host in arbuscular mycorrhizal symbiosis: application of virusinduced gene silencing. The New Phytologist. 2016;**211**:1202-1208

[127] Tsuzuki S, Handa Y, Takeda N, Kawaguchi M. Strigolactone-induced putative secreted protein 1 is required for the establishment of symbiosis by the arbuscular mycorrhizal fungus Rhizophagus irregularis. Molecular Plant-Microbe Interactions. 2016;**29**:1-59

[128] Voß S, Betz R, Heidt S, Corradi N, Requena N. RiCRN1, a crinkler effector from the arbuscular mycorrhizal fungus Rhizophagus irregularis, functions in arbuscule development. Frontiers in Microbiology. 2018;**9**:1-18

[129] Silvestri A, Fiorilli V, Miozzi L, Accotto GP, Turina M, Lanfranco L. In silico analysis of fungal small RNA accumulation reveals putative plant mRNA targets in the symbiosis between an arbuscular mycorrhizal fungus and its host plant. BMC Genomics. 2019;**20**(1):1-8

[130] García-Garrido JM, Ocampo JA. Regulation of the plant defence response in arbuscular mycorrhizal symbiosis. Journal of Experimental Botany. 2002;**53**:1377-1386

[131] Siebers M, Brands M, Wewer V, Duan Y, Hölzl G, Dörmann P. Lipids in plant–microbe interactions. Biochimica et Biophysica Acta-Molecular Cell Biology Lipids. 2016;**1861**:1379-1395

[132] Feddermann N, Reinhardt D. Conserved residues in the ankyrin domain of VAPYRIN indicate potential protein-protein interaction surfaces. Plant Signaling & Behavior. 2011;**6**:680-684

[133] Pumplin N, Mondo SJ, Topp S, Starker CG, Gantt JS, Harrison MJ. Medicago truncatula Vapyrin

is a novel protein required for arbuscular mycorrhizal symbiosis. The Plant Journal. 2010;**61**:482-494

[134] Zhang X, Pumplin N, Ivanov S, Harrison MJ. EXO70I is required for development of a sub-domain of the periarbuscular membrane during arbuscular mycorrhizal symbiosis. Current Biology. 2015;**25**:2189-2195

[135] Aloui A, Recorbet G, Lemaître-Guillier C, Mounier A, Balliau T, Zivy M, et al. The plasma membrane proteome of *Medicago truncatula* roots as modified by Arbuscular Mycorrhizal symbiosis. Mycorrhiza. 2018;**28**:1-16

[136] Silvestri A, Turina M, Fiorilli V, Miozzi L, Venice F, Bonfante P, et al. Different genetic sources contribute to the small RNA population in the Arbuscular Mycorrhizal fungus Gigaspora margarita. Frontiers in Microbiology. 2020;**11**:395

[137] Huang CY, Wang H, Hu P, Hamby R, Jin H. Small RNAs-big players in plant-microbe interactions. Cell Host & Microbe. 2019;**26**(2):173-182

[138] Roth R, Hillmer S, Funaya C, Chiapello M, Schumacher K, Lo Presti L, et al. Arbuscular cell invasion coincides with extracellular vesicles and membrane tubules. Nature Plants. 2019;**5**:204-211

[139] Wong-Bajracharya J, Singan VR, Monti R, Plett KL, Ng V, Grigoriev IV, et al. The ectomycorrhizal fungus Pisolithus microcarpus encodes a microRNA involved in cross-kingdom gene silencing during symbiosis. Proceedings of the National Academy of Sciences of the United States of America. 2022;**119**(3):e2103527119

Chapter 13

Receptor Kinases and Signal Pathway in the Arbuscular Mycorrhizal Symbiosis

Jiashan Wu, Weiyun Wang, Hui Zhu and Yangrong Cao

Abstract

Most terrestrial plants establish symbiotic interactions with arbuscular mycorrhizal fungi (AMF) to acquire phosphorus and nitrogen nutrients. The current understanding regarding how plants recognize symbiotic signals has now been updated. Plant Lysin-Motif receptor kinases, that is, rice OsCERK1 and OsMYR1 or orthologs from other plants, perceive Myc factor, a lipochitooligosaccharide from AMF, to initiate symbiotic signaling pathway. The Myc factor receptor model is quite similar to the known Nod factor receptors required for rhizobial symbiosis and chitin receptors for chitin-triggered immunity. Thus, the open question is how plants use similar receptor complexes to recognize structurally similar molecules to induce different signaling pathways. Upon recognition of Myc/Nod factors signaling, LysM receptors could activate the symbiosis receptor kinase (SymRK), which is an essential component of common symbiotic signaling pathway (CSSP) for both mycorrhizal symbiosis and rhizobial symbiosis. Downstream of SymRK, a clear module in the CSSP by CCaMK-CYCLOPS-DELLA was identified to promote both mycorrhizal symbiosis by activating the expression of *RAM1*, and rhizobial symbiosis by forming a complex with NSP1/NSP2 to regulate the expression of *NIN*. In this chapter, we discussed the roles of receptor kinases and CSSP in mycorrhizal symbiosis, as well as in rhizobial symbiosis.

Keywords: symbiosis, LysM-receptor kinases, root nodule symbiosis, SymRK, transcription factor, common symbiosis signal pathway

1. Introduction

Arbuscular mycorrhizal symbiosis (AMS) is a mutualistic interaction formed between more than 80% of terrestrial plants and members of the Glomeromycotina fungi, referred to as the arbuscular mycorrhizal fungi (AMF) [1]. It was proposed that AMS evolved approximately 400–450 million years ago, while root nodule symbiosis (RNS) originated about 60 million years ago [2, 3]. Thus, it was consistent with the generally accepted theory that RNS might be a result of a gradual attenuation of AMS, and both of them might evolve from the ancient plant-pathogens interaction [4, 5]. In the AMS, AMF could help plants to absorb more phosphorus and nitrogen nutrients from environment, and in return, plants provide carbohydrates mainly in the form of lipids for AMF [6–9]. As one of the pivotal nutrients for host plants' growth, phosphate is known negatively correlate with AMS [10–12]. Meanwhile, AMF could help host plants adapt to stressful environmental conditions [13, 14]. The development of AMS is a highly dynamic process, including presymbiotic communication between both symbionts, colonization of AMF in the plant root cortex and highly branched structures called arbuscules formation [15, 16], vesicle and spore maturation. Before making physical contact, phytohormones strigolactones (SLs) secreted by the roots of plants into the rhizosphere under Pi-deficient conditions [17], promote the branching of mycorrhizal hyphae. Simultaneously, the secreted cutin monomer promotes the colonization of AMF in the host roots. But the molecular basis of SL perception by AMF spores has not yet been elucidated.

Similar to the process of rhizobial symbiosis, branched hyphae of AMF secrete mycorrhizal (Myc) factors, a mixture of short-chain chitooligosaccharides (CO4/CO5) and lipochitooligosaccharides (LCOs), both of which are similar structures with rhizobial Nod factors [18, 19], to activate the common symbiosis signaling pathway (CSSP), required for both AMS and RNS [2, 20]. Over the past two decades, advances have been made in the areas of symbiotic signals perception, including the identification of LysM receptor-like kinases in non-legumes, especially in rice, and symbiotic signaling transduction in plants. Downstream of LysM receptors, several key common components were identified as shared components for both AMS and RNS, that is, Symbiosis receptor kinase (SymRK), also called Nodulation Receptor Kinase (NORK) or Does not Make Infection 2 (DMI2) in other plant species [21, 22], calcium and calmodulin-dependent kinase (CCaMK) [23], CYCLOPS [24]. GRAS transcription factor (TF) family proteins such as DELLAs [25], Nodulation Signaling Pathway 1 (NSP1), NSP2 [26], Reduced Arbuscular Mycorrhiza 1 (RAM1) [27, 28] are also involved in AMS and/or RNS. In addition, some downstream components, such as TFs like Nodulation Inception (NIN) [29, 30], Phosphate Starvation Responses (PHRs) [31], SYG1/Pho81/XPR1 (SPX) [32], etc., participate in AMS and/or RNS. Overall, the current study has provided a rough linear pathway of AMS/RNS in plants.

2. Roles of LysM-RLKs in perceiving symbiotic signals in AMS and RNS

In nature, only a few species of microbes can establish compatible interactions with host plants to cause either pathogenic or mutualistic symbiosis. More and more data suggest that plant innate immunity plays a key role in distinguishing invading microbes to establish different interactions. Hence, how plants recognize and distinguish signals from different microbes could be precisely regulated. The existing data indicate that LysM-RLKs play such roles in distinguishing different microbes and initiating different physiological responses in plants. N-Acetylglucosamine (GlcNAc)-containing molecules are conserved components of cell walls for different microbes. For example, chitin, the major component of fungal cell wall, and bacterial peptidoglycan (PGN), function as microbe-associated molecular patterns (MAMPs) perceived by LysM-containing proteins to trigger plant immunity against invading pathogens.

Whereas lipo-chitooligosaccharides (LCOs), for example, rhizobial NFs and mycorrhizal Myc factors are key signals recognized by two LysM-RLKs to induce symbiotic signaling transduction in plants [33]. Rhizobial NF, a short-chain of chitin with different modifications at the terminal residues, plays an important role in specific recognition between rhizobial and legumes [34]. In AMS, Myc factors that contain Myc-LCOs, and Myc-COs can activate the CSSP with resultant calcium oscillations in root epidermal cells [35–37]. Myc factors were proposed to be mixtures of CO4/5 and LCO, while the function of CO4 appears to be the predominant molecule activating symbiotic responses in rice. Thus, symbiotic signaling pathways induced by Myc-LCOs and COs seem to be a little bit different since AMF produces a mixture of molecules during the symbiotic interaction with hosts [18, 37, 38]. In this process, a class of LysM receptor kinases (LYKs) participate in the discrimination of these GlcNAc molecules and determine the outcomes of the downstream signaling pathway to immunity or symbiosis [39–41].

In the establishment of symbiosis between rhizobial and legume host plants, LjNFR1 (Nod Factor Receptor 1) and LjNFR5 in *Lotus japonicus*, also named MtLYK3 (LysM containing receptor Kinase 3) and MtNFP (Nod Factor Perception) in *Medicago truncatula* are two essential LYKs regulating NFs specific perception [42–44]. In non-legume species, for example, rice OsCERK1 (Chitin-elicitor receptor kinase 1), a LysM-RLK, is a necessary receptor involved in COs-induced immunity and mycorrhizal symbiotic responses. CERK1 was originally identified as an essential receptor for chitin elicitor signaling in *Arabidopsis thaliana*, and the KO mutant for *AtCERK1* completely lost the ability to respond to chitin [45]. Subsequent studies have demonstrated that OsCERK1 and a LysM protein OsCEBiP (Chitin Elicitor binding protein) could cooperatively regulate chitin elicitor signaling in rice [39, 46]. But, interestingly, OsCERK1 plays a dual role in mediating both AMS and immunity [47, 48]. The AMS in the rice *Oscerk1* mutant plants was severely diminished but was normal in the *Oscebip* mutant, the function of OsCERK1 in mediating symbiosis or immune responses seems to be dependent on specific interaction with different receptors in response to either symbiotic signals or pathogenic signals [19]. Recent breakthroughs have revealed $OsCERK1^{DY}$ from Dongxiang wild rice with two amino acids substitutions in the second LysM domain exhibited stronger colonization with AMF than the rice cultivar Zhongzao 35 (ZZ35), as well as promoting phosphorus acquisition [49].

CO4/CO5 are necessary signals for symbiotic interactions between AMF and host plants, however, rice OsCERK1 does not seem to bind to CO4 directly [39, 50]. It was implied that another component was needed to perceive these signals, just like the sandwich models of OsCERK1/OsCEBiP in mediating chitin-triggered immunity and LjNFR1/LjNFR5 complex for rhizobial symbiosis in *L. japonicus* [44, 46]. OsMYR1/OsLYK2/OsNFR5/OsRLK2, grouped in the same clade as LjNFR5/MtNFP/SlLYK10, is the co-receptor of OsCERK1 required for AMS. OsMYR1 directly binds to CO4 but not Nod factors or lipopolysaccharides (LPS) [20]. Significant reduction of AMF colonization, as well as transcription levels of AM-specific marker genes and calcium spiking, were observed in the *Osmyr1-1/Oslyk2-1* mutant compared to wide type (WT) inoculated with *Rhizophagus irregularis* spores [20]. However, the AM colonization in a high dosage did not show too many differences between *Osmyr1-1/Osnfr5* and WT plants, but a significant decrease in transcript levels of AM-responsive gene was detected in the *Osmyr1-1/Osnfr5* mutant plants [51]. Whether OsMYR1/OsNFR5 responds differentially to different dosages of *R. irregularis* is unclear. But all these data indicated that OsMYR1 seems to be a binding receptor for sensing CO4, and the subsequent dimerization and phosphorylation between OsMYR1 and OsCERK1 activate symbiotic signaling pathway [20].

It was identified that only the long chain of COs with polymer of degree between 6 and 8 (CO6, CO7, and CO8) but not CO4 or CO5 could trigger a plant immunity [52, 53]. A recent study has shown that CO8 has a similar function as CO4 to induce symbiotic nuclear calcium oscillations and activates some of the symbiosis-related genes expression [18], raising a question that nuclear calcium oscillation might not be a specific signal representing symbiosis. In *Medicago truncatula*, MtCERK1 and MtLYR4 can bind both CO4 and CO8, in which MtCERK1 is required in both COs induced immune and symbiotic signaling pathways [18]. Interestingly, the plant defense triggered by CO8 could be suppressed by additional LCOs in both legumes and non-legumes, and CO4 could reduce CO8-triggered ROS generation via OsMYR1 in rice or AtLYK3 in Arabidopsis [18, 19, 54], suggesting that non-legumes could still respond to LCOs. Hence, a Single Pole Double Throw (SPDT) switch model in symbiosis and defense signaling pathways was proposed [19]. In this model, OsMYR1 can recognize CO4 from symbiotic fungi and then associates with OsCERK1, which suppresses the formation of chitin-induced OsCEBiP/OsCERK1 complex and trans-phosphorylation of immune-associated substrates, then activate mycorrhizal colonization. When OsCEBiP senses CO8, it competes with OsMYR1 to bind OsCERK1, which induces immunity and negatively regulates AMS [19].

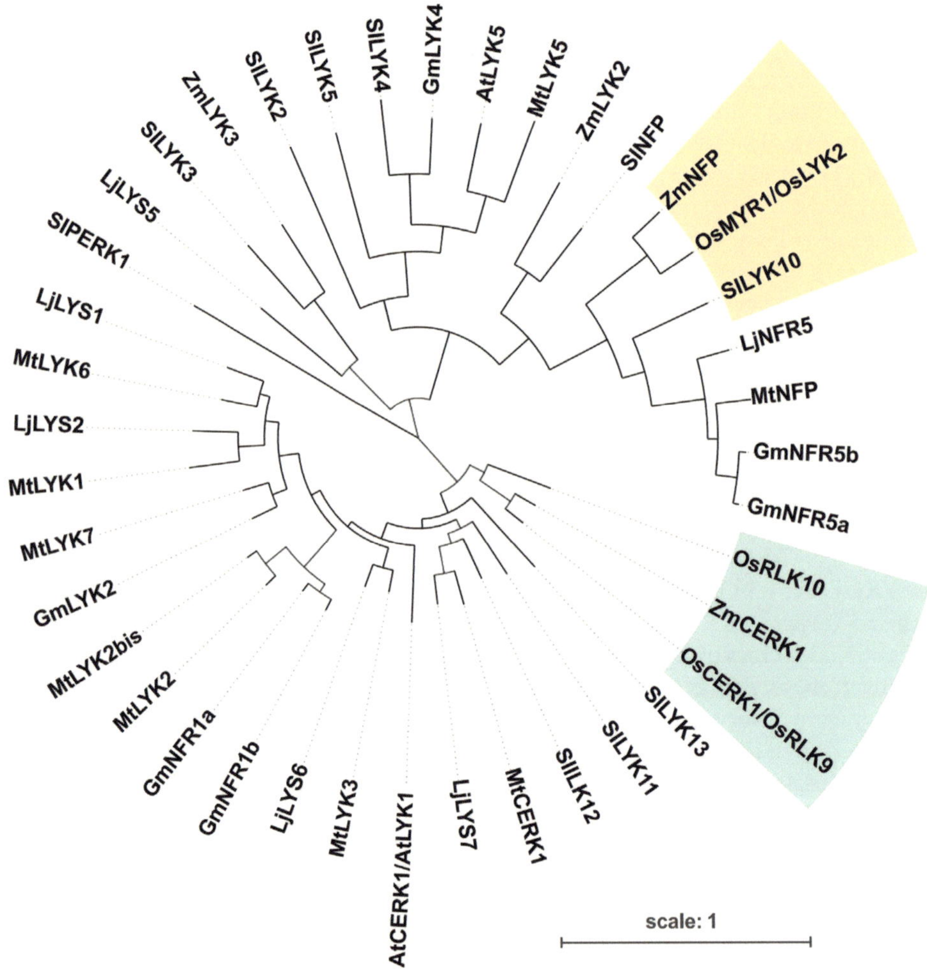

Figure 1.
Phylogenetic tree of LysM-containing receptor-like kinases. Molecular phylogeny of OsCERK1 and OsMYR1 with homologs from Oryza sativa (Os), Medicago truncatula (Mt), Glycine max (Gm), Lotus Japonicus (Lj), Arabidopsis thaliana (At), Zea mays (Zm), Solanum lycopersicum (SI) was constructed using a maximum-likelihood algorithm in MEGA-X with 1000 bootstraps value.

Rice is a very-well studied model species used for AMS study. The dual function of OsCERK1 homologs in both symbiosis and immunity was also studied in other plant species. Similar to rice OsCERK1 in symbiosis and immunity, OsCERK1 homologs in leguminous plants also play a dual role in both symbiosis and immunity (**Figure 1**). For example, MtLYK9 in *M. truncatula* and PsLYK9, a close ortholog of CERK1 in *Pisum sativum* regulate both plant immunity and AMS [55, 56]. Likewise, in *Parasponia andersonii*, the only known non-legume plant with an ability of RNS, PanLYK1, and PanLYK3 were essential for intracellular arbuscule formation, while PanLYK3 also acted as a chitin receptor for innate immunity signaling [57]. Nonetheless, LYKs in some species, such as the CERK1 homologs in tomato (SlLYK1, SlLYK11, SlLYK12, and SlLYK13) are also functional in AMS and chitin-triggered defense. It has been found that knockdown of *SILYK12* would significantly reduce AMF colonization, but the chitin-induced defense response was unaffected, whereas SILYK1 and SILYK13 only participate in immune signaling but not AMS [58]. Hence, plant CERK1 is such a receptor that works as a shared kinase mediating both chitin-triggered immunity and AMF symbiotic signals. The LYKs receptor model might at least support the hypothesis that symbiotic interaction might be a result of a graduate attenuation from plant-pathogens interaction until a condition that both symbionts and plants could benefit.

3. Common signaling pathway in mycorrhizal and rhizobial symbiosis

In legumes, the establishment and development of AMS and RNS require a set of common symbiosis genes [59, 60], including a conserved SymRK protein from different species and several essential TFs. When LCOs and COs from bacteria and/or fungi are recognized, SymRK is activated to associate with a set of essential proteins like HMGR1 to regulate both AMS and RNS or interact with SymRK-interacting protein 2 (SIP2) which is specifically involved in RNS. Currently, some interacting proteins of SymRK have been confirmed to participate in RNS, but whether they also take part in AMS remains unknown. As critical components of CSSP, SymRK and other receptor complexes could promote the signaling pathway downstream by triggering nuclear calcium spiking and activating CCaMK. CCaMK could interact with and phosphorylate the downstream transcription factor CYCLOPS [24, 61]. Meanwhile, DELLAs bind the CCaMK-CYCLOPS complex to promote the expression of *RAM1* and regulate AMF colonization [59]. On the other hand, DELLAs associate CCaMK-CYCLOPS with NSP1-NSP2 to enhance the expression of *NIN* which regulates RNS positively [62, 63]. Recently, several other important transcription factors, for example, PHRs and SPXs have been identified to involve in AMS under different conditions of Pi [31, 32, 64].

3.1 SymRK and its interacting proteins involved in AMS and RNS

As a typical LRR-RLK, SymRK was identified as an important membrane-localized receptor kinase required for activating a series of physiological responses in the symbiosis between AMF, rhizobial, Frankia bacteria, and their corresponding host plants [65, 66]. Studies have found only SymRK but not Nod factor receptors (NFRs) overexpression triggers the expression of AM-related genes, and the *symrk* mutant fails to form arbuscule, indicating that SymRK plays a crucial role in the exchange of signals and in the decision between the development of AMS or RNS [67]. According to polymorphisms and amino acid length of SymRK in both legumes and non-legumes, SymRK exists in at least three different structural versions. Rice and tomato, two plant species widely used for AMS study, have a shorter form of SymRK that is sufficient for AMS but cannot fully complement *symrk* mutant in legumes for rhizobial endosymbiosis [68]. Based on the sequence comparison, the extracellular domain of SymRK homologs might play important roles in determining

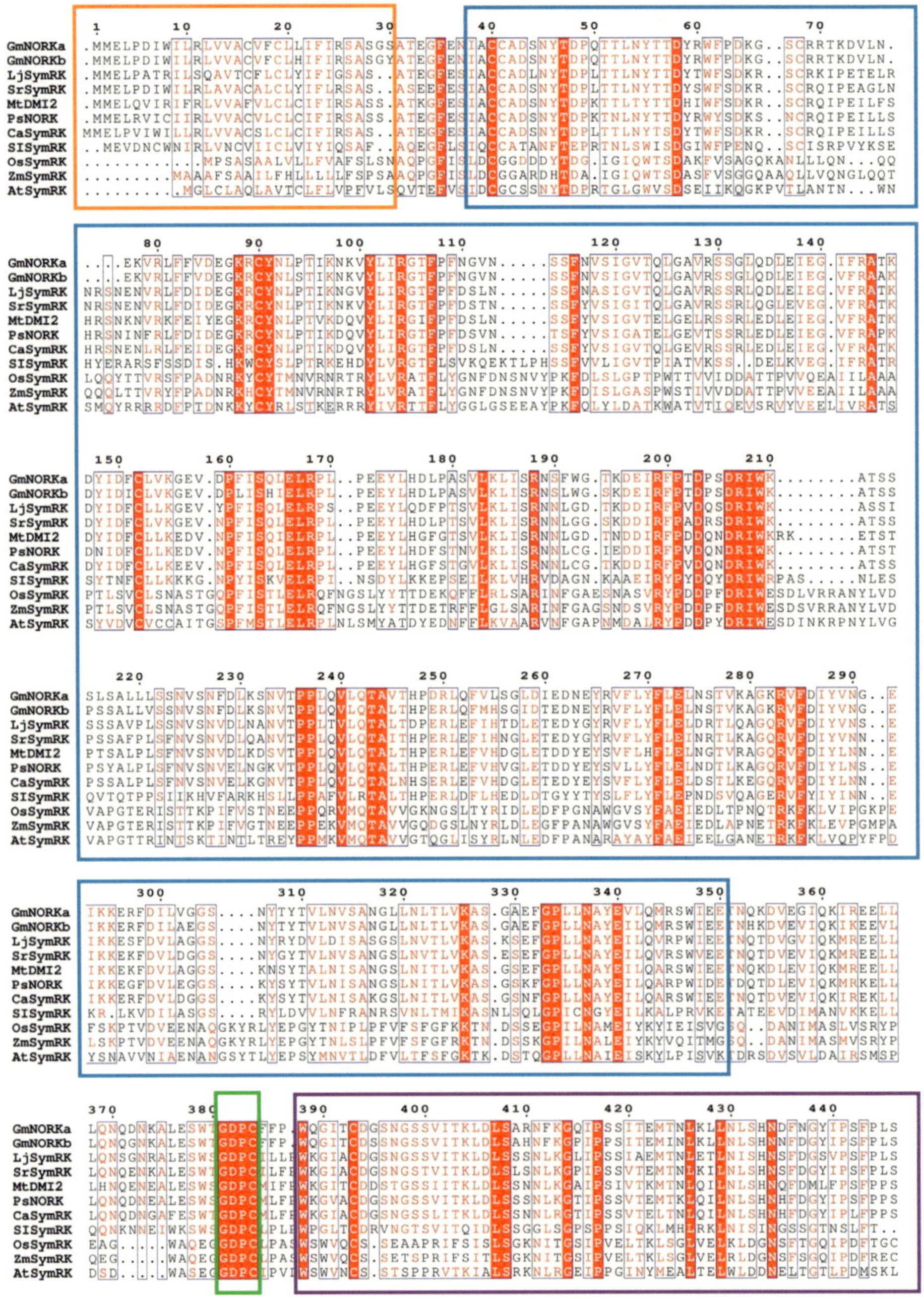

Figure 2.
Multiple sequence alignment of extracellular domain of SymRK from legumes and non-legumes. The orange box represents the signal peptide, the blue box represents malectin-like domain, the green box represents the conserved Gly-Asp-Pro-Cys (GDPC) sequence, and the purple box represents leucine-rich repeat (LRR) domain.

their specific functions (**Figure 2**). In legumes, such as *L. japonicus*, *M. truncatula,* and *Glycine max*, great conservation with about 75% identities is found in the extracellular domain of SymRK homologs. Such conservation might pinpoint its specific role in RNS. However, the sequence identities of SymRK homologs between legumes and non-legumes are reduced to about 40–50%. Althoug, SymRK is a central player in CSSP required for both RNS and AMS, the sequence differences at its extracellular region might give direct evidence that responds to either mycorrhizal signals or rhizobial signals. However, what determines the functional difference between legumes and non-legumes is of great interest to be studied in future, and the functional difference might be related to the evolution of AMS and RNS.

Although the *symrk* knock-out mutant completely loses the ability to allow rhizobial infection, the root hairs of *symrk* mutant plants were observed to be exaggerated after rhizobial attachments [21, 69], indicating that SymRK plays an essential role in determining rhizobial infection but not rhizobial attachment. Due to the key role of SymRK in both AMS and RNS, several SymRK-interacting proteins and protein modifications of SymRK have been studied to elucidate the precise function of SymRK in RNS and/or AMS (summarized in **Table 1**). HMGR1 (3-Hydroxy-3-Methylglutaryl Coenzyme A Reductase1) was identified as an interacting protein of MtDMI2 (SymRK homolog in *M. truncatula*), suggesting that mevalonate biosynthesis was involved in mediating the function of SymRK to initiate calcium spiking and symbiotic gene expression in response to both rhizobia and AMF [70]. Transcriptome expression analysis characterized that SYMREM1 (Symbiotic Remorin 1) from *M. truncatula* could interact with MtDMI2 and may act as a scaffold protein for assembly of receptor complexes involved in rhizobial infection [71, 81]. In addition to interacting with DMI2/SymRK, SYMREM1 was also identified to interact with other receptor kinases in legumes, such as MtNFP/LjNFR5 and MtLYK3/LjNFR1 [81]. Thus, the function of SYMREM1 might provide corresponding structural support for the molecular network of NFR1/5-SymRK receptor complexes.

SymRK-interacting protein 1 (SIP1) is a major AT-rich sequence binding (ARID) transcription factor [72]. Two major splicing forms, SIP1 and SIP1L (a longer variant of the SIP1 transcripts) in *L. japonicus* were characterized. Interestingly, SIP1 was found to interact with SymRK, while SIP1L could not. Both SIP1L and SIP1 could specifically bind to the promoter of *NIN* to positively regulate symbiosis. Knockdown of SIP1using RNAi technology in transgenic hairy roots resulted in impairment in the nodule and arbuscular development, suggesting an important role of SIP1 in the CSSP [72, 73]. As a typical mitogen-activated protein kinase (MAPKK), SymRK-interacting protein 2 (SIP2) could specifically interact with SymRK homologs from different legumes [74]. Although SymRK and SIP2 have a strong interaction, the inter-phosphorylation between them was not detected [74, 82]. Additional studies identified that LjMPK6 is a phosphorylation target of SIP2 [83], and SymRK could inhibit the phosphorylation activity of SIP2, therefore SymRK might negatively regulate the SIP2-MPK6 signaling cascade. Like SIP1, SIP2 also functions as a positive regulator in RNS conformed by RNA interference methods. However, SIP2 is not required for AMF colonization, suggesting that SymRK-SIP2 interaction might be specific for rhizobium infection [74]. Interestingly, SymRK was also identified to directly associate with and suppress the kinase activity of LjBAK1, a homolog protein of Arabidopsis AtBAK1 that functions as a coreceptor for multiple MAMP receptors. The function of SymRK in suppressing LjBAK1-mediated immunity is required for rhizobial infection, but whether this suppression favors AMF infection is to be determined [80].

Interaction protein	Summary	AMS	RNS
HMGR1	As an interacting protein of MtDMI2, HMGR1 is involved in the synthesis of isoprenoid compounds and mevalonate (MVA) pathway, and positively regulates AMS and RNS [70].	Green	Green
SYMREM1	Symbiotic Remorin 1 (SYMREM1) has been shown to interact with SymRK and may act as a scaffold protein for assembly of signaling complexes involved in rhizobial infection [71].	Orange	Green
SIP1	SIP1 is a major AT-rich sequence binding (ARID) transcription factor, and two major splicing forms: SIP1 and SIP1L both can specifically bind to the promoter of *NIN*, but only SIP1 can interact with SymRK [72, 73].	Green	Green
SIP2	SIP2 is a typical mitogen protein kinase (MPKK), which can specifically interact with legume NORK/SymRK, and SymRK can inhibit its kinase activity. It has been shown that SIP2 was positively regulating RNS but was not involved in AMS [74].	Orange	Green
SINA4	SINA4 is induced by Nod factors and could promote the ubiquitination degradation of SymRK, which negatively regulate nodulation [75].	Orange	Green
SIE3	As a newly discovered E3 ubiquitin ligase, SIE3 could interact with SymRK and enhance the degree of ubiquitination of SymRK in vivo. The homodimerization of SIE3 is essential for ubiquitin-related degradation of SIP1 [76–78].	Orange	Green
PUB1	PUB1 is an E3 ubiquitin ligase and another interacting protein of DMI2, which could be phosphorylated by DMI2 and LYK3. Although PUB1 plays a negative role in RNS and AMS, DMI2 is not the ubiquitinated substrate of PUB1, only LYK3 could be degraded [79].	Green	Green
PUB2	MtPUB2 could be activated by MtDMI2 via phosphorylation and activated MtPUB2 directly targets MtDMI2 through ubiquitination-mediated degradation. In addition, MtDMI2-MtPUB2 negative feedback loop plays a role in symbiosis homeostasis [79].	Orange	Green
BAK1	SymRK could directly interact with and suppress the kinase activity of LjBAK1 in *L. japonicus*, which is a well-characterized positive regulator of plant innate immunity, then suppress the plant immunity during rhizobial infection [80].	Orange	Orange

Table 1.
Summary of interacting proteins of SymRK. The green dot represents the confirmed function of SymRK in AMS or RNS, while the orange dot represents the function of SymRK in AMS/RNS that was not confirmed or studied.

SymRK-Interacting E3 ligase (SIE3) is a protein containing CTLH/CRA/RING domains, which mediates the ubiquitination of SymRK, but does not mediate the protein degradation of SymRK in an *in vitro* ubiquitination assay [76]. It is possible that ubiquitinated SymRK may allow sustained signal transduction to downstream host responses. In addition, SIE3 plays a positive role in the SymRK-mediated signaling pathway in RNS. Further study showed that SIE3 can interact with SIP1 and form a homodimer via Cys266 residue [77, 78]. SEVEN IN ABSENTIA4 (SINA4) was identified as another E3 ligase that could interact with SymRK, and coexpression of

SymRK and SINA4 caused SymRK relocalization [75]. On the contrary, SINA4 is able to mediate the degradation of SymRK and plays a negative role in RNS by inhibiting the development of infection threads [75]. It seems like SINA4 might be a key protein working in a negative feedback loop by suppressing excessive symbiotic signal responses. However, whether SIE3 and SINA4 also regulate AMS needs to be further explored.

At present, two E3 ubiquitin ligases of Plant U-Box (PUB) family, PUB1 and PUB2, have also been proved that could interact with DMI2 in *M. truncatula* [79, 84]. PUB1 could interact with two significant symbiotic receptors (LYK3 and DMI2) and modulate the establishment of both AMS and RNS [38]. In addition, PUB1 has been shown to be directly phosphorylated by LYK3 and acts as a negative regulator to inhibit rhizobial infection and nodulation [85]. Although the E3 ligase activity of PUB1 is necessary for negative regulation in RNS and AMS, DMI2 is not a ubiquitination substrate of PUB1 [79], only LYK3 could be degraded by PUB1. It is possible that the involvement of PUB1 in the early symbiosis signal pathway might be a strategy used by plants to actively suppress the excessive infection by rhizobia and AMF. The other PUB-type E3 ubiquitin ligase, MtPUB2, was identified to be a direct regulator of DMI2 in *M. truncatula* and could enhance the ligase activity of MtPUB2 via phosphorylation at Ser421, then the activated MtPUB2 directly ubiquitinates MtDMI2 for degradation in vitro. These studies demonstrated that MtDMI2-MtPUB2 forms a negative feedback loop that displays an important role in nodulation homeostasis [79]. As a key receptor kinase involved in both RNS and AMS, the direct phosphorylation target and protein modification of SymRK need to be further elucidated.

3.2 Transcription factor complexes regulate arbuscule branching

Nuclear calcium oscillations are essential components of signals leading to AMS and RNS in host plant root cells. A couple of proteins and GRAS domain TFs cooperatively mediate calcium signals and induce symbiotic process. CCaMK (as known as DMI3 in *M. truncatula*), the initiation of calcium spiking perception in nucleus, is required in both AMS and RNS [86]. A gain-of-function of CCaMK leads to exaggerated symbiosis response by forming spontaneous nodules in the absence of rhizobial in *L. japonicus*. Interestingly, the gain-of-function of CCaMK could induce rhizobial and AMF infection, as well as calcium spiking even in the mutant plants, that is, *symrk*, *castor* or *pollux*, suggesting that calcium oscillations mediated by CCaMK are a downstream response of symbiosis pathway [61, 87, 88]. The activation of CCaMK by calcium always needs two steps, direct calcium binding to three EF-hand motifs, and calmodulin (CaM) binding to the kinase domain [89]. Thus, different levels of CCaMK activity are needed in AMS and RNS. As it has been confirmed that Thr265 residue is an essential autophosphorylation site for CCaMK activation [90]. CCaMK autophosphorylated at Thr265 and disrupts the hydrogen bonds network with residues of side chains, leading to nodule organogenesis and AMF infection after two EF-hand motifs bind to calcium. CaM binds the kinase domain and stimulates the activity of CCaMK, which is only required for rhizobial infection but not for AMF colonization. Consistent with that CCaMK with kinase domain only could restore the symbiotic entry of AMF in *ccamk* mutant in *L. japonicus* [88, 91]. In addition, rice OsDMI3 is able to functionally complement the AMS in *M. truncatula*, but partially restore the nodulation phenotype in legumes [92].

Downstream and phosphorylated by CCaMK [24], CYCLOPS/IPD3 is required for rhizobial infection, nodule development [93, 94], AMF infection, and arbuscule formation [24, 95]. Acting as a member in CSSP, OsCYCLOPS could complement the AMS and RNS phenotype of *cyclops-3* mutant in *L. japonicus*, indicating functional conservation of CYCLOPS in legumes and non-legumes [24]. Recently, IPD3 and IPD3-LIKE are identified functional redundancy in AMS in *M. truncatula*, the double mutant of *IPD3* and *IPD3L* can form arbuscular but hyphal entry into epidermis cells is impaired [96, 97]. What's more, the development of AMS is remarkably reduced in *ipd3/ipd3l* under a high concentration of Pi treatment [97]. Recent research has demonstrated that other important regulators in Pi uptake, SPXs, and PHRs also take part in regulating AMF colonization under different degrees of phosphate [31, 32, 64]. The formation of arbuscule is reduced in the absence of PHRs. In low Pi conditions, PHRs could bind to P1BS cis-element in promotor of AM-associated genes and induce AM-mediated Pi uptake of host plants; however, SPXs could bind to PHRs, leading to the suppression of transcription of genes downstream in Pi-sufficient conditions. Besides, PHRs could not only regulate genes including *RAM1, PT11, and WRI5A*, which are required for arbuscule formation or nutrient exchange [31], but they also target genes like *CERK1*, *SymRK*, *NSP2*, etc., that are essential for signal perception of AMF entry in the early stage of symbiosis [64].

The CSSP plays a conserved role in regulating AMS and RNS, and plants discriminate between such processes by CCaMK-CYCLOPS complex promoting different GRAS domain TFs through DELLA proteins. Exogenous GA treatment could inhibit infection threads formation and nodule development, as well as hyphal entry and arbuscule formation in *L. japonicus* [98]. As integrators of GA signaling, DELLAs positively regulate rhizobial infection and arbuscule formation, acting as a bridge to complex CCaMK/DMI3-CYCLOPS/IPD3 with NSP1-NSP2 or RAM1 [25, 98]. In RNS, DELLAs work downstream of DMI3-IPD3, enhancing the intensity of phosphorylation of IPD3 by DMI3, then interact with NSP2 to promote the DMI3-IPD3-NSP1-NSP2 complex formation and induce the expression of symbiotic associated genes like *NIN* [99]. While in AMS, DELLAs interact with CCaMK-CYCLOPS complex, activating RAM1 via DIP1 (DELLA Interacting Protein 1) in rice, and RAD1 (Required for Arbuscule Development) in *M. truncatula* and *L. japonicus* [28, 59, 100, 101]. NSP1 specifically functions in Nod factor signaling, and NSP2 may have a minor role in AMS [35]. By contrast, RAM1 is only required in AMS to support arbuscule branching and has no role in Nod factor signaling [27]. Interestingly, the experiment in *Nicotiana benthamiana* showed that RAM1 may compete NSP2 with NSP1, indicating that RAM1 and NSP1 may be the first step downstream of CSSP to distinct Myc factors and NFs signaling [102]. Meanwhile, other GRAS-type transcription factors, maybe RAD1, are involved in Myc factors signaling, for NSP2 only has a weak function in AMS [27, 101, 102].

4. Conclusions and future perspectives

Plants establish mutualistic symbiosis with AMF and rhizobial for nutrient uptake. In AMS, mineral nutrients, especially Pi, are supplied by AMF via AM fungal hyphae, and host plants mainly concurrently transfer fatty acids to fungi as carbon resources. Although AMS facilitates the uptake of Pi, the concentration of Pi, in turn, impairs the colonization of AMF. Thus, the status of Pi is essential in AMS establishment, and some TFs involved in CSSP like PHRs play important roles in the regulation of this process. As a central regulator of Pi homeostasis, PHR2 is required for the activation of AMS-associated genes under Pi-deficient conditions. While in RNS, legumes interact with rhizobial to fix N, and N homeostasis is closely related to NIN, a member of NIN-like protein (NLP) family.

The opening question is how plants discriminate chitin, NFs and Myc factors signals that are structurally similar and then promote different signaling pathways. Receptor kinases play critical roles in primary signal recognition in immunity and AMS and/or RNS. A part of LYKs like OsCERK1 play a dual role in both immunity and symbiosis, while some others show subfunctional in regulating these two signaling pathways. What's more, duplication of LysM genes has happened, compared with 10 LysM genes in rice, *M. truncatula* has 21 and *L. japonicus* has 26 genes that are predicted to encode LYKs. Therefore, the versatile combination of LYKs receptors plays an essential role in sensing structurally similar polysaccharides to initiate symbiosis or immunity thereby discriminating symbionts or pathogens.

SymRK functions as a vital component of the genetic basis for both plant-fungal and plant-bacterial endosymbiosis. It perceives signals dependent on extracellular malectin domain and LRR domain, whose sequences vary wildly between non-legumes and legumes, but are much more conserved in legumes. It is suggested that the diversity among them may be one of the reasons for different responses to Myc factors or NFs. SymRK activates signaling pathways downstream mediating the

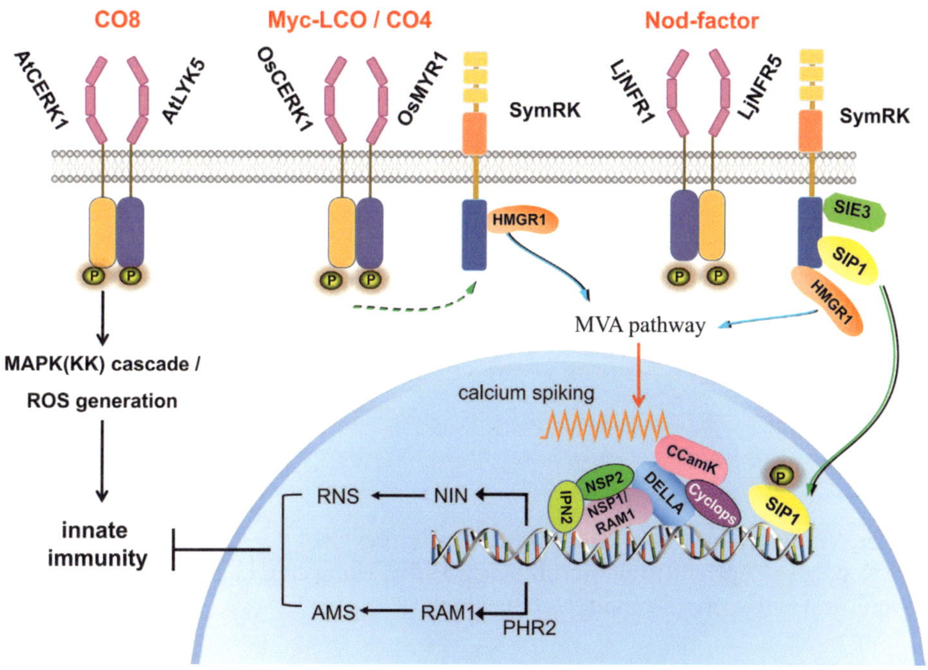

Figure 3.
The chitin-triggered immunity and common symbiotic signaling pathway. AtCERK1 and AtLYK5 form a receptor complex perceiving CO8 to induce innate immunity in A. thaliana. In AMS, the complex of OsCERK1/OsMYR1 receives Myc factors in rice. And LjNFR1/LjNFR5 receptor complex in L. japonicus perceive Nod factors (structurally similar to Myc factors) in RNS. Upon recognition of symbiotic signals of AMS/RNS, SymRK, as the coreceptor of LysM receptors, then interacts with some key proteins such as SIP1 and SIE3. Phosphorylated SIP1 could specifically bind to the promoter of NIN gene to positively regulate symbiosis. Another SymRK interacting protein, HMGR1 is a common signaling component in both AMS/RNS, which participates in MVA pathway and triggers calcium spiking in nucleus. CCaMK-CYCLOPS-DELLA complex interacts with RAM1 or NSP1/NSP2, inducing the expression of RAM1 or NIN to regulate AMS and RNS. The activation of symbiosis signaling inhibits innate immunity signaling at certain levels.

post-translational modification of the interacting proteins; however, whether these interacting proteins function in RNS also participate in AMS remains to be further investigated.

It is probable that a single pathway mediating both AMS and RNS after the recognition of signal molecules, for some homolog proteins such as CCaMK and CYCLOPS play the same role in both AMS and RNS. There may also be some parallel signaling pathways to regulate the TFs in nucleus. The complex of CCaMK-CYCLOPS directly regulates at least 3 genes in different pathways: *NIN*(RNS), *RAM1*(AMS), *CBP1*(RNS and AMS) [103]. In addition, NIN and RAM1 may be the first divergence for AMS or RNS in CSSP, and we wonder whether there are other TFs involved or not.

In summary, receptor kinases are essential in the specifical recognition of signals, and OsCERK1/OsMYR1 were confirmed to be the receptor complex perceiving Myc factors in recent research, but the pivotal receptors for Myc factors in other species remain to be studied. As SymRK could receive symbiotic signals from NFRs, it is of great interest whether SymRK could receive Myc-factor signals to participate in symbiotic signal transduction between AMF and plants. Activated by the calcium spiking, CCaMK-CYCLOPS-DELLA complex could regulate TFs which promote the expression of AMS-related or RNS-related genes (**Figure 3**). Therefore, further study on the difference in signal recognition and signaling pathways between AMS and RNS may help us to apply RNS in non-legumes.

Acknowledgements

This work was supported by the National Natural Science Foundation of China (32090063) and the Natural Science Foundation of Hubei Province (2020CFA008).

Author details

Jiashan Wu†, Weiyun Wang†, Hui Zhu and Yangrong Cao*
State Key Lab of Agricultural Microbiology, Hubei Hongshan Laboratory, Huazhong Agricultural University, Wuhan, Hubei, China

*Address all correspondence to: yrcao@mail.hzau.edu.cn

† Contributed equally to this chapter.

References

[1] Spatafora JW, Chang Y, Benny GL, Lazarus K, Smith ME, Berbee ML, et al. Mint: A phylum-level phylogenetic classification of zygomycete fungi based on genome-scale data. Mycologia. 2016;**108**(5):1028-1046. DOI: 10.3852/16-042

[2] Choi J, Summers W, Paszkowski U. Mint: Mechanisms underlying establishment of arbuscular mycorrhizal symbioses. Annual Review of Phytopathology. 2018;**56**:135-160. DOI: 10.1146/annurev-phyto-080516-035521

[3] Young ND, Debelle F, Oldroyd GE, Geurts R, Cannon SB, Udvardi MK, et al. Mint: The Medicago genome provides insight into the evolution of rhizobial symbioses. Nature. 2011;**480**(7378): 520-524. DOI: 10.1038/nature10625

[4] Parniske M. Mint: Arbuscular mycorrhiza: The mother of plant root endosymbioses. Nature Reviews. Microbiology. 2008;**6**(10):763-775. DOI: 10.1038/nrmicro1987

[5] Markmann K, Parniske M. Mint: Evolution of root endosymbiosis with bacteria: How novel are nodules? Trends in Plant Science. 2009;**14**(2):77-86. DOI: 10.1016/j.tplants.2008.11.009

[6] Hodge A, Campbell CD, Fitter AH. Mint: An arbuscular mycorrhizal fungus accelerates decomposition and acquires nitrogen directly from organic material. Nature. 2001;**413**(6853):297-299. DOI: 10.1038/35095041

[7] Smith SE, Smith FA. Mint: Roles of arbuscular mycorrhizas in plant nutrition and growth: New paradigms from cellular to ecosystem scales. Annual Review of Plant Biology. 2011;**62**:227-250. DOI: 10.1146/annurev-arplant-042110-103846

[8] Rich MK, Nouri E, Courty PE, Reinhardt D. Mint: Diet of arbuscular mycorrhizal fungi: Bread and butter? Trends in Plant Science. 2017;**22**(8):652-660. DOI: 10.1016/j.tplants.2017.05.008

[9] Bravo A, Brands M, Wewer V, Dormann P, Harrison MJ. Mint: Arbuscular mycorrhiza-specific enzymes FatM and RAM2 fine-tune lipid biosynthesis to promote development of arbuscular mycorrhiza. The New Phytologist. 2017;**214**(4):1631-1645. DOI: 10.1111/nph.14533

[10] Nouri E, Surve R, Bapaume L, Stumpe M, Chen M, Zhang Y, et al. Mint: Phosphate suppression of arbuscular mycorrhizal symbiosis involves gibberellic acid signaling. Plant & Cell Physiology. 2021;**62**(6):959-970. DOI: 10.1093/pcp/pcab063

[11] Ma X, Li X, Ludewig U. Mint: Arbuscular mycorrhizal colonization outcompetes root hairs in maize under low phosphorus availability. Annals of Botany. 2021;**127**(1):155-166. DOI: 10.1093/aob/mcaa159

[12] Wang W, Shi J, Xie Q, Jiang Y, Yu N, Wang E. Mint: Nutrient exchange and regulation in arbuscular mycorrhizal symbiosis. Molecular Plant. 2017;**10**(9):1147-1158. DOI: 10.1016/j.molp.2017.07.012

[13] Jung SC, Martinez-Medina A, Lopez-Raez JA, Pozo MJ. Mint: Mycorrhiza-induced resistance and priming of plant defenses. Journal of Chemical Ecology. 2012;**38**(6):651-664. DOI: 10.1007/s10886-012-0134-6

[14] Bahadur A, Batool A, Nasir F, Jiang S, Mingsen Q, Zhang Q, et al. Mint:

Mechanistic insights into arbuscular mycorrhizal fungi-mediated drought stress tolerance in plants. International Journal of Molecular Sciences. 2019;**20**(17):4199. DOI: 10.3390/ijms20174199

[15] Jarratt-Barnham E, Zarrabian D, Oldroyd GED. Mint: Symbiotic regulation: How plants seek salvation in starvation. Current Biology. 2022;**32**(1):R46-R48. DOI: 10.1016/j.cub.2021.11.059

[16] Javot H, Penmetsa RV, Terzaghi N, Cook DR, Harrison MJ. Mint: A *Medicago truncatula* phosphate transporter indispensable for the arbuscular mycorrhizal symbiosis. Proceedings of the National Academy of Sciences of the United States of America. 2007;**104**(5):1720-1725. DOI: 10.1073/pnas.0608136104

[17] Kretzschmar T, Kohlen W, Sasse J, Borghi L, Schlegel M, Bachelier JB, et al. Mint: A petunia ABC protein controls strigolactone-dependent symbiotic signalling and branching. Nature. 2012;**483**(7389):341-344. DOI: 10.1038/nature10873

[18] Feng F, Sun J, Radhakrishnan GV, Lee T, Bozsoki Z, Fort S, et al. Mint: A combination of chitooligosaccharide and lipochitooligosaccharide recognition promotes arbuscular mycorrhizal associations in *Medicago truncatula*. Nature Communications. 2019;**10**(1):5047. DOI: 10.1038/s41467-019-12999-5

[19] Zhang C, He J, Dai H, Wang G, Zhang X, Wang C, et al. Mint: Discriminating symbiosis and immunity signals by receptor competition in rice. Proceedings of the National Academy of Sciences of the United States of America. 2021;**118**(16):e2023738118. DOI: 10.1073/pnas.2023738118

[20] He J, Zhang C, Dai H, Liu H, Zhang X, Yang J, et al. Mint: A LysM receptor Heteromer mediates perception of arbuscular mycorrhizal symbiotic signal in rice. Molecular Plant. 2019;**12**(12):1561-1576. DOI: 10.1016/j.molp.2019.10.015

[21] Stracke S, Kistner C, Yoshida S, Mulder L, Sato S, Kaneko T, et al. Mint: A plant receptor-like kinase required for both bacterial and fungal symbiosis. Nature. 2002;**417**(6892):959-962. DOI: 10.1038/nature00841

[22] Endre G, Kereszt A, Kevei Z, Mihacea S, Kalo P, Kiss GB. Mint: A receptor kinase gene regulating symbiotic nodule development. Nature. 2002;**417**(6892):962-966. DOI: 10.1038/nature00842

[23] Miller JB, Pratap A, Miyahara A, Zhou L, Bornemann S, Morris RJ, et al. Mint: Calcium/calmodulin-dependent protein kinase is negatively and positively regulated by calcium, providing a mechanism for decoding calcium responses during symbiosis signaling. The Plant Cell. 2013;**25**(12):5053-5066. DOI: 10.1105/tpc.113.116921

[24] Yano K, Yoshida S, Muller J, Singh S, Banba M, Vickers K, et al. Mint: CYCLOPS, a mediator of symbiotic intracellular accommodation. Proceedings of the National Academy of Sciences of the United States of America. 2008;**105**(51):20540-20545. DOI: 10.1073/pnas.0806858105

[25] Floss DS, Levy JG, Levesque-Tremblay V, Pumplin N, Harrison MJ. Mint: DELLA proteins regulate arbuscule formation in arbuscular mycorrhizal symbiosis. Proceedings of the National Academy of Sciences of the United States of America. 2013;**110**(51):E5025-E5034. DOI: 10.1073/pnas.1308973110

[26] Liu W, Kohlen W, Lillo A, Op den Camp R, Ivanov S, Hartog M, et al.

Mint: Strigolactone biosynthesis in *Medicago truncatula* and rice requires the symbiotic GRAS-type transcription factors NSP1 and NSP2. The Plant Cell. 2011;**23**(10):3853-3865. DOI: 10.1105/tpc.111.089771

[27] Park HJ, Floss DS, Levesque-Tremblay V, Bravo A, Harrison MJ. Mint: Hyphal branching during Arbuscule development requires reduced arbuscular Mycorrhiza1. Plant Physiology. 2015;**169**(4):2774-2788. DOI: 10.1104/pp.15.01155

[28] Yu N, Luo D, Zhang X, Liu J, Wang W, Jin Y, et al. Mint: A DELLA protein complex controls the arbuscular mycorrhizal symbiosis in plants. Cell Research. 2014;**24**(1):130-133. DOI: 10.1038/cr.2013.167

[29] Guillotin B, Couzigou JM, Combier JP. Mint: NIN is involved in the regulation of arbuscular mycorrhizal symbiosis. Frontiers in Plant Science. 2016;**7**:1704. DOI: 10.3389/fpls.2016.01704

[30] Kumar A, Cousins DR, Liu CW, Xu P, Murray JD. Mint: Nodule inception is not required for arbuscular mycorrhizal colonization of *Medicago truncatula*. Plants (Basel). 2020;9(1):71. DOI: 10.3390/plants9010071

[31] Shi J, Zhao B, Zheng S, Zhang X, Wang X, Dong W, et al. Mint: A phosphate starvation response-centered network regulates mycorrhizal symbiosis. Cell. 2021;**184**(22):5527-40 e18. DOI: 10.1016/j.cell.2021.09.030

[32] Wang P, Snijders R, Kohlen W, Liu J, Bisseling T, Limpens E. Mint: Medicago SPX1 and SPX3 regulate phosphate homeostasis, mycorrhizal colonization, and arbuscule degradation. The Plant Cell. 2021;**33**(11):3470-3486. DOI: 10.1093/plcell/koab206

[33] Limpens E, van Zeijl A, Geurts R. Mint: Lipochitooligosaccharides modulate plant host immunity to enable endosymbioses. Annual Review of Phytopathology. 2015;**53**:311-334. DOI: 10.1146/annurev-phyto-080614-120149

[34] D'Haeze W, Holsters M. Mint: Nod factor structures, responses, and perception during initiation of nodule development. Glycobiology. 2002;**12**(6):79R-105R. DOI: 10.1093/glycob/12.6.79r

[35] Maillet F, Poinsot V, Andre O, Puech-Pages V, Haouy A, Gueunier M, et al. Mint: Fungal lipochitooligosaccharide symbiotic signals in arbuscular mycorrhiza. Nature. 2011;**469**(7328):58-63. DOI: 10.1038/nature09622

[36] Sun J, Miller JB, Granqvist E, Wiley-Kalil A, Gobbato E, Maillet F, et al. Mint: Activation of symbiosis signaling by arbuscular mycorrhizal fungi in legumes and rice. The Plant Cell. 2015;**27**(3):823-838. DOI: 10.1105/tpc.114.131326

[37] Genre A, Chabaud M, Balzergue C, Puech-Pages V, Novero M, Rey T, et al. Mint: Short-chain chitin oligomers from arbuscular mycorrhizal fungi trigger nuclear Ca2+ spiking in *Medicago truncatula* roots and their production is enhanced by strigolactone. The New Phytologist. 2013;**198**(1):190-202. DOI: 10.1111/nph.12146

[38] Girardin A, Wang T, Ding Y, Keller J, Buendia L, Gaston M, et al. Mint: LCO receptors involved in arbuscular mycorrhiza are functional for rhizobia perception in legumes. Current Biology. 2019;**29**(24):4249-59 e5. DOI: 10.1016/j.cub.2019.11.038

[39] Kaku H, Nishizawa Y, Ishii-Minami N, Akimoto-Tomiyama C,

Dohmae N, Takio K, et al. Mint: Plant cells recognize chitin fragments for defense signaling through a plasma membrane receptor. Proceedings of the National Academy of Sciences of the United States of America. 2006;**103**(29):11086-11091. DOI: 10.1073/pnas.0508882103

[40] Broghammer A, Krusell L, Blaise M, Sauer J, Sullivan JT, Maolanon N, et al. Mint: Legume receptors perceive the rhizobial lipochitin oligosaccharide signal molecules by direct binding. Proceedings of the National Academy of Sciences of the United States of America. 2012;**109**(34):13859-13864. DOI: 10.1073/pnas.1205171109

[41] Gust AA, Willmann R, Desaki Y, Grabherr HM, Nurnberger T. Mint: Plant LysM proteins: Modules mediating symbiosis and immunity. Trends in Plant Science. 2012;**17**(8):495-502. DOI: 10.1016/j.tplants.2012.04.003

[42] Limpens E, Franken C, Smit P, Willemse J, Bisseling T, Geurts R. Mint: LysM domain receptor kinases regulating rhizobial nod factor-induced infection. Science. 2003;**302**(5645):630-633. DOI: 10.1126/science.1090074

[43] Madsen EB, Madsen LH, Radutoiu S, Olbryt M, Rakwalska M, Szczyglowski K, et al. Mint: A receptor kinase gene of the LysM type is involved in legume perception of rhizobial signals. Nature. 2003;**425**(6958):637-640. DOI: 10.1038/nature02045

[44] Radutoiu S, Madsen LH, Madsen EB, Felle HH, Umehara Y, Gronlund M, et al. Mint: Plant recognition of symbiotic bacteria requires two LysM receptor-like kinases. Nature. 2003;**425**(6958):585-592. DOI: 10.1038/nature02039

[45] Miya A, Albert P, Shinya T, Desaki Y, Ichimura K, Shirasu K, et al. Mint: CERK1, a LysM receptor kinase, is essential for chitin elicitor signaling in Arabidopsis. Proceedings of the National Academy of Sciences of the United States of America. 2007;**104**(49):19613-19618. DOI: 10.1073/pnas.0705147104

[46] Shimizu T, Nakano T, Takamizawa D, Desaki Y, Ishii-Minami N, Nishizawa Y, et al. Mint: Two LysM receptor molecules, CEBiP and OsCERK1, cooperatively regulate chitin elicitor signaling in rice. The Plant Journal. 2010;**64**(2):204-214. DOI: 10.1111/j.1365-313X.2010.04324.x

[47] Miyata K, Kozaki T, Kouzai Y, Ozawa K, Ishii K, Asamizu E, et al. Mint: The bifunctional plant receptor, OsCERK1, regulates both chitin-triggered immunity and arbuscular mycorrhizal symbiosis in rice. Plant & Cell Physiology. 2014;**55**(11):1864-1872. DOI: 10.1093/pcp/pcu129

[48] Zhang X, Dong W, Sun J, Feng F, Deng Y, He Z, et al. Mint: The receptor kinase CERK1 has dual functions in symbiosis and immunity signalling. The Plant Journal. 2015;**81**(2):258-267. DOI: 10.1111/tpj.12723

[49] Huang R, Li Z, Mao C, Zhang H, Sun Z, Li H, et al. Mint: Natural variation at OsCERK1 regulates arbuscular mycorrhizal symbiosis in rice. The New Phytologist. 2020;**225**(4):1762-1776. DOI: 10.1111/nph.16158

[50] Carotenuto G, Chabaud M, Miyata K, Capozzi M, Takeda N, Kaku H, et al. Mint: The rice LysM receptor-like kinase OsCERK1 is required for the perception of short-chain chitin oligomers in arbuscular mycorrhizal signaling. The New Phytologist. 2017;**214**(4):1440-1446. DOI: 10.1111/nph.14539

[51] Miyata K, Hayafune M, Kobae Y, Kaku H, Nishizawa Y, Masuda Y, et al.

Mint: Evaluation of the role of the LysM receptor-like kinase, OsNFR5/OsRLK2 for AM Symbiosis in rice. Plant & Cell Physiology. 2016;**57**(11):2283-2290. DOI: 10.1093/pcp/pcw144

[52] Cao Y, Liang Y, Tanaka K, Nguyen CT, Jedrzejczak RP, Joachimiak A, et al. Mint: The kinase LYK5 is a major chitin receptor in Arabidopsis and forms a chitin-induced complex with related kinase CERK1. eLife. 2014;**3**:e03766. DOI: 10.7554/eLife.03766

[53] Bozsoki Z, Cheng J, Feng F, Gysel K, Vinther M, Andersen KR, et al. Mint: Receptor-mediated chitin perception in legume roots is functionally separable from nod factor perception. Proceedings of the National Academy of Sciences of the United States of America. 2017;**114**(38):E8118-E8E27. DOI: 10.1073/pnas.1706795114

[54] Liang Y, Cao Y, Tanaka K, Thibivilliers S, Wan J, Choi J, et al. Mint: Nonlegumes respond to rhizobial nod factors by suppressing the innate immune response. Science. 2013;**341**(6152):1384-1387. DOI: 10.1126/science.1242736

[55] Gibelin-Viala C, Amblard E, Puech-Pages V, Bonhomme M, Garcia M, Bascaules-Bedin A, et al. Mint: The *Medicago truncatula* LysM receptor-like kinase LYK9 plays a dual role in immunity and the arbuscular mycorrhizal symbiosis. The New Phytologist. 2019;**223**(3):1516-1529. DOI: 10.1111/nph.15891

[56] Leppyanen IV, Shakhnazarova VY, Shtark OY, Vishnevskaya NA, Tikhonovich IA, Dolgikh EA. Mint: Receptor-like kinase LYK9 in Pisum sativum L. is the CERK1-like receptor that controls both plant immunity and AM symbiosis development. International Journal of Molecular Sciences 2017;19(1):8. DOI: 10.3390/ijms19010008

[57] Rutten L, Miyata K, Roswanjaya YP, Huisman R, Bu F, Hartog M, et al. Mint: Duplication of symbiotic Lysin motif receptors predates the evolution of nitrogen-fixing nodule symbiosis. Plant Physiology. 2020;**184**(2):1004-1023. DOI: 10.1104/pp.19.01420

[58] Liao D, Sun X, Wang N, Song F, Liang Y. Mint: Tomato LysM receptor-like kinase SlLYK12 is involved in arbuscular mycorrhizal symbiosis. Frontiers in Plant Science. 2018;**9**:1004. DOI: 10.3389/fpls.2018.01004

[59] Pimprikar P, Carbonnel S, Paries M, Katzer K, Klingl V, Bohmer MJ, et al. Mint: A CCaMK-CYCLOPS-DELLA complex activates transcription of RAM1 to regulate arbuscule branching. Current Biology. 2016;**26**(8):987-998. DOI: 10.1016/j.cub.2016.01.069

[60] Oldroyd GE. Mint: Speak, friend, and enter: Signalling systems that promote beneficial symbiotic associations in plants. Nature Reviews. Microbiology. 2013;**11**(4):252-263. DOI: 10.1038/nrmicro2990

[61] Kistner C, Winzer T, Pitzschke A, Mulder L, Sato S, Kaneko T, et al. Mint: Seven *Lotus japonicus* genes required for transcriptional reprogramming of the root during fungal and bacterial symbiosis. The Plant Cell. 2005;**17**(8):2217-2229. DOI: 10.1105/tpc.105.032714

[62] Hayashi T, Shimoda Y, Sato S, Tabata S, Imaizumi-Anraku H, Hayashi M. Mint: Rhizobial infection does not require cortical expression of upstream common symbiosis genes responsible for the induction of Ca(2+) spiking. The Plant Journal. 2014;**77**(1):146-159. DOI: 10.1111/tpj.12374

[63] Montiel J, Reid D, Gronbaek TH, Benfeldt CM, James EK, Ott T, et al. Mint: Distinct signaling routes mediate intercellular and intracellular rhizobial infection in *Lotus japonicus*. Plant Physiology. 2021;**185**(3):1131-1147. DOI: 10.1093/plphys/kiaa049

[64] Das D, Paries M, Hobecker K, Gigl M, Dawid C, Lam HM, et al. Mint: PHOSPHATE STARVATION RESPONSE transcription factors enable arbuscular mycorrhiza symbiosis. Nature Communications. 2022;**13**(1):477. DOI: 10.1038/s41467-022-27976-8

[65] Gherbi H, Markmann K, Svistoonoff S, Estevan J, Autran D, Giczey G, et al. Mint: SymRK defines a common genetic basis for plant root endosymbioses with arbuscular mycorrhiza fungi, rhizobia, and Frankia bacteria. Proceedings of the National Academy of Sciences of the United States of America. 2008;**105**(12):4928-4932. DOI: 10.1073/pnas.0710618105

[66] Abdel-Lateif K, Bogusz D, Hocher V. Mint: The role of flavonoids in the establishment of plant roots endosymbioses with arbuscular mycorrhiza fungi, rhizobia and Frankia bacteria. Plant Signaling & Behavior. 2012;**7**(6):636-641. DOI: 10.4161/psb.20039

[67] Ried MK, Antolin-Llovera M, Parniske M. Mint: Spontaneous symbiotic reprogramming of plant roots triggered by receptor-like kinases. eLife. 2014;**3**:e03891. DOI: 10.7554/eLife.03891

[68] Markmann K, Giczey G, Parniske M. Mint: Functional adaptation of a plant receptor-kinase paved the way for the evolution of intracellular root symbioses with bacteria. PLoS Biology. 2008;**6**(3):e68. DOI: 10.1371/journal.pbio.0060068

[69] Indrasumunar A, Wilde J, Hayashi S, Li D, Gresshoff PM. Mint: Functional analysis of duplicated Symbiosis receptor kinase (SymRK) genes during nodulation and mycorrhizal infection in soybean (*Glycine max*). Journal of Plant Physiology. 2015;**176**:157-168. DOI: 10.1016/j.jplph.2015.01.002

[70] Kevei Z, Lougnon G, Mergaert P, Horvath GV, Kereszt A, Jayaraman D, et al. Mint: 3-hydroxy-3-methylglutaryl coenzyme a reductase 1 interacts with NORK and is crucial for nodulation in *Medicago truncatula*. The Plant Cell. 2007;**19**(12):3974-3989. DOI: 10.1105/tpc.107.053975

[71] Liang P, Stratil TF, Popp C, Marin M, Folgmann J, Mysore KS, et al. Mint: Symbiotic root infections in *Medicago truncatula* require remorin-mediated receptor stabilization in membrane nanodomains. Proceedings of the National Academy of Sciences of the United States of America. 2018;**115**(20):5289-5294. DOI: 10.1073/pnas.1721868115

[72] Zhu H, Chen T, Zhu M, Fang Q, Kang H, Hong Z, et al. Mint: A novel ARID DNA-binding protein interacts with SymRK and is expressed during early nodule development in *Lotus japonicus*. Plant Physiology. 2008;**148**(1):337-347. DOI: 10.1104/pp.108.119164

[73] Wang C, Zhu H, Jin L, Chen T, Wang L, Kang H, et al. Mint: Splice variants of the SIP1 transcripts play a role in nodule organogenesis in *Lotus japonicus*. Plant Molecular Biology. 2013;**82**(1-2):97-111. DOI: 10.1007/s11103-013-0042-3

[74] Chen T, Zhu H, Ke D, Cai K, Wang C, Gou H, et al. Mint: A MAP kinase kinase interacts with SymRK and regulates nodule organogenesis in *Lotus japonicus*. The Plant Cell. 2012;**24**(2):823-838. DOI: 10.1105/tpc.112.095984

[75] Den Herder G, Yoshida S, Antolin-Llovera M, Ried MK, Parniske M. Mint: *Lotus japonicus* E3 ligase SEVEN IN ABSENTIA4 destabilizes the symbiosis receptor-like kinase SYMRK and negatively regulates rhizobial infection. The Plant Cell. 2012;**24**(4):1691-1707. DOI: 10.1105/tpc.110.082248

[76] Yuan S, Zhu H, Gou H, Fu W, Liu L, Chen T, et al. Mint: A ubiquitin ligase of symbiosis receptor kinase involved in nodule organogenesis. Plant Physiology. 2012;**160**(1):106-117. DOI: 10.1104/pp.112.199000

[77] Feng Y, Wu P, Fu W, Peng L, Zhu H, Cao Y, et al. Mint: The *Lotus japonicus* ubiquitin ligase SIE3 interacts with the transcription factor SIP1 and forms a homodimer. Frontiers in Plant Science. 2020;**11**:795. DOI: 10.3389/fpls.2020.00795

[78] Wu P, Feng Y, Zou Z, Cao Y, Yuan S. Mint: Critical role of cysteine-266 of SIE3 in regulating the ubiquitination and degradation of SIP1 transcription factor in *Lotus japonicus*. Planta. 2021;**253**(6):126. DOI: 10.1007/s00425-021-03647-8

[79] Liu J, Deng J, Zhu F, Li Y, Lu Z, Qin P, et al. Mint: The MtDMI2-MtPUB2 negative feedback loop plays a role in nodulation homeostasis. Plant Physiology. 2018;**176**(4):3003-3026. DOI: 10.1104/pp.17.01587

[80] Feng Y, Wu P, Liu C, Peng L, Wang T, Wang C, et al. Mint: Suppression of LjBAK1-mediated immunity by SymRK promotes rhizobial infection in *Lotus japonicus*. Molecular Plant. 2021;**14**(11):1935-1950. DOI: 10.1016/j.molp.2021.07.016

[81] Lefebvre B, Timmers T, Mbengue M, Moreau S, Herve C, Toth K, et al. Mint: A remorin protein interacts with symbiotic receptors and regulates bacterial infection. Proceedings of the National Academy of Sciences of the United States of America. 2010;**107**(5):2343-2348. DOI: 10.1073/pnas.0913320107

[82] Capoen W, Goormachtig S, De Rycke R, Schroeyers K, Holsters M. Mint: SrSymRK, a plant receptor essential for symbiosome formation. Proceedings of the National Academy of Sciences of the United States of America. 2005;**102**(29):10369-10374. DOI: 10.1073/pnas.0504250102

[83] Yan Z, Cao J, Fan Q, Chao H, Guan X, Zhang Z, et al. Mint: Dephosphorylation of LjMPK6 by phosphatase LjPP2C is involved in regulating nodule organogenesis in *Lotus japonicus*. International Journal of Molecular Sciences. 2020;**21**(15). DOI: 10.3390/ijms21155565

[84] Vernie T, Camut S, Camps C, Rembliere C, de Carvalho-Niebel F, Mbengue M, et al. Mint: PUB1 interacts with the receptor kinase DMI2 and negatively regulates Rhizobial and arbuscular mycorrhizal symbioses through its ubiquitination activity in *Medicago truncatula*. Plant Physiology. 2016;**170**(4):2312-2324. DOI: 10.1104/pp.15.01694

[85] Mbengue M, Camut S, de Carvalho-Niebel F, Deslandes L, Froidure S, Klaus-Heisen D, et al. Mint: The *Medicago truncatula* E3 ubiquitin ligase PUB1 interacts with the LYK3 symbiotic receptor and negatively regulates infection and nodulation. The Plant Cell. 2010;**22**(10):3474-3488. DOI: 10.1105/tpc.110.075861

[86] Levy J, Bres C, Geurts R, Chalhoub B, Kulikova O, Duc G, et al. Mint: A putative Ca2+ and calmodulin-dependent protein kinase required for bacterial and fungal symbioses. Science. 2004;**303**(5662):1361-1364. DOI: 10.1126/science.1093038

[87] Gleason C, Chaudhuri S, Yang T, Munoz A, Poovaiah BW, Oldroyd GE. Mint: Nodulation independent of rhizobia induced by a calcium-activated kinase lacking autoinhibition. Nature. 2006;**441**(7097):1149-1152. DOI: 10.1038/nature04812

[88] Takeda N, Maekawa T, Hayashi M. Mint: Nuclear-localized and deregulated calcium- and calmodulin-dependent protein kinase activates rhizobial and mycorrhizal responses in *Lotus japonicus*. The Plant Cell. 2012;**24**(2):810-822. DOI: 10.1105/tpc.111.091827

[89] Ramachandiran S, Takezawa D, Wang W, Poovaiah BW. Mint: Functional domains of plant chimeric calcium/calmodulin-dependent protein kinase: Regulation by autoinhibitory and visinin-like domains. Journal of Biochemistry. 1997;**121**(5):984-990. DOI: 10.1093/oxfordjournals.jbchem.a021684

[90] Tirichine L, Imaizumi-Anraku H, Yoshida S, Murakami Y, Madsen LH, Miwa H, et al. Mint: Deregulation of a Ca2+/calmodulin-dependent kinase leads to spontaneous nodule development. Nature. 2006;**441**(7097):1153-1156. DOI: 10.1038/nature04862

[91] Shimoda Y, Han L, Yamazaki T, Suzuki R, Hayashi M, Imaizumi-Anraku H. Mint: Rhizobial and fungal symbioses show different requirements for calmodulin binding to calcium calmodulin-dependent protein kinase in *Lotus japonicus*. The Plant Cell. 2012;**24**(1):304-321. DOI: 10.1105/tpc.111.092197

[92] Godfroy O, Debelle F, Timmers T, Rosenberg C. Mint: A rice calcium- and calmodulin-dependent protein kinase restores nodulation to a legume mutant. Molecular Plant-Microbe Interactions. 2006;**19**(5):495-501. DOI: 10.1094/MPMI-19-0495

[93] Messinese E, Mun JH, Yeun LH, Jayaraman D, Rouge P, Barre A, et al. Mint: A novel nuclear protein interacts with the symbiotic DMI3 calcium- and calmodulin-dependent protein kinase of *Medicago truncatula*. Molecular Plant-Microbe Interactions. 2007;**20**(8):912-921. DOI: 10.1094/MPMI-20-8-0912

[94] Horvath B, Yeun LH, Domonkos A, Halasz G, Gobbato E, Ayaydin F, et al. Mint: *Medicago truncatula* IPD3 is a member of the common symbiotic signaling pathway required for rhizobial and mycorrhizal symbioses. Molecular Plant-Microbe Interactions. 2011;**24**(11):1345-1358. DOI: 10.1094/MPMI-01-11-0015

[95] Chen C, Ane JM, Zhu H. Mint: OsIPD3, an ortholog of the *Medicago truncatula* DMI3 interacting protein IPD3, is required for mycorrhizal symbiosis in rice. The New Phytologist. 2008;**180**(2):311-315. DOI: 10.1111/j.1469-8137.2008.02612.x

[96] Jin Y, Chen Z, Yang J, Mysore KS, Wen J, Huang J, et al. Mint: IPD3 and IPD3L function redundantly in Rhizobial and mycorrhizal symbioses. Frontiers in Plant Science. 2018;**9**:267. DOI: 10.3389/fpls.2018.00267

[97] Lindsay PL, Williams BN, MacLean A, Harrison MJ. Mint: A phosphate-dependent requirement for transcription factors IPD3 and IPD3L during arbuscular mycorrhizal Symbiosis in *Medicago truncatula*. Molecular Plant-Microbe Interactions. 2019;**32**(10):1277-1290. DOI: 10.1094/MPMI-01-19-0006-R

[98] Foo E, Ross JJ, Jones WT, Reid JB. Mint: Plant hormones in arbuscular mycorrhizal symbioses: An emerging role for gibberellins. Annals of Botany. 2013;**111**(5):769-779. DOI: 10.1093/aob/mct041

[99] Jin Y, Liu H, Luo D, Yu N, Dong W, Wang C, et al. Mint: DELLA proteins

are common components of symbiotic rhizobial and mycorrhizal signalling pathways. Nature Communications. 2016;7:12433. DOI: 10.1038/ncomms12433

[100] Floss DS, Levesque-Tremblay V, Park HJ, Harrison MJ. Mint: DELLA proteins regulate expression of a subset of AM symbiosis-induced genes in *Medicago truncatula*. Plant Signaling & Behavior. 2016;**11**(4):e1162369. DOI: 10.1080/15592324.2016.1162369

[101] Xue L, Cui H, Buer B, Vijayakumar V, Delaux PM, Junkermann S, et al. Mint: Network of GRAS transcription factors involved in the control of arbuscule development in *Lotus japonicus*. Plant Physiology. 2015;**167**(3):854-871. DOI: 10.1104/pp.114.255430

[102] Gobbato E, Marsh JF, Vernie T, Wang E, Maillet F, Kim J, et al. Mint: A GRAS-type transcription factor with a specific function in mycorrhizal signaling. Current Biology. 2012;**22**(23):2236-2241. DOI: 10.1016/j.cub.2012.09.044

[103] Gong X, Jensen E, Bucerius S, Parniske M. Mint: A CCaMK/cyclops response element in the promoter of *Lotus japonicus* calcium-binding protein 1 (CBP1) mediates transcriptional activation in root symbioses. The New Phytologist. 2022;**235**(3):1196-1211. DOI: 10.1111/nph.18112

Chapter 14

Diversity of Arbuscular Mycorrhizal Fungi in the Rhizosphere of *Argania spinosa* in Morocco

Zineb Sellal, Amina Ouazzani Touhami, Jamila Dahmani, Soukaina Maazouzi, Najoua Mouden, Mohamed Chliyeh, Karima Selmaoui, Rachid Benkirane, Cherkaoui El Modafar and Allal Douira

Abstract

Despite the importance of arbuscular mycorrhizal fungi (AMF) within forest and agroecosystems, few data are available about how AMF communities are structured in the root zone of the argan tree. Some studies have characterized endomycorrhizal fungi population occurring in rhizosphere soils of argan trees grown in southwest of Morocco, numerous sites in this area harbored unexplored communities. The endomycorrhizae diversity of rhizosphere soils collected from 15 argan forest stands located in Lakhssas, Smimou, Ait Baha, Tamanar, Essaouira, Taroudant (Elkodya), Irherm, Guelmim, Imsouane, Anzi, Tiznit, Taghazoute, Ait Melloul, Bouizakarne, and Oulad Teima have revealed the presence of different AMF communities sharing some species but dissimilar AMF community compositions are noted according to sampling time and site. Additionally, the diverse AMF structures detected such as vesicles, arbuscules and hyphae reflect implicitly the germination of AMF propagules in the rhizospheric area of the Argan tree. The pre-evaluation of AMF in the soil through spores' density can indicate AMF community dynamics, signaling either the adaptability of mycorrhizal symbionts to the local conditions or its decline. In total, 39 morphotypes of endomycorrhizal fungal spores were identified and described, representing seven genera: *Glomus* (15 species), *Scutellospora* (3 species), *Entrophospora* (4 species), *Pacispora* (2 species), *Gigaspora* (4 species), *Acaulospora* (10 species), and *Ambispora* (1 species). The genus *Glomus* has a wide occurrence and had the largest number of species. This chapter gives a great overview of the mycorrhizal status of argan trees in their natural habitats of the main Moroccan argan forests.

Keywords: Morocco, *Argania spinosa*, arbuscular mycorrhizal fungi, diversity

1. Introduction

In their natural environment, plants are part of a rich ecosystem, including numerous and diverse microorganisms in the soil and the arbuscular mycorrhizal fungi (AMF), which represent the main component of the soil microbiota in most agroecosystems. Arbuscular mycorrhizal fungi (AMF) are obligate biotrophs and rely on their autotrophic host to complete their life cycle and produce the next generation of spores [1]. These symbionts colonize the roots of the vast majority of plants, either the roots of 86% of terrestrial plants [2] and most crop plants [3]. By forming an extended, intricate hyphal network, AMF can efficiently absorb mineral nutrients from the soil and deliver them to their host plants in exchange for carbohydrates. They play an important role in soil fertility, the acquisition of mineral nutrients, especially immobile nutrients, such as phosphorus [4, 5]. AMF can also enhance tolerance or resistance to root pathogens [6] or abiotic stresses, such as metal toxicity [7]. Yet another benefit conferred by the mycorrhizal fungi is plant growth increase under water deficit conditions. It does so by aiding drought avoidance, enhancing mineral nutrition, improvement in soil physicochemical and biological properties [8].

AMF protects the plant health against other environmental stresses [9, 10] and improves the soil structure by the formation of stable soil aggregates, building up a macroporous structure of soil that allows penetration of water and air and prevents erosion, which results in promoting root system development [11].

Due to all of these advantageous attributes of AMF related to the extended absorptive root surface and the available soil volume by hyphae mycelium of mycorrhizal fungi, some ecological scientists have advocated their use in the regeneration of tropical forests and the restoration of degraded soil in arid and semi-arid areas. In Morocco, there are many representative areas where potential resources are affected by the grazing pressure, arid climate, and anthropogenic activities, such as the northwest palm grove [12, 13], Thuya [14], and argan forest [15]. Of these latter, the argan-ecosystem, suffers from an increase in the deterioration of its various components and needs rehabilitation and reforestation programs to restore a sustainable natural environment.

The use of AMF is one of the natural processes that gains an increasing interest. Its success depends on the knowledge of the diversity and richness of AMF as probable indicators of adaptation in certain environments and the setting of symbiosis with plants [16]. In this context, the study of the diversity of AMF in argan tree rhizosphere through the isolation, identification, and quantification of the number of spores constitute the key step to the characterization of the native AMF associated with this plant species before using as inoculants with a better chance of adapting to particular soil, climate conditions [17].

Several works have shown that the argan tree benefits from a symbiotic association established between the roots of the plant with mycorrhizal fungi [18–21]. Indeed, in semi-arid and arid seeded areas, soils are deficient in nutrients and subject to long periods of drought, hence the need for such root symbiosis [22]. Describing the diversity of the community of AMF at numerous sites from the same area can be useful tool awarding eventual changes that can occur in the course of years before undertaking preservation strategies of this endemic tree, such as incorporating AMF-based biotechnology to cope with stressful conditions that threaten both the perennity and production of this agroforestry system.

2. Argan stands in southwestern Morocco

Argan tree forest covers an area of 3,976,000 ha, spanning from the city of Safi in northeastern Morocco to the Saharan fringe in the south, where the argan tree

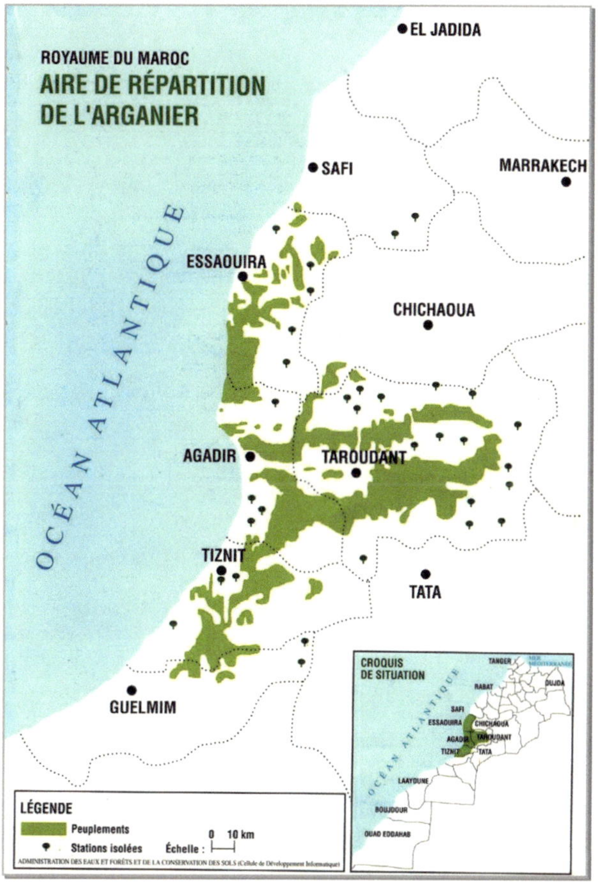

Figure 1.
Distribution area of the argan tree in Morocco [23].

(a) (b)

Figure 2.
Argan tree from Tamanar (a) and Oulad Teima (b) regions.

Figure 3.
Fruits of the argan trees of southwestern Morocco (A) Smimou, (B) Ait Baha, (C) Tiznit, and (D) Bouizakarne.

occupies about 70% of the woodland area [23]. The most important stands extend mainly from the Northeast of Essaouira to the valley of Souss (**Figure 1**).

This locality constitutes the central area of the argan grove and this is because of the state of development and the exceptional vigor that this species presents as shown in 15 sites covering areas of Lakhssas, Smimou, Ait Baha, Tamanar, Essaouira, Taroudant, Irherm, Guelmim, Imsouane, Anzi, Tiznit, Taghazoute, Ait Melloul, Bouizakrane, and Oulad Teima (**Figures 2** and **3**).

3. Physicochemical properties of soil—AMF community

Soil properties are critical in determining the fertility of soils, and some parameters can define the composition and species richness of AMF communities. Hazard *et al.* [24] stated that soil pH has a stronger effect than land use itself on AMF communities in agroecosystems and crops. Alguacil *et al.* [25] suggested that three soil properties related to microbial activity, that is, pH and levels of two micronutrients (Mn and Zn) also determined the distribution of AMF communities in soils. Differences in soil have been found to be key factors in determining AMF community composition [26], and this is particularly relevant in stressed environments. Soil properties have been found to affect the AMF community [27], especially in terms of the availability of nutrients

[28] and variations in pH [29, 30]. Moreover, the structure and dynamic of the AMF community can be influenced by edaphic features, including soil texture and structure, organic matter content, the pH, and macro and micronutrient levels [31, 32]. As pH increases above 7.0 in aqueous solutions, most of the dissolved phosphorus reacts with calcium forming calcium phosphates resulting in a decrease in solubility and availability of phosphate [33, 34]. Indeed, external abiotic factors, such as precipitation or edaphic characteristics, can directly influence the available habitat for a species, which affects an organism's ability to survive in a given location [35]. Furthermore, soil characteristics, such as pH, electrical conductivity, and assimilable phosphorus levels, may also affect the spore number of endomycorrhizal fungi [36].

4. AMF community composition associated with Argan trees

The AM fungi are the important rhizospheric microorganisms whose diversity can be decisive for both plant community structure and ecosystem productivity. Studies on AMF occurrence and distribution have been made by spore extraction from soil and identification based on the morphology of the spores. Thus, the identification of spores has also been widely used to characterize AMF communities in soil [26, 37, 38].

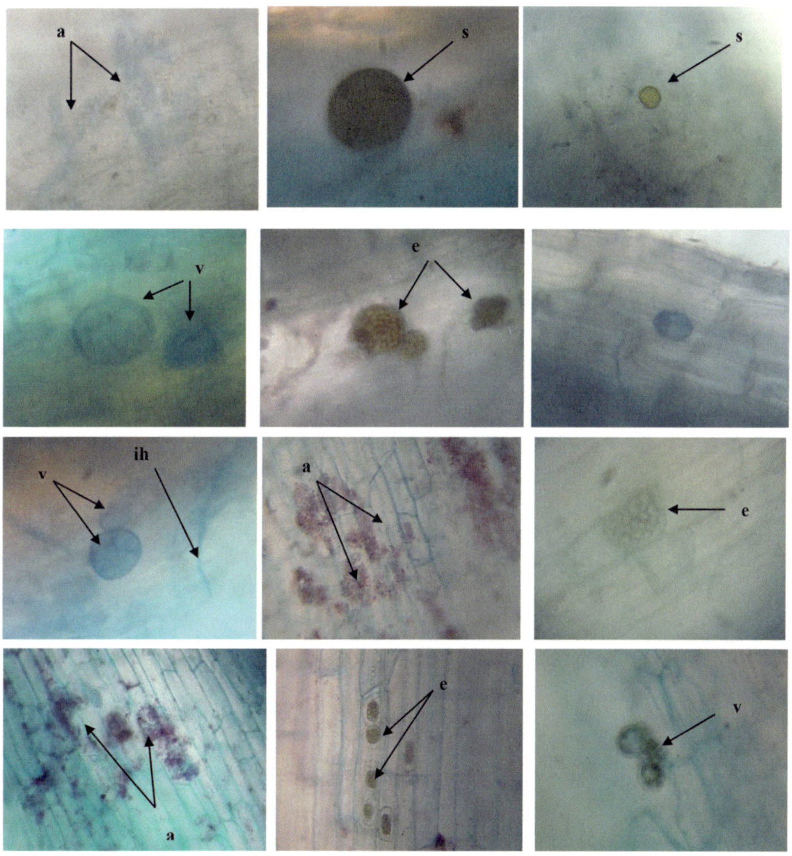

Figure 4.
Different structures of endomycorrhizal fungi colonizing the roots of Argania spinosa. Arbuscules (a); intra hyphae (ih), spores (s); vesicles (v) and endophytes (e). (G. × 400).

4.1 Root colonization with AM fungi

The root colonization by AM fungi relies on the presence of microscopic structures, such as external and internal hyphae, vesicles and arbuscules, as well as endophytes (**Figure 4**).

4.2 AMF spore density

According to Morton *et al.* [39] and Sturmer and Bellei [40], spore density is the common tool for quantifying the AMF population in the soil. The highlighting of the structure mycorrhizal community consists of spores' number enumeration and abondance of each one. The communities of these fungi present in soil can be estimated in terms of the number of species observed and the abundance of each of them in the community. In *Argania spinosa* rhizosphere soil gathers 561 spores/100 g of soil (**Figure 5**) [41].

Oliveira and Oliveira [42] have revealed significant variations in spore density between the soil samples collected in August (dry season) obviously lower than in the sampling performed during the rainy season. Likewise, Khaekhum *et al.* [43] noted a higher number of AMF spores in the rainy season than in the dry season. The changes in spore densities are probably attributable to annual variations in climatic and edaphic conditions, especially as spore density increases in dry climates [44] reflecting adaptability to temperate, dry, and arid ecosystems [45, 46]. It is well known that edaphoclimatic conditions, such as pluvial precipitation can influence AMF spore density [47]. According to Pringle and Bever [48], fungal species sporulate differently on the season. For these authors, the seasonal variations in spore densities probably reflect seasonal differences in spore formation. Smith [49] showed that maximum spore densities are noted in the spring and decline in the summer.

The variation of spore density of AMF is directly related to the plant growth stage [50]. Various medicinal plants have displayed the highest intensity of AMF colonization and spore population in the flowering stage [51]. Hatimi and Tahrouch [52] have demonstrated that mycorrhization is nutrient level-dependent, and the spore production of AMF tends to be significant at the flowering stage and then decreased at the end of the growing season when the physiological cycle of plant roots changed. Indeed, disturbance of semi-arid ecosystems decreased mycorrhizal spore density and nutrient availability.

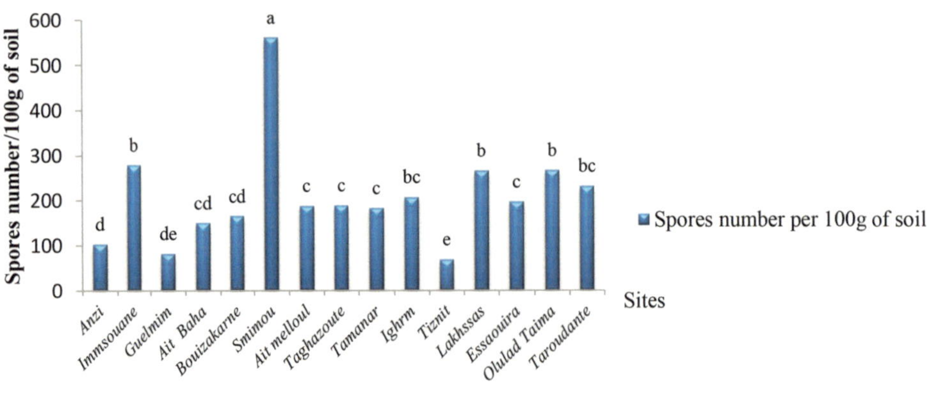

Figure 5.
Average of AMF spore density according to soils of sampled sites from Argania spinosa distribution areas [41].

Figure 6.
Some AMF species and morphotypes isolated from the rhizospheric soil of argan tree [41].

4.3 AMF Community and species richness

As all natural plant communities, the argan tree contains arbuscular mycorrhizal fungi at rhizospheric soil level. The total number of AMF morphotypes was 35 in 2016 [53] and 39 in 2021 [41] illustrated in **Figure 6**.

The specific richness of this assembly of community attains 18, 14, and 9 species in some sites (**Figure 7**). Almost the same number of AMF spore morphotypes (31) was found in the rhizosphere of *Ceratonia siliqua* developing in different ecological zones (Afourar, Ksiba Khénifra, Taroudant, and Nador) [54]. El Maati *et al.* [55] detail a low specific richness (nine species) of native AMF communities from *Argania spinosa*, *Acacia gummifera,* and *C. siliqua* in southwest Morocco, 11 morphotypes from the argan tree in northwest Morocco [56]. Several factors can explain these

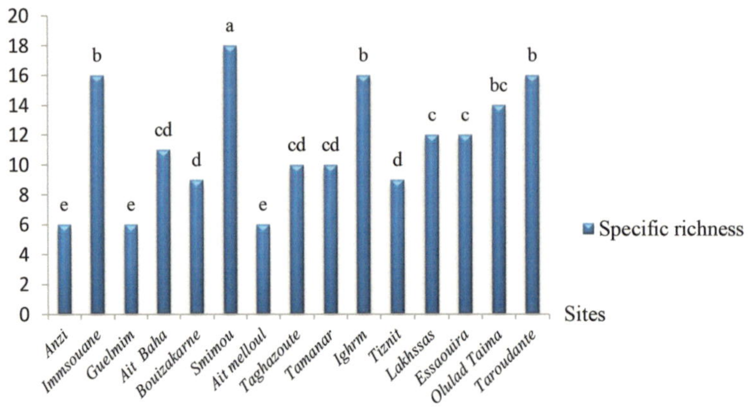

Figure 7.
Specific richness of mycorrhizal species in the rhizosphere of argan tree according to studied sites.

disparities. Relative air humidity and rainfall are significant drivers for AMF spore density, especially for members of the families Acaulosporaceae, Diversisporaceae, and Glomeraceae, which were positively correlated with these abiotic factors [57]. The precipitation and water availability could drive the changes in AMF communities at a regional scale [58]. Spore abundance and species richness can also be influenced by elevation gradients [59, 60] and mycorrhizal fungi pH tolerance [61], plant density [62], and productivity and land-use intensity [63].

Regarding the dominance of genera *Glomus* and *Acaulospora* in the rhizospheric soil of argan tree, it was also cited by El Maati *et al.* [55], in the rhizosphere of diverse plant species [64–69], in soil from different ecosystems, in Senegal [70], in China [71], Burkina Faso [72], Kenya [73], Sudan [74], and in central Europe [75]. The high occurrence of the *Glomus* genus is due to its ability to produce more spores in a shorter time than other genera, such as *Gigaspora* and *Scutellospora,* and its adaptation to drought and soil salinity [76]. In disturbed habitats, the high abundance of Glomeraceae is related to the considerable capacity of some of its most frequently found members, for example, *Rhizophagus irregularis*, to sporulate [77]. *Acaulosporaceae* members may be confined to the harsh environmental conditions of uplands [78] and are dominant in protected areas. In fact, the high anthropic impact may modify the AMF community and cause decreased AMF biodiversity, root colonization, and sporulation [79]. It was emphasized that degraded lands harbor low levels of AMF abundance and diversity [80]. Several studies found that disturbance of semi-arid ecosystems decreased mycorrhizal spore density and root colonization [81]. It was also reported that livestock and human disturbances decreased AMF spore density, root colonization, and nutrient availability [82].

5. Conclusion

Mycorrhizal fungi play a complex role in ecosystem function, so knowledge of their distributional patterns is important, especially in view of the current environmental threats to AMF diversity and plant productivity under climate changes. The present study provides useful information about the composition of the AMF community associated with *Argania spinose* tree within its natural environment where some conditions exert strong pressure leading to the appearance, dominance of AMF

DOI: http://dx.doi.org/10.5772/intechopen.106162

type or disappearance of other AMF species, and replaced by others. Thus, we can expect the success of restoration programs if the suited AMF is used for multiplication in soil with plants displaying great mycorrhizal capacity.

Author details

Zineb Sellal[1], Amina Ouazzani Touhami[1], Jamila Dahmani[1], Soukaina Maazouzi[1], Najoua Mouden[2], Mohamed Chliyeh[1], Karima Selmaoui[1], Rachid Benkirane[1], Cherkaoui El Modafar[3] and Allal Douira[1*]

1 Faculté des Sciences, Laboratoire des Productions Végétales, Animales et Agro industrie, Equipe de Botanique, Biotechnologie et Protection des Plantes, Université Ibn Tofail, Morocco

2 Faculté Pluridisciplinaire de Nador, Laboratoire de Chimie Moléculaire et molécules de l'environnement LCM2E, Université Mohammed 1er Oujda, Maroc

3 Faculté des Sciences et Techniques Guéliz, Centre d'Agrobiotechnologie et Bioingénierie, Unité de Recherche Labellisée CNRST (URL-CNRST 05), Université Cadi Ayyad, Marrakech, Maroc

*Address all correspondence to: labobotanique@gmail.com

References

[1] Garbaye J. La symbiose mycorhizienne. Une association entre les plantes et les champignons. Editions QUAE. 2013:280

[2] Newman EI, Reddell P. The distribution of mycorrhizas among families of vascular plants. The New Phytologist. 1987;**106**:745-751

[3] Smith SE, Read DJ. Mycorrhizal Symbiosis. 2nd ed. London, England: Academic Press Ltd.; 1997. p. 1997

[4] Clark RB, Zeto SK. Mineral acquisition by arbuscular mycorrhizal plants. Journal of Plant Nutrition. 2000;**23**:867-902. DOI: 10.1080/01904160009382068

[5] Li XL, Marschner H, George E. Acquisition of phosphorus and copper by VA-mycorrhizal hyphae and root-to-shoot transport in white clover. Plant and Soil. 1991;**136**:49-57

[6] Whipps JM. Prospects and limitations for mycorrhizas in biocontrol of root pathogens. Canadian Journal of Botany. 2004;**82**:1198-1227

[7] Meharg AA, Cairney JWG. Co-evolution of mycorrhizal symbionts and their hosts to metal-contaminated environments. Advances in Ecological Research. 2000;**30**:69-112

[8] Meyer J, Oehl F. Integration of mycorrhizal inoculum in high alpine revegetation. mycorrhiza works. In: Feldmann F, Kapulnik Y, Baar J, editors. Proceedings of the International Symposium "Mycorrhiza for Plant Vitality" and the Joint Meeting of Working Groups 1 4 of COST Action 870. Braunschweig: Deutsche Phytomedizinische Gesellschaft; 2008. pp. 278-288

[9] Martínez-Garcia LB. Micorrizas arbusculares en ecosistemas semiáridos. Respuesta a factores de estrés ambiental. Ecosistemas. 2011;**20**(2-3):117-120

[10] Liu S, Guo X, Feng G, Maimaitiaili B, Fan J, He X. Indigenous arbuscular mycorrhizal fungi can alleviate salt stress and promote growth of cotton and maize in saline fields. Plant and Soil. 2016;**398**:195-206

[11] Gutjahr C, Paszkowski U. Multiple control levels of root system remodeling in arbuscular mycorrhizal symbiosis. Frontiers in Plant Science. 2013;**4**:204. DOI: 10.3389/fpls.2013.00204

[12] Botes A, Zaid A. The economic importance of date production and international trade. In: Zaid A, Arias-Jimenez EJ, editors. Date Palm Cultivation. FAO Plant Production and Protection Paper. Rome: FAO; 2002

[13] Meddich A, Ait El Mokhtar M, Bourzik W, Mitsui T, Baslam M, Hafidi M. Optimizing growth and tolerance of date palm (Phoenix dactylifera L.) to drought, salinity, and vascular fusarium induced wilt (*fusarium oxysporum*) by application of arbuscular mycorrhizal fungi (AMF). Root Biology, Soil Biology. 2018;**52**:239-258. DOI: 10.1007/978-3-319-75910-4_9

[14] Khaddari AE, Gabardi SE, Touhami AO, Aoujdad J, Ouajdi M, Antry SE, et al. Diversity of endomycorrhizal fungi in the thuya rhizosphere, sefrou region (middle eastern atlas, Morocco). Plant Cell Biotechnology And Molecular Biology. 2019;**20**(23-24):1143-1159

[15] M'hirit O. L'arganier: une espèce fruitière à usage multiple, Formation

forestière continue, thème « l'arganier ». In: Division de recherche et d'expérimentations forestières. Rabat; 1989. pp. 59-64

[16] Van der Heijden MGA, Klironomos JN, Ursic M. Mycorrhizal fungal diversity determines plant biodiversity, ecosystem variability and productivity. Nature. 1998;**396**:69-72. DOI: 10.1038/23932

[17] López-Gómez BF, Alarcón A, Quintero-Lizaola R, Lara-Herrera A. Selection of strains of arbuscular mycorrhizal fungi in two Chile production systems. Mexican Journal of Agricultural Sciences. 2015;**6**:1203-1214

[18] Nouaim R, Chaussod R. Mycorrhizal dependency of two clones of micro-propagated argan tree (*Argania spinosa*). I. Growth and biomass production. Agroforestry Systems. 1994;**27**(1):53-65

[19] Nouaim R, Chaussod R. Rôle des mycorhizes dans l'alimentation hydrique et minérale des plantes notamment des ligneux de zones arides. Cahiers options mediterraniennes. 1996;**20**:9-26

[20] Sellal Z, Ouazzani Touhami A, Mouden N, El Ouarraqi M, Selmaoui K, Dahmani J, et al. Effect of an endomycorrhizal inoculum on the growth of Argan tree. International Journal of Environmental and Agricultural Biotechnology. 2017;**2**: 928-939. DOI: 10.22161/ijeab/2.2.47

[21] Sellal Z, Ouazzani Touhami A, Mouden N, Chliyeh M, Selmaoui K, Dahmani J, et al. The effectiveness of seed coating with composite endomycorrhizal inoculum for establishing mycorrhization and growing Argan seedlings. Interciencia. 2019;**4**(5):24-37

[22] Nouaim R, Chaussod R. Effet de la mycorhization contrôlée sur la croissance de l'arganier (*Argania spinosa*) après sa transplantation en sol non désinfecté. Al Awamia. 1997;**96**:65-76

[23] M'Hirit O, Benzyane M, Benchekroun F, el Yousfi M, Bendaanoun M. L'arganier: une espèce fruitière-forestière à usages multiples. Belgique: Mardaga; 1998. p. 144

[24] Hazard C, Gosling P, Van Der Gast CJ, Mitchell DT, Doohan FM, Bending GD. The role of local environment and geographical distance in determining community composition of arbuscular mycorrhizal fungi at the landscape scale. The ISME Journal. 2013;7:498-508

[25] Alguacil MDM, Torres MP, Montesinos-Navarro A, Roldán A. Soil characteristics driving arbuscular mycorrhizal fungal communities in semiarid Mediterranean soils. Applied and Environmental Microbiology. 2016;**82**:3348-3356

[26] Oehl F, Laczko E, Bogenrieder A, Stahr K, Bösch R, van der Heijden M, et al. Soil type and land use intensity determine the composition of arbuscular mycorrhizal fungal communities. Soil Biology and Biochemistry. 2010;**42**:724-738

[27] Johnson NC, Tilman D, Wedin D. Plant and soil controls on mycorrhizal fungal communities. Ecology. 1992;**73**:2034-2042

[28] Johnson NC, Wilson GWT, Bowker MA, Wilson JA, Miller RM. Resource limitation is a driver of local adaptation in mycorrhizal symbioses. Proceedings of the National Academy of Science USA. 2010;**107**:2093-2098

[29] Dumbrell AJ, Nelson M, Helgason T, Dytham C, Fitter AH. Relative roles of niche and neutral processes in

structuring a soil microbial community. The ISME Journal. 2009;**4**:337-345

[30] Bueno CG, Marín C, Silva-Flores P, Aguilera P, Godoy R. Think globally, research locally: Emerging opportunities for mycorrhizal research in South America. The New Phytologist. 2017;**215**:1306-1309

[31] Mohammad MJ, Hamad SR, Malkawit HI. Population of arbuscular mycorrhizal fungi in semi-arid environment of Jordan as influenced by biotic and abiotic factors. Journal of Arid Environments. 2003;**53**:409-417

[32] Rivaton D. Étude des champignons mycorhiziens arbusculaires des sols en systèmes de grandes cultures biologiques sans élevage : application à la nutrition phosphatée. Mémoire de Master. France: Agrocampus Ouest Renne; 2016. p. 68

[33] Christophe D. Les racines face cachée des arbres. Institut pour le développement forestier CNPPF. Paris; 2006

[34] Siebielec G, Ukalska-Jaruga A, Kidd P. Bioavailability of trace elements in soils amended with high-phosphate materials. In: Phosphate in Soils: Interaction with Micronutrients, Radionuclides and Heavy Metals. Boca Raton, FL, USA: CRC Press; 2014. pp. 237-268

[35] Chaudhary VB, Lau MK, Johnson NC. Macroecology of Microbes Biogeography of the *Glomeromycota*. Mycorrhiza. Berlin, Heidelberg: Springer; 2008. pp. 529-563

[36] Radi M. Thèse de Doctorat, Université Cadi Ayyad de Marrakech (Maroc). 2014:143

[37] Robinson-Boyer L, Grzyb I, Jeffries P. Shifting the balance from qualitative to quantitative analysis of arbuscular mycorrhizal communities in field soils. Fungal Ecology. 2009;**2**:1-9. DOI: 10.1016/j.funeco.2008.11.001

[38] Tchabi A, Coyne D, Hountondji F, LawouinL, WiemkenA, OehlF. Arbuscular mycorrhizal fungal communities in sub-Saharan savannas of Benin, West Africa, as affected by agricultural land use intensity and ecological zone. Mycorrhiza. 2008;**18**:181-195

[39] Morton JB, Franke M, Bentivenga SP. Developmental foundations for morphological diversity among endomycorrhizal fungi in Glomales. In: Varma A, Hock B, editors. Mycorrhiza Structure, Function, Molecular Biology andBiotechnology. Heidelberg: Springer; 1995. pp. 669-683

[40] Stürmer SL, Bellei MM. Composition and seasonal variation of spore populations of arbuscular mycorrhizal fungi in dune soils on the island of Santa Catarina, Brazil. Canadian Journal of Botany. 1994;**72**:359-363

[41] Sellal Z, OuazzaniTouhami A, Dahmani J, Maazouzi S, Mouden N, ChliyehM,et al.Distributionandabundance of arbuscularmycorrhizal fungi of *Argania spinosa* tree and mycorrhizal infectious potential of rhizospheric soil of 15 argania groves in southwestern Morocco. Plant Cell Biotechnology and Molecular Biology. 2021;**22**(3&4):1-29

[42] Oliveira AN, Oliveira LA. Seasonal dynamics of arbuscular mycorrhizal fungi in plantas of *Teobroma grandiflorum* Schum and *Paullinia cupana* Mart. Of an agroforestry system in central Amazonía, Amazonas state, Brazil. Brazilian Journal of Microbiology. 2005;**36**:262-270

[43] Khaekhum S, Lumyong S, Kuyper TW, Boonlue S. Species richness and composition of arbuscular

mycorrhizal fungi occurring on eucalypt trees (*Eucalyptus camaldulensis* Dehnh.) in rainy and dry season. Current Research in Environmental & Applied Mycology. 2017;7(4):282-292

[44] Uhlmann E, Gorke C, Petersen A, Oberwinkler F. Arbuscular mycorrhizae from arid parts of Namibia. Journal of Arid Environments. 2006;**6**:221-237

[45] Mangan SA, Eom AH, Adler GH, Yavitt JB, Herre EA. Diversity of mycorrhizal fungi across a fragmented forest in Panama: Insular spore communities differ from mainland communities. Oecologia. 2004;**141**:687-700

[46] Tao L, Zhiwei Z. Arbuscular mycorrhizas in a hot and arid ecosystem in Southwest China. Applied Soil Ecology. 2005;**29**:135-141

[47] Johnson NC, Rowland DL, Corkidi L, Egerton-Warburton LM, Allen EB. Nitrogen enrichment alters mycorrhizal allocation at five Mesic to semiarid grasslands. Ecology. 2003;**84**:1895-1908

[48] Pringle A, Bever JD. Divergent phenologies may facilitate the coexistence of arbuscular mycorrhizal fungi in a North Carolina grassland. American Journal of Botany. 2002;**89**:1439-1446

[49] Smith TF. The effect of season and crop rotation on the abundance of spores of vesiculararbuscular (VA) mycorrhizal endophytes. Plant and Soil. 1980;**57**(2):475-479

[50] Moreira-Souza M, Trufem SFB, Gomes-daCosta SM, Cardoso EJBN. Arbuscular mycorrhizal fungi associated with *Araucaria angustifolia* (Bert.) O. Ktze. Mycorrhiza. 2003;**13**:211-215

[51] Yaseen ZM, Jaafar O, Deo RC, Kisi O, Adamowski J, Quilty J, et al. Stream flow forecasting using extreme learning machines: A case study in a semi-arid region in Iraq. Journal of Hydrology. 2016;**542**:603-614

[52] Hatimi A, Tahrouch S. Caractérisations chimique, botanique et microbiologique du sol des dunes littorales du Souss-Massa. Biomatec Echo. 2007;**2**(5):85-97

[53] Sellal Z, Ouazzani Touhami A, Chliyeh M, Dahmani J, Benkirane R, Douira A. Arbuscular mycorrhizal fungi species associated with rhizosphere of *Argania spinosa* (L.) Skeels in Morocco. International Journal of Pure and Applied Bioscience. 2016;**4**(1):82-99

[54] Talbi Z, El Asri A, Jihane Touati J, Chliyeh M, Ait Aguil F, Selmaoui K, et al. Morphological characterization and diversity of endomycorrhizae in the rhizosphere of Carob tree (*Ceratonia siliqua*) in Morocco. Biolife. 2015;**3**(1):196-211

[55] El Maati Y, Msanda F, el Hamdaoui A, el Mrabet S, Ouahmane L. contribution to the characterization of mycorrhizae in the south west of Morocco and their effect on growth parameters of *Argania spinosa*. The American Journal of Innovative Research and Applied Sciences. 2015;**1**(7):235-243

[56] Maazouzi S, Aoujdad J, Selmaoui K, El Gabardi S, Artib M, Elantry S, et al. Mycorrhizal status and mycorrhizal colonization potential of Rhizospheric soils around introduced and natural Argan trees in Northwest Morocco. Tree Planters' Notes. 2021;**64**(1):62-71

[57] Melo CD, Walker C, Krüger C, Borges PAV, Luna S, Mendonça D, et al. Environmental factors driving arbuscular mycorrhizal fungal communities associated with endemic woody plant *Picconia azorica* on native

forest of Azores. Annals of Microbiology. 2019;**69**:1309-1327

[58] Zhang J, Wang F, Che R, Wang P, Liu H, Ji B, et al. Precipitation shapes communities of arbuscular mycorrhizal fungi in Tibetan alpine steppe. Scientific Reports. 2016;**6**:23488

[59] Li X, Gai J, Cai X, Li X, Christie P, Zhang F, et al. Molecular diversity of arbuscular mycorrhizal fungi associated with two co-occurring perennial plant species on a Tibetan altitudinal gradient. Mycorrhiza. 2014;**24**(2):95-107

[60] Coutinho ES, Fernandes GW, Berbara RL, Valério HM, Goto BT. Variation of arbuscular mycorrhizal fungal communities along an altitudinal gradient in rupestrian grasslands in Brazil. Mycorrhiza. 2015;**25**(8):627-638

[61] Bainard LD, Bainard JD, Hamel C, Gan Y. Spatial and temporal structuring of arbuscular mycorrhizal communities is differentially influenced by abiotic factors and host crop in a semi-arid prairie agroecosystem. FEMS Microbiology Ecology. 2014;**88**:333-344

[62] Burrows RL, Pfleger FL. Arbuscular mycorrhizal fungi respond to increasing plant diversity. Canadian Journal of Botany. 2002;**80**:120-130

[63] Verbruggen E, Röling WFM, Gamper HA, Kowalchuk GA, Verhoef HA, van der Heijden MGA. Positive effects of organic farming on below-ground mutualists: Large-scale comparison of mycorrhizal fungal communities in agricultural soils. The New Phytologist. 2010;**186**:968-979

[64] Sghir F, Touati J, Chliyeh M, Ouazzani Touhami A, Filali Maltouf A, Cherkaoui E, et al. Diversity of arbuscular mycorrhizal fungi in the rhizosphere of date palm tree (*Phoenix dactylifera*) in Tafilalt and Zagora regions (Morocco). International Journal of Pure and Applied Bioscience. 2014;**2**(6):1-11

[65] Chliyeh M, Ouazzani Touhami A, Filali-Maltouf A, El Modafar C, Moukhli A, Oukabli A, et al. Effect of a composite endomycorrhizal inoculum on the growth of olive trees under nurseries conditions in Morocco. International Journal of Pure and Applied Bioscience. 2014;**2**(2):1-14

[66] Touati J, Chliyeh M, Ouazzani Touhami A, Benkirane R, Douira A. Effect of arbuscular mycorrhizal fungi on plant growth and root development of the boxthorn tree (*Lycium europaeum*) under greenhouse conditions. International Journal of Pure and Applied Bioscience. 2015;**2**(6):84-91

[67] Hibilik N, Selmaoui K, Touati J, Chliyeh M, Ouazzani Touhami A, Benkirane R, et al. Mycorrhizal status of Eryngium maritimum in the mobile dunes of Mehdia (Northwest of Morocco). International Journal of Pure and Applied Bioscience. 2016;**4**(1):35-44

[68] Artib M, Chliyeh M, Touati J, Talbi Z, Selmaoui K, Ouazzani Touhami A, et al. Study of arbuscular mycorrhizal fungi diversity in the rhizosphere of Citrus grown in Morocco. International Journal of Advances in Pharmacy, Biology and Chemistry. 2016;**5**(3):2277-4688

[69] Maazouzi S, Aoujdad J, Selmaoui K, El Gabardi S, Artib M, Elantry S, et al. Evaluation of the mycorrhizal status of aAcacia in the Rhamna-Sidi Bouathman and the Haha regions in Morocco. Plant Cell Biotechnology and Molecular Biology. 2020;**21**(1&2):1-18

[70] Manga A, Diop TA, Tuinen DV, Neyra M. Variabilité Génétique des champignons mycorhiziens associés

à *Acacia seyal* en zone semi-aride du sénégal. Secheresse. 2007;**18**(2):129-133

[71] Zhao D, Zhao Z. Biodiversity of arbuscular mycorrhizal fungi in the hot-dry valley of the Jinsha River, Southwest China. Applied Soil Ecology. 2007;**37**:118-128

[72] Bâ AM, Dalpé Y, Guissou T. Les Glomales d'*Acacia holosericea* et d'*Acacia mangium*. Bois et Forêt des Tropiques. 1996;**250**:5-18

[73] Jefwa JM, Mungatu J, Okoth P, Muya E, Roimen H, Njuguini S. Influence of land use types on occurrence of arbuscular mycorrhizal fungi in the high altitude regions of MT, Kenya. Tropical and Subtropical Agroecosystems. 2009;**11**:277-290

[74] Abdelhalim TS, Finckh MR, Babiker AG, Oehl F. Species composition and diversity of arbuscular mycorrhizal fungi in White Nile state, Central Sudan. Archives of Agronomy and Soil Science. 2013;**60**:377-391

[75] Oehl F, Sieverding E, Ineichen K, Mäder P, Boller T, Wiemken A. Impact of land use intensity on the species diversity of arbuscular mycorrhizal fungi in agroecosystems of Central Europe. Applied and Environmental Microbiology. 2003;**69**:2816-2824

[76] Brito I, Goss MJ, De Carvalho M, Chatagnier O, van Tuinen D. Impact of tillage system on arbuscular mycorrhiza fungal communities in the soil under Mediterranean conditions. Soil and Tillage Research. 2012;**121**:63-67

[77] Moora M, Davison J, Öpik M, et al. Anthropogenic land use shapes the composition and phylogenetic structure of soil arbuscular mycorrhizal fungal communities. FEMS Microbiology Ecology. 2014;**90**:609-621

[78] Senés-Guerrero C, Schüßler A. A conserved arbuscular mycorrhizal fungal core-species community colonizes potato roots in the Andes. Fungal Diversity. 2016;**77**:317-333

[79] Moreira M, Nogueira MA, Tsai SM, Gomes-da-Costa SM, Cardoso EJBN. Sporulation and diversity of arbuscular mycorrhizal fungi in Brazil Pine in the field and in the greenhouse. Mycorrhiza. 2007;**17**:519-526

[80] Asmelash F, Bekele T, Birhane E. The potential role of arbuscular mycorrhizal fungi in the restoration of degraded lands. Frontiers in Microbiology. 2016;**7**:1-15

[81] Barea JM, Palenzuela J, Cornejo P, Sanchez-Castro I, Navarro-Fernandez C, Lopez-Garcea AZC, et al. Ecological and functional roles of mycorrhizas in semi-arid ecosystems of Southeast Spain. Journal of Arid Environments. 2011;**75**(12):1292-3011

[82] Birhane E, Sterck FJ, Fetene M, Bongers F, Kuyper TW. Arbuscular mycorrhizal fungi enhance photosynthesis, water use efficiency, and growth of frankincense seedlings under pulsed water availability conditions. Oecologia. 2012;**169**:895-904